ROADSIDE GEOLOGY

OF TEXAS

Third Edition

ROADSIDE GEOLOGY OF TEXAS

Third Edition

PAUL BRANDES AND DARWIN SPEARING

Illustrated by Chelsea M. Feeney

2025

THE GEOLOGICAL
SOCIETY OF AMERICA
GSA

Boulder, Colorado

Photos by author unless otherwise credited.

Illustrations constructed by Chelsea M. Feeney (cmcfeeney.com)
using data from the Texas Water Science Center (USGS TWSC)
Geologic Database of Texas, 2014-02-01.

ROADSIDE
GEOLOGY

Roadside Geology is a registered trademark
of The Geological Society of America.

Library of Congress Cataloging-in-Publication Data

Names: Brandes, Paul, 1974- author. | Spearing, Darwin, author. | Feeney,
  Chelsea McRaven, 1980- illustrator.
Title: Roadside Geology of Texas / Paul Brandes and Dar Spearing ;
  illustrated by Chelsea M. Feeney.
Other titles: Roadside Geology series.
Description: Third edition. | [Boulder, Colorado] : The
  Geological Society of America, [2025] | Series: Roadside geology |
  Includes bibliographical references and index. | Summary: "This new
  edition updates the previous version's road guides with new information
  and color photos. After an introduction to Texas geology, the book is
  subdivided into four unique geographic regions: the Gulf Coast, Central
  Texas/Hill Country, the High Plains, and West Texas. The back of the
  book features a glossary of terms and references for further reading" —
  Provided by publisher.
Identifiers: LCCN 2025018745 | ISBN 9780813741284 (paperback)
Subjects: LCSH: Geology—Texas—Guidebooks. | Texas—Guidebooks. | Gulf
  Coast (Tex.)—Guidebooks. | Texas Hill Country (Tex.)—Guidebooks. |
  High Plains (U.S.)—Guidebooks. | Texas, West—Guidebooks. | LCGFT:
  Guidebooks.
Classification: LCC QE167 .B74 2025 | DDC 557.64—dc23/eng/20250512
LC record available at https://lccn.loc.gov/2025018745

PRINTED IN THE UNITED STATES BY VERSA PRESS, INC.

**THE GEOLOGICAL
SOCIETY OF AMERICA**

GSA

3300 Penrose Place · P.O. Box 9140 · Boulder, CO 80301-9140
303-357-1000, option 3 · 800-472-1988
gsaservice@geosociety.org · www.geosociety.org

To my mother Nadine, who never doubted my ability as a scientist

and

to the many geologists before me who have studied Texas geology. May their legacies live on forever.

—Paul Brandes

Roads and sections of Roadside Geology of Texas.

CONTENTS

ACKNOWLEDGMENTS

When I was asked by Mountain Press to work on the revision of *Roadside Geology of Texas*, I knew Texas was a big, diverse state with many different geologic features; however, it is even bigger and greater when you drive every road and stop to examine the hundreds of outcrops and roadcuts that are present. As I pressed forward with this project, it became clear early on that this was to be a daunting task, but one that became very interesting and enjoyable to complete. This project could not have been possible without a great group of individuals to assist me at a moment's notice.

While I cannot mention everyone by name here, many people were integral to this book coming together. I also want to give thanks to the two previous authors of *Roadside Geology of Texas*: Robert Sheldon and Darwin Spearing. Without their efforts in writing the first two editions, this new color edition would not have been possible. Sadly, Dar passed away as the process of writing this revision and gathering photographs was just beginning. Many of the new figures are based on his original sketches, and some of his original text was included, particularly where he introduced geological regions with his colorful and whimsical descriptions.

This book would not be possible without the invaluable help I received from Mountain Press Publishing Company and the Geological Society of America. Their assistance throughout this project was priceless. I also want to thank every national, state, and local park, every historical society, public library, museum, and numerous other organizations that added important information to this book. I especially want to thank the Texas Parks and Wildlife Department for their willingness in taking me to places so I didn't have to walk miles to reach a feature, and for revealing information about their parks that otherwise would not have been known without their welcoming demeanor and expertise.

Most of the photographs used in this book are my own, but in some cases, it was impossible to get that perfect shot. I want to thank Tama Higuchi, Rob Lavinsky, Kelly Nash, Jonathan Woolley, and Amir Akhaven for their gracious use of photographs and artwork. In addition, I must also thank Chelsea Feeney for creating the maps, graphs, and figures seen throughout this book. Satellite images were obtained from Google Earth and the US Geological Survey Earth Explorer sites.

Lastly, I must thank my wife, Nathalie, who provided much needed suggestions and recommendations to make this book the best it could be. Not only did she travel to every site, ride shotgun and take notes about the geology on every road guide, and provide guidance along the way, but she also did it with a smile on her face and a willingness to learn about the geology of Texas, even while writing her own book on Texas geology, *Texas Rocks!*

NOTE TO THE READER

More than forty-five years have passed since the first *Roadside Geology of Texas* was published. In that time, roads have been rerouted, outcrops that were visible when the first two books were written are now either gone or have been covered up and are no longer visible, and cities have grown and engulfed entire mountains of once beautiful rock layers. While it might seem like a lot of Texas geology is no longer available with all the "improvements," nothing could be further from the truth. Many spectacular roadcuts remain, new road guides have been added that were left out of the older editions, and state parks provide access to many amazing geological sites.

This new edition of *Roadside Geology of Texas* updates the original road guides with new information and color photos. The book begins with an introduction to Texas geology to give an overall view of what the reader will see before heading out on the highway. Following the introduction, the book is subdivided into four unique geographic regions: the Gulf Coastal Plain, Central Plains/Hill Country, the High Plains, and West Texas. Each region begins with an introduction to its geology, followed by road guides with figures, maps, and photos. The back of the book features a glossary of terms used throughout, followed by a list of references for further reading should the reader want to know more.

All the road guides follow paved roads, and most of the outcrops featured are publicly accessible with no fee required. Some of the sites are in state or federal parks where a small admission fee is required for entry. A few of the sidebars require travel on a gravel road with one (El Solitario in Big Bend Ranch State Park) requiring a high-clearance, four-wheel drive vehicle. One thing to note: about 95 percent of all land in Texas is privately owned. It is imperative to know where you are and observe all private property signs when venturing off the main roads on your own. You don't have to leave the main roads, however, to learn almost everything you need to know about Texas geology.

GPS (Global Positioning System) points are included at the end of each photo caption to indicate where the photo was taken. You can type these numbers directly into a portable GPS unit or the Google Maps search bar, and it will take you to that spot.

The Gulf of Mexico was renamed to the Gulf of America in the United States in 2025 by the US Department of the Interior, but it retained its original name internationally and during the writing of this work. It appears as the Gulf of Mexico throughout this publication, following the naming standards for international bodies of water set by the International Hydrographic Organization.

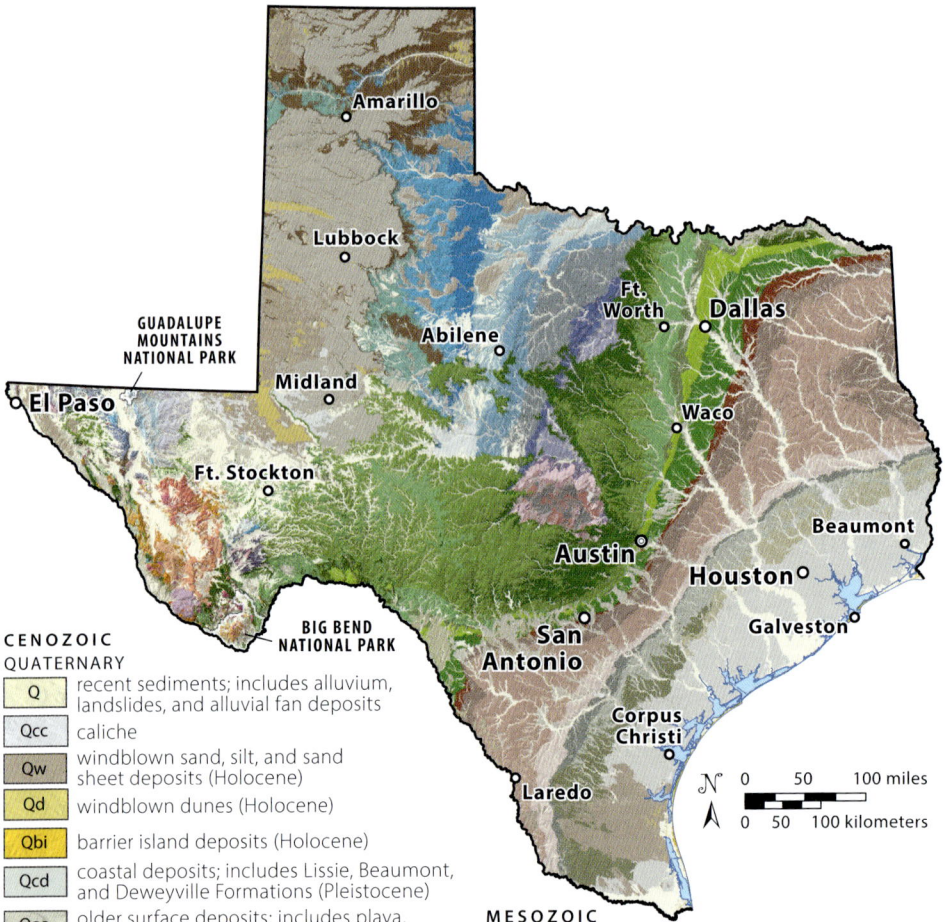

Map labels (cities and parks):
Amarillo, Lubbock, Abilene, Midland, Ft. Stockton, El Paso, GUADALUPE MOUNTAINS NATIONAL PARK, BIG BEND NATIONAL PARK, Ft. Worth, Dallas, Waco, Austin, San Antonio, Houston, Beaumont, Galveston, Corpus Christi, Laredo

Scale:
0   50   100 miles
0   50   100 kilometers

CENOZOIC

QUATERNARY

Q	recent sediments; includes alluvium, landslides, and alluvial fan deposits
Qcc	caliche
Qw	windblown sand, silt, and sand sheet deposits (Holocene)
Qd	windblown dunes (Holocene)
Qbi	barrier island deposits (Holocene)
Qcd	coastal deposits; includes Lissie, Beaumont, and Deweyville Formations (Pleistocene)
Qoa	older surface deposits; includes playa, terrace, and high gravel deposits (Pleistocene)
Qbd	Blackwater Draw and Tule Formations and windblown cover sand (Pleistocene)

QUATERNARY–NEOGENE

| QTb | Blanco Formation (Pliocene?–Pleistocene) |
| QTs | sedimentary deposits; includes Willis Formation (Miocene–Pleistocene) |

NEOGENE and PALEOGENE

To	Ogallala Formation (middle Miocene to early Pliocene)
Tg	Goliad Sand (Pliocene)
Tf	Fleming Group; includes Oakville Sandstone (Miocene)
Tcf	Catahoula and Frio Formations; includes Anahuac Formation (in subsurface only) (Oligocene–Miocene)
Tj	Jackson Group (Eocene and Oligocene)
Tc	Claiborne Group; includes Laredo, Queen City, Reklaw, Weches, and Yegua Formations, and Sparta Sand (Eocene)
Tw	Wilcox Group; includes Carrizo Sand and Hooper Formation (Eocene)
Tm	Midway Group (Paleocene–Eocene)
Tbp	Black Peaks Formation (Paleocene)

MESOZOIC

LATE CRETACEOUS

K	sedimentary rocks, undivided
Kt	Tornillo Group; includes Aguja and Javelina Formations
Knt	Navarro and Taylor Groups; includes Corsicana Formation (Navarro Group) Anacacho Limestone, Pecan Gap Chalk, (Taylor Group)
Kau	Austin Group; includes Austin Chalk
Ktb	Terlingua Group; includes Boquillas and Pen Formations
Kef	Eagle Ford Group

EARLY–MIDDLE CRETACEOUS

Kww	Woodbine Formation and Washita Group; includes Salmon Peak and Santa Elena Limestones, Buda Formation and Boracho Limestone of the Sixshooter Group, Del Rio Clay, and Sue Peaks Formation
Kf	Fredericksburg Group; includes Devils River, Comanche Peak, Del Carmen, Edwards, and Goodland Limestones, Segovia and Kiamichi Formations, and Walnut Clay
Ktr	Trinity Group; includes Antlers, Presidio, Shafter, Travis Peak, Twin Mountains and Glen Rose Formations, Cow Creek Limestone, and Hensell, Hosston, and Paluxy Sands
Kbi	Bissett Conglomerate of the Coahuila Group

The geologic map of Texas. Within each legend description, the rocks listed as "included" are those mentioned in the text. Additional rocks, unnamed in the legend, may also be part of the unit. Question marks mean the geologic age is uncertain. —Map constructed from the Geologic Database of Texas

MESOZOIC (continued)

JURASSIC

Jcv/Jm Cotton Valley Formation (in subsurface only) and Malone Formation

Jl Louann Salt (in subsurface only)

TRIASSIC

ℝ Dockum Group; includes Chinle, Trujillo, and Tecovas Formations

PALEOZOIC

PERMIAN

P sedimentary rocks, undivided; includes Cutoff Shale

Pqr Quartermaster Formation; includes Alibates Dolomite Member of the Quartermaster Formation, Cloud Chief Formation, and Whitehorse Sandstone

Pg Guadalupe Group; includes Capitan and Goat Seep Limestones and Mina Grande Formation

Po Ochoa Group; includes Castile Formation

Ppr Pease River Group; includes Blaine and San Angelo Formations

Pcf Clear Fork Group

Pp Patrolia Formation

Pwa Wichita Albany Group; includes Lueders, Elm Creek, Admiral, Grape Creek, and Talpa Formations

Pwh Whitehorse Group; includes Tansill, Yates, and Seven Rivers Formations

Pdm Delware Mountain Group; includes Brushy Canyon, Bell Canyon, and Cherry Canyon Formations, and Carlsbad Group

Pbs Bone Spring Group; includes Victorio Peak Limestone and Bone Spring Formation

Ph Hueco Limestone

Pz sedimentary rocks, undivided; includes Magdalena Formation at south end of Franklin Mountains (Paleozoic)

PENNSYLVANIAN–EARLY PERMIAN

PIPcb Cisco-Bowie Group

PENNSYLVANIAN

IP sedimentary rocks, undifferentiated

IPc Canyon Group

IPs Strawn Group; includes Haymond Formation, Mineral Wells Formation, and Mingus Shale

IPb Bend Group; includes Smithwick Shale, Marble Falls Limestone, and Dimple Limestone

MISSISSIPPIAN–PENNSYLVANIAN

IPMt Tesnus Formation of the Morrow Group

SILURIAN–MISSISSIPPIAN

MD sedimentary rocks, undifferentiated (Devonian–Mississippian); includes Barnett and Helms Shales (Mississippian) and Woodford Shale (in subsurface only) (Devonian–Mississippian)

Sf Fusselman Dolomite (Silurian)

MDO Caballos Novaculite (Mississippian); includes Maravillas Chert (Ordovician)

PALEOZOIC (continued)

ORDOVICIAN

O sedimentary rocks, undivided; includes Montoya Dolomite

Oel Ellenburger Group; includes Honeycut Formation, Gorman Limestone, and Tanyard Formation

Oe El Paso Group

Om Marathon Limestone

CAMBRIAN–ORDOVICIAN

COb Bliss Sandstone (late Cambrian–Ordovician)

CAMBRIAN

Moore Hollow Group

€mw Wilberns Formation; includes San Saba Dolomite, Point Peak Siltstone, Morgan Creek Limestone, and Welge Sandstone

€mr Riley Formation; includes Lion Mountain Sandstone, Cap Mountain Limestone, and Hickory Sandstone

€vh Van Horn Sandstone

PROTEROZOIC

p€mm metamorphic rocks; includes Packsaddle Schist, Coal Creek Serpentinite, and Valley Spring and Big Branch Gneisses

p€ms sedimentary and metasedimentary rocks; includes quartzite, Carrizo Mountain Group, Mundy Breccia, Castner Marble, and Hazel and Allamoore Formations

---

VOLCANIC and IGNEOUS ROCKS

Tv volcanic rocks, undifferentiated (Neogene and Paleogene)

Tr Las Burras Basalt Member of the Rawls Formation (Oligocene–Miocene)

Tgb Garren Group; includes Bell Valley Andesite of the Jones Formation (Oligocene)

Tfr Fresno Formation (Oligocene)

Tmr Morita Ranch Formation (Oligocene)

Tpc Perdiz Conglomerate (Oligocene)

Tdm Davis Mountains volcanic complex; includes Barrel Springs Rhyolite, Sheep Pasture, Mount Locke, Adobe Canyon, Frazier Canyon, and Sleeping Lion Formations, and Gomez Tuff (Eocene–Oligocene)

Tbh Buck Hill Group; includes Duff (includes Decie Member), Pruett (includes Crossen Trachyte), and Tascotal Formations, and Mitchell Mesa Rhyolite (Eocene–Oligocene)

Tbb Big Bend Park Group (Eocene–Oligocene); includes Canoe and South Rim Formations, Burro Mesa Rhyolite, and Chisos Group (includes Mule Ear Spring Tuff Member)

Ti igneous rocks; includes stocks, vent rocks, laccoliths, sills, and dikes; includes Government Spring laccolith (Neogene and Paleogene)

Ki igneous rocks, undivided (Cretaceous)

p€g granitic rocks; includes Red Bluff, Town Mountain, and Oatman Creek Granites

p€m metavolcanic and meta-igneous rocks; includes Hackett Peak Formation, mafic igneous, metarhyolite, and Ilanite dike

EON	ERA	PERIOD	EPOCH	AGE (mya)*	MAJOR GEOLOGIC EVENTS IN TEXAS
PHANEROZOIC	CENOZOIC	QUATERNARY	HOLOCENE		Galveston and Padre Islands begin forming as sea level rises at end of ice ages
				0.01	rivers incise channels during lowered sea levels of Pleistocene ice ages
			PLEISTOCENE		
				2.58	Basin and Range extension from 25 to 2 million years ago
		NEOGENE	PLIOCENE		
				5.33	modern Rockies are uplifted; eroded debris spreads east, becoming Ogallala Formation from 18 to 5 million years ago
			MIOCENE		Balcones fault zone active from 25 to 8 million years ago
				23.03	
		PALEOGENE	OLIGOCENE		caldera volcanism in West Texas; Trans-Pecos volcanic field from 48 to 27 million years ago
				33.9	
			EOCENE		
				56	Laramide mountain building from 80 to 40 million years ago folds rocks in West Texas
			PALEOCENE		
				66	extinction of dinosaurs
	MESOZOIC	CRETACEOUS			two magma pulses at 80 and 72 million years ago
					Western Interior Seaway floods continent about 95 to 79 million years ago; limestones deposited
					rivers deposit sediments to fill deepening Gulf of Mexico
				145	
		JURASSIC			deposition of Louann Salt from 166 to 162 million years ago
				201	Pangea begins to break up; Gulf of Mexico begins to open
		TRIASSIC			red beds are deposited on land
					dinosaurs appear
				252	
	PALEOZOIC	PERMIAN			climate dries, and evaporites are deposited as shallow seas dry
				299	basins form west and north of Ouachita Mountains; Capitan Reef forms around edge of Delaware Basin
		PENNSYLVANIAN			Ouachita Mountains form when North America collides with South America during assembly of Pangea 323 to 290 million years ago
				323	shallow sea covers much of continent
		MISSISSIPPIAN			
				359	episodes of sea level rise and fall
		DEVONIAN		419	erosion over Texas Arch from 400 to 380 million years ago
		SILURIAN		444	Great American carbonate bank includes El Paso Group and Ellenburger Group; invertebrate organisms and fish abundant in warm, shallow seas
		ORDOVICIAN		485	
		CAMBRIAN			
				541	sandstones deposited at edge of sea
PRECAMBRIAN	PROTEROZOIC				rifting of Rodinia about 700 to 520 million years ago and long period of erosion at end of Proterozoic time
					Llano mountain building 1,150 to 1,070 million years ago as supercontinent Rodinia forms; granites intrude
					Llano sedimentary and volcanic rocks deposited 1,300 to 1,200 million years ago
					oldest rock in Texas is 1,382-million-year-old Carrizo Mountain Group; part of the Southern Granite–Rhyolite province
				2,500	
	ARCHEAN				
				4,000	
	HADEAN				approximate formation of the Earth, 4,540 mya

*mya = millions of years ago

Major events in the geologic history of Texas.

THE BIG PICTURE

The geologic panorama of Texas is as wide as the big state itself, sweeping from volcanic mesas and thrusted mountains in the west across central limestone plateaus and hard granitic terrain to tropical sand barriers of the Gulf Coast. Rocks of all ages and types, from ancient Precambrian gneiss to weakly consolidated, young sandstone are found at the surface in the state, covered in places by loose sediments. Moreover, Texas is blessed with an array of natural geologic resources, including oil and gas fields, salt, sulfur, lignite, and building stone. For Texans who live in large cities or for the out-of-staters who think the stereotypical Texas is endless, featureless, windswept prairies covered by cattle and oil derricks, it may come as a surprise to learn of deep canyons, extensive caves, dinosaur footprints, coral reefs, and exfoliating granite knobs.

This diverse geology has been shaped over millions of years of Earth's history. When the planet formed 4.54 billion years ago, it was an unpleasant and uninhabitable place. Small blocks of solid rock floated on a magma ocean. Meteorites pummeled the surface. The atmosphere held no oxygen but was a mix of water vapor, carbon dioxide, nitrogen, ammonia, and traces of methane. This time period, from Earth's formation to 4 billion years ago, is aptly named the Hadean Eon. By the end of this eon, Earth had cooled enough for a solid crust to form, but no rocks from this time are found in Texas, and they are extremely rare worldwide.

During the Archean Eon, the next subdivision of time, plate tectonics began. Tectonic plates are pieces of the lithosphere, Earth's outermost shell of rigid rock. Lithosphere comes in two varieties: the thin, dense variety beneath the oceans and the thick, buoyant rock of the continents. Where oceanic and continental lithosphere push against each other, the denser oceanic lithosphere sinks beneath the continent, a process called subduction. Subducting plates often carry small pieces of lithosphere that cannot subduct and are accreted, or added, to the edge of the continent. The early continent of North America, known as Laurentia, grew as islands and other pieces of less-dense lithosphere were added to it.

As plates move, sometimes two continental plates collide, an event that forms tall mountains like the Himalayas. Continents can also split apart along a rift. When this begins to occur, a rift valley is created. If the pull-apart movement continues, the valley will transform into a new ocean basin where new oceanic crust is created. Although the movement of tectonic plates is slow, it can cause profound changes as the plates interact with each other.

No rocks from the Archean Eon are found in Texas, but rocks from the next eon, the Proterozoic, are exposed in the Central Texas Uplift, also called the Llano Uplift. The old rocks appear as a circular region in the middle of the geologic map of the state. Proterozoic rocks also appear in the Van Horn area and in the Franklin Mountains near El Paso. Other patterns on the geologic map reveal important chapters in the state's geologic history.

On the geologic map of Texas, notice the curving S-shaped line through the center of the state, running from Dallas–Ft. Worth, south past Austin and San Antonio, and westward toward the Big Bend of the Rio Grande. Note how rocks east of the line are

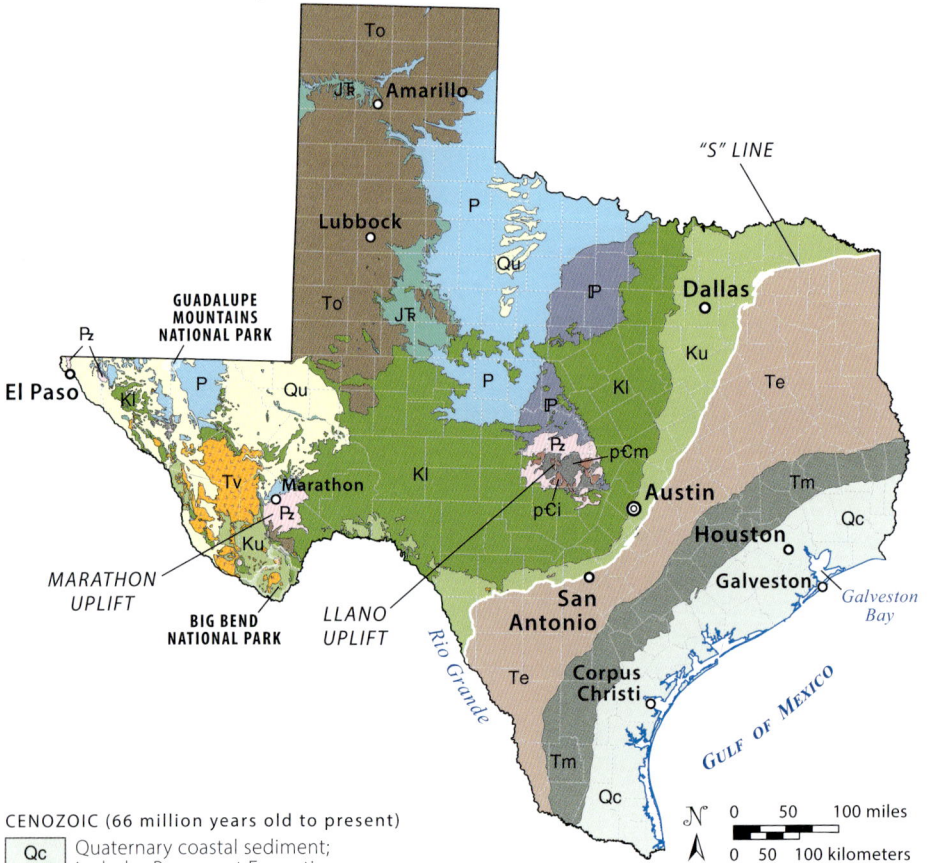

CENOZOIC (66 million years old to present)

Qc	Quaternary coastal sediment; includes Beaumont Formation
Qu	Quaternary sediment, undivided
Tv	Oligocene and Eocene volcanic rocks and conglomerates in West Texas
To	Pliocene and Miocene rocks; includes Blackwater Draw (Quaternary) and Ogallala Formations (Miocene to early Pliocene)
Tm	Pliocene, Miocene, Oligocene coastal rocks; includes Goliad, Fleming, Catahoula, and Frio Formations
Te	Eocene and Paleocene rocks; includes Jackson, Claiborne, Wilcox, and Midway Groups

MESOZOIC (252–66 million years old)

Ku	Cretaceous rocks, undivided; includes Navarro, Taylor, Austin, Eagle Ford, Woodbine, and Terlingua Groups
Kl	Cretaceous rocks, undivided; includes Washita, Fredericksburg, and Trinity Groups
JꝨ	Jurassic–Triassic rocks, undivided; includes Dockum Group

"S" line shown in white

national park boundaries shown in white

county boundaries

PALEOZOIC (541–252 million years old)

P	Permian rocks, undivided
℗	Pennsylvanian rocks, undivided
Pz	older Paleozoic rocks (Cambrian, Ordovician, Silurian, Devonian, and Mississippian rocks)

PRECAMBRIAN (2,000–541 million years old)

pЄm	metamorphic rocks
pЄi	igneous rocks

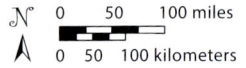

General geologic map of Texas with units organized by age.

all younger than Cretaceous in age and get progressively younger toward the Gulf of Mexico. The S line marks the edge of the North American continent in Cretaceous time when the Gulf of Mexico was forming. This curved line also follows quite closely to a tightly folded mountain belt, which is buried across much of Texas but extends in the subsurface from the Ouachita Mountains of southern Oklahoma to southwest Texas, where the range reappears in the Marathon Uplift around the town of Marathon, north of Big Bend National Park. All the bands to the east of the S line are younger sediment layers deposited into the Gulf of Mexico over the past 66 million years or so. The Paleozoic rocks west of the S line were deposited in basins west of the ancient mountains. Cenozoic rocks in the High Plains of the Panhandle eroded from the rising Rocky Mountains to the west and spread east over the basin rocks.

The Caballos Novaculite, a chert of Paleozoic age exposed near Marathon, shows the contorted nature of rocks associated with the Ouachita mountain building. (30.1588, -103.2367)

GEOLOGIC TIME

There is nothing magic or overly complicated about geologic time. Geologists have simply divided up the geologic history of the Earth into convenient packages and called it the geologic time scale. The more-or-less official geologic time scale, used throughout the world, has been around for more than one hundred years and was developed quite some time before anybody had any idea of how long, in years, the Earth has been around.

Thus, the original geologic time scale was based on relative time. For example, geologists originally recognized a group of rocks with a distinctive set of fossils that occurred in England, and then they saw similar rocks with similar fossils in Russia and named the collective assemblage the Permian Period. In other words, they interpreted that the two sets of similar rocks, though quite far apart, were deposited at the same time in the geologic past. But then these geologists, or maybe other geologists, recognized this Permian package of rocks always lay under another set of distinctive rocks that had a different fossil assemblage. They named this second group of rocks Triassic, and so the time scale developed. A stack of rocks representing the very oldest to the very youngest was recognized, named, and ordered, and by about 1900, geologists everywhere agreed on this relative age scale.

But nobody knew how old any of these rocks were. Geologists blithely talked about Permian rocks and could say they were older than the Triassic ones, and younger than the Devonian ones, but nobody knew whether these Permian rocks were formulated 50 years ago, 50,000 years ago, 50 million years ago, or 500 million years ago. For the most part, the nineteenth-century geologic fraternity thought, through rather inexact measuring techniques, that the age of the Earth was probably a few million years. Early in the twentieth century, uranium clocks, based on the constancy of radioactive decay, were discovered in minerals. It is difficult for us today to understand the immense psychic impact on the human mind that must have occurred when the first age dates produced several hundred million years for Paleozoic rocks and a few billion-year dates for the really old Precambrian rocks. Since then, radiometric age dates have been combined with the relative geologic time scale. Geologists now know that Permian rocks were deposited between 299 and 252 million years ago. Efforts over the last one hundred years continue to refine and upgrade the radiometric age dates.

PRECAMBRIAN ERA
4,500 to 541 million years ago

Several rocks in Texas are vying to be the oldest. Dates of 1,360 to 1,250 million years have been obtained from the Valley Spring Gneiss in the Llano Uplift, putting these old rocks in the middle of the Proterozoic Eon. The Hackett Peak Formation of the Carrizo Mountain Group, a stack of metamorphic rocks of the Southern Granite–Rhyolite Province of West Texas, has yielded a 1,382-million-year date and a 1,333-million-year date. This ancient province, part of the North American craton, is mostly in the subsurface beneath parts of Texas, Oklahoma, and points farther east. A large, unexposed body of igneous rock of this province is thought to produce a gravity anomaly, the Abilene gravity minimum, in the subsurface from Van Horn to Ft. Worth.

The oldest publicly accessible rock in Texas can be found at Inks Lake State Park in the Llano Uplift. Here, the Valley Spring Gneiss is around 1.3 billion years old. (30.7479, -98.3592)

About 1,300 to 1,200 million years ago in the Proterozoic Eon, thick sequences of sediment were dumped into an ancient sea bordering Laurentia, the ancient core of North America. Mostly coarse-grained sediments were deposited first, followed by finer-grained sediment. Volcanic materials also accumulated in the sediment pile. Most of the radiometric dates for this rock sequence have been obtained from metamorphosed basalts, metamorphosed rhyolites, and volcanic ashes, indicating active volcanoes existed along the continental margin.

Later, Laurentia collided with either another continent or an ocean margin in a plate tectonic event known as the Llano mountain building that buried, squeezed, and heated the borderlands, including the sediment piles. The Llano deformation, from 1,150 to 1,070 million years ago, was part of the larger Grenville mountain building that occurred during the formation of the supercontinent Rodinia. The collision built mountains, created the metamorphic Packsaddle Schist and Valley Spring Gneiss out of the deeply buried sediments, and generated magmas that cooled to form granite stocks, batholiths, and dikes. The compression of the collision shoved rocks to the north over other rocks along thrust faults. Areas of Texas west of the curving S on the geologic map were part of the Laurentian continent by 1 billion years ago.

In the Franklin Mountains near El Paso, stromatolites (mounded forms created by cyanobacteria) are present in the Castner Marble, which was originally limestone deposited in shallow seas covering this ancient continent. These early life forms tell us something about the environment in which they formed but not much else. Stromatolites existed for millions of years before organisms with shells evolved, and they still exist today, most notable at Shark Bay, Australia.

All these Proterozoic rocks are displayed in the Llano country of central Texas. Proterozoic rocks are also found in the Franklin Mountains near El Paso, and in West Texas mountain ranges near Van Horn. Proterozoic rocks have been reached by oil drilling over much of the western half of Texas, though little is known about ancient rocks beneath the Gulf Coastal Plain because they are buried so deeply there.

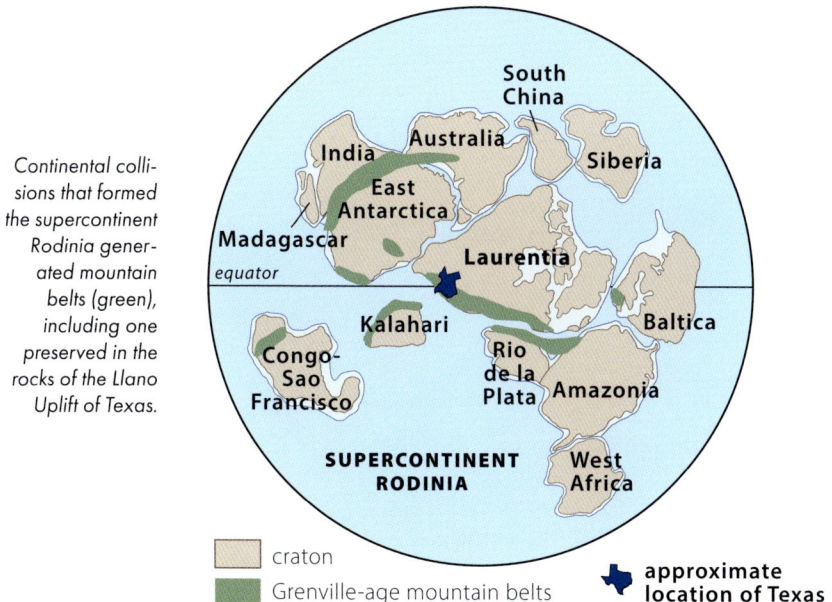

Continental collisions that formed the supercontinent Rodinia generated mountain belts (green), including one preserved in the rocks of the Llano Uplift of Texas.

South China

India   Australia   Siberia

East Antarctica

Madagascar

equator   Laurentia

Kalahari   Baltica

Congo-Sao Francisco   Rio de la Plata   Amazonia

SUPERCONTINENT RODINIA   West Africa

craton

Grenville-age mountain belts

approximate location of Texas

We know Rodinia began rifting apart about 700 million years ago, and the southern edge of North America had rifted from South America by Cambrian time, the first period of the Paleozoic Era. There's not much evidence for the rifting in Texas, at the surface anyway, but rift-related rocks are exposed in Oklahoma.

Erosion flattened the Proterozoic mountains to a tabletop by Cambrian time. The contact between the long-eroded metamorphic rocks and granite and the unmetamorphosed sandstones from the end of Proterozoic time and into Cambrian time is known as the Great Unconformity. It is present throughout western North America and can be found in a few places in Texas, although most are on private property and thus not accessible. For more on the rocks of the Llano region, see the Llano Uplift section in the Central Plains and Hill Country chapter.

PALEOZOIC ERA
541 to 252 million years ago

In Cambrian time, shallow marine seas lapped against the low, erosion-worn central core of the North American continent. Streams carrying sediment eroded from the low continental terrain to the north and west deposited sandy sediments at the margin of these seas. Cambrian sandstones, which originated as beach sand and near-shore sand, are found around the Llano Uplift in central Texas and in the Marathon region of West Texas.

Farther from shore in clear marine water, dolomite and limestone accumulated from the shells of Cambrian organisms. Outcrops of Cambrian limestones are also found in the Llano Uplift. Cambrian time is noteworthy because it represents the appearance, rather suddenly in the geologic record, of abundant fossils. Trilobites,

This close-up of the Hickory Sandstone, photographed along TX 16 south of Cherokee, shows the sand to granule-sized nature of the sediments that were washed into the shallow Cambrian seas by continental rivers from the north and west. The large, gray clast at the bottom of the specimen measures 1 inch across. (30.9340, -98.6910)

brachiopods, sponges, snails, clams, and bryozoans were all present by Cambrian time, whereas rocks from the end of the Precambrian Era display only rare fossils of algae and a few soft-bodied marine animals.

Extensive dolomite and limestone deposits, with less extensive chert deposits, were laid down in shallow seas that spread north and covered Texas in Ordovician time, a rise in sea level known as the Sauk transgression. The seas were very shallow, less than a few hundred feet deep, because they were covering part of a continent. The land remained mostly underwater even though sedimentation into the sea was continuous because the seafloor subsided under the weight of the sediments. Evidence of Ordovician limestones overlying Cambrian sandstones are seen in the Franklin Mountains, where the Ordovician carbonates overlie the Bliss Sandstone, and in the Beach Mountains near Van Horn, where the El Paso Group overlies the Van Horn Sandstone.

The Ordovician carbonates of Texas are great sources and reservoirs of oil and gas. The Ellenburger Group, a major remnant of a large carbonate bank from Early Ordovician time, is especially noted for its gas production from great depths in basins of West Texas. Ordovician limestones yield brachiopods, corals, cephalopods, and gastropods. The first primitive land plants appeared in this period, colonizing areas above sea level.

Limestone, dolomite, and chert were laid down in shallow marine waters in West Texas during the Silurian Period. Silurian limestones in the Franklin Mountains north of El Paso bear rare brachiopods and corals. Corals, first seen in Late Ordovician time, exploded in numbers and types during the Silurian.

Shale, sandstone, limestone, and chert of shallow marine origin from Devonian time are well known in the subsurface in Texas and appear in outcrops in the Llano and Marathon Uplifts. A major unconformity in the Texas geologic record occurs from 400 to 380 million years ago in the middle of the Devonian Period. An unconformity is where rocks are missing, either because they were never deposited or because they were eroded. In this case, part of Texas was uplifted above sea level in the Middle Devonian, subjecting it to erosion. Deposited above the unconformity is the Woodford Shale, an important source of hydrocarbons in West Texas. Amphibians first appear in the Devonian, and this was the golden age for the development of fishes. On land, ferns, seedferns, and huge trees related to present-day horsetail rushes appeared in the Devonian Period but exploded in numbers during the Mississippian Period, when shallow marine seas still covered Texas.

In the Mississippian seas, brachiopods, bryozoans, trilobites, and corals were common. Mississippian shales and limestones are found around the Llano Uplift, and folded, upended Mississippian rocks occur in roadcuts east of Marathon. In Late Mississippian time, the basins deepened, and deposition resulted in the Barnett Shale. This organic-rich unit is a source of natural gas exploited with fracking technology in the Ft. Worth Basin.

THE RISE AND FALL OF THE OUACHITA MOUNTAINS

About 323 to 290 million years ago, the North American continent, already joined to Eurasia, collided with South America and Africa to become part of the supercontinent Pangea. The Appalachian Mountains formed along the east coast, the Arbuckles formed in Oklahoma, and the Ouachita Mountains formed from Texas to Oklahoma. The Ouachita Mountains and their western end, the Marathon Uplift, followed the zone of weakness in the Earth's crust that was established during the Llano collision of Proterozoic time and the later rifting at the end of Proterozoic time. In Colorado, large blocks of crust were uplifted as the Ancestral Rocky Mountains, and between them lay deep basins. Uplifted blocks extended south into Texas but are deeply buried today. These uplifts, however, affected the geometry of the basins in Texas west of the Ouachita Mountains, creating subsurface arches between basins.

Although most of the rocks involved in the Ouachita mountain building are hidden in Texas, buried by younger rocks, the rocks exposed near Marathon display the folds, reverse faults, and thrust faults one would expect from the collision of tectonic plates. Turbidites of the Haymond and Tesnus Formations also provide evidence of the collision. These layered sandstones and mudstones were deposited from turbidity

Turbidites of the Tesnus Formation, exposed along US 90 east of Marathon, formed as sediments raced down the steep continental slope prior to the collision of plates that produced the Ouachita mountain building. (30.2045, -102.9672)

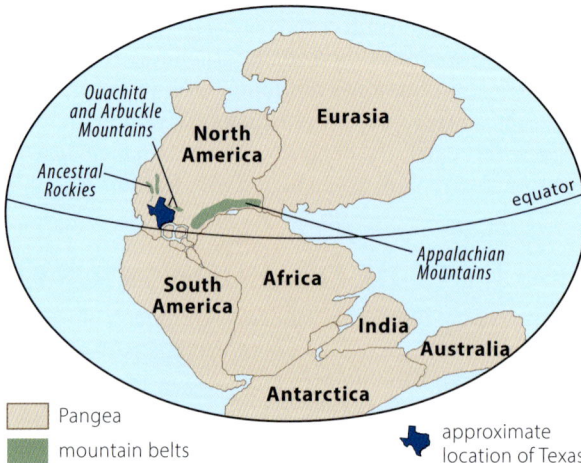

Mountain belts formed during the assembly of the supercontinent Pangea.

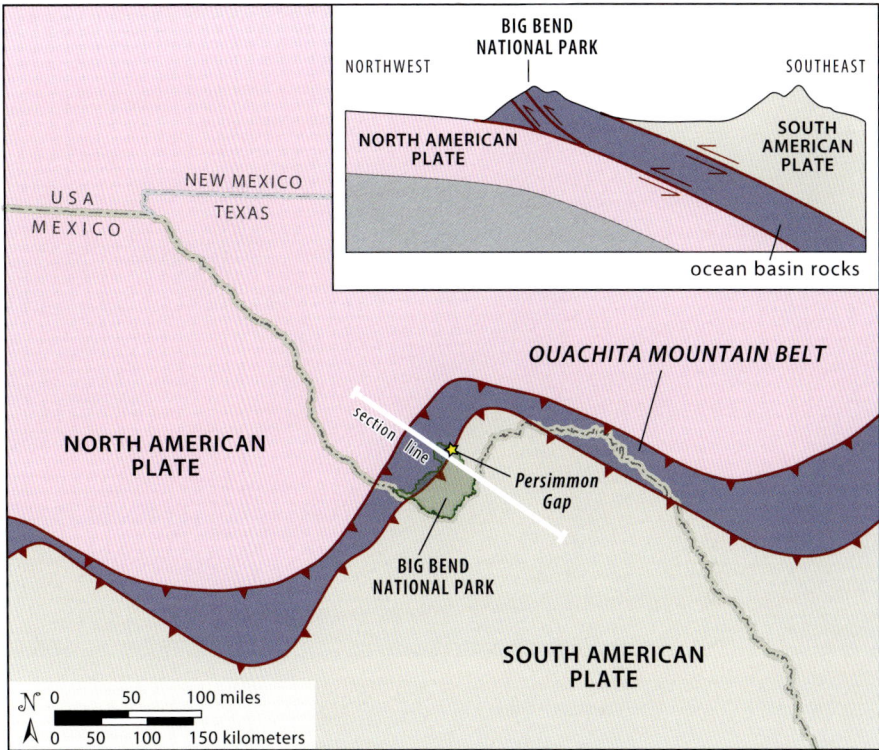

In Pennsylvanian time, South America collided with the North American coast at Texas and formed the Ouachita Mountains. —Modified from Gray and Page, 2008

currents that flowed down the steep, underwater slopes of the approaching thrust belt in Pennsylvanian time, prior to the final collision.

To the west of the growing mountain belt, the crust sagged in response, and several downwarps, or basins, formed and deepened. The basins continued to settle over several millions of years, receiving organic-rich deposits of calcareous mud from marine organisms living in shallow seas. Reefs and banks of limestones formed around the edges of the basins from the growth of abundant reef-building animals. The organic-rich mudstones, and even the basin-edge limestones, became the source for much of the oil in West Texas, while the cavernous limestone reefs and banks became the reservoirs to store the oil. The limestones are best seen in the Guadalupe and Delaware Mountains of West Texas. The Ft. Worth Basin received sediment from both the Ouachita Mountains to the east and the Arbuckle Mountains to the north.

Extensive forests of conifers, ferns, seed-ferns, and horsetail trees in the Pennsylvanian Period gave rise to coal deposits, and reptiles first roamed these forests at the end of the period. The Mississippian and Pennsylvanian Periods are together called the Carboniferous Period across much of the world because when the abundant organic matter was eventually buried and compressed, the heat and pressure turned it into coal.

Colorful Permian-aged red beds of sandstone, reminiscent of what might be seen in Utah, are well displayed in Caprock Canyons State Park. (34.4202, -101.0600)

Erosion wore the mountains down during Permian time. Being landlocked in the center of the supercontinent Pangea, the climate was dry. While marine reefs and banks prevailed in shallow marine waters of the inland basin of West Texas, nearshore evaporation flats and isolated pockets of seawater in the Panhandle area produced deposits of bright-red shales and salt and gypsum deposits. Nonetheless, a thick accumulation of marine sediment accumulated in what is known as the Permian Basin. Colorful red beds at the edge of the High Plains (Caprock Canyons State Park, Palo Duro Canyon) are Permian, as are the magnificent reef limestones on El Capitan in Guadalupe Mountains National Park, as well as oil-rich limestones in the subsurface of West Texas. For more on the Permian Basin, see the introduction to the West Texas chapter.

Vertebrates flourished during the Permian Period, including the fin-backed predator *Dimetrodon*. Thecodonts (the precursors to dinosaurs and crocodiles) and therapsids (stem mammals) had their beginnings in the Permian Period.

The boundary between the Permian and Triassic Periods, 251.9 million years ago, is defined by the greatest mass extinction to ever occur on Earth. Known as the Great Dying, it is thought to have been caused by carbon dioxide released by massive volcanic eruptions in Siberia. Trilobites and horn corals went extinct, along with thousands of other marine organisms, including many that built reefs. The Capitan Reef in Guadalupe Mountains National Park serves as a catalog of all the life forms that existed in Permian time prior to the extinction event. A majority of those life forms went extinct in the Great Dying.

MESOZOIC ERA
252 to 66 million years ago

The Mesozoic Era is divided into three periods: Triassic, Jurassic, and Cretaceous. Rumbles of change in the configuration of continents were seen in Texas as the supercontinent Pangea began to split apart in the Middle Triassic, beginning about 240 million years ago. As the Gulf of Mexico began to shudder open, red shale, siltstone, and sandstone were the first deposits to be shed into the downwarping on the southeast and east side of the mostly eroded Ouachita Mountains. Colorful shales and sandstones were still being deposited in the west, although in more restricted areas of the Panhandle. The Dockum Group, deposited in a nonmarine basin west of the Ouachita highlands in the Late Triassic, is well exposed at Palo Duro Canyon and Caprock Canyons State Parks and in the breaks along the Canadian River north of Amarillo. A wide patch of Triassic rocks also occurs east of Big Spring. On land, ferns began anew, and cycads appeared, while dinosaurs, pterosaurs, crocodiles, ichthyosaurs, turtles, and mammals arrived on the scene. In the sea, modern scleractinian corals developed, and there was a reawakening of bryozoans.

South of Post along US 84 are wonderful roadcuts of the Triassic-aged Dockum Group sediments. (33.1583, -101.3456)

During Early and Middle Jurassic time, Pangea broke up in earnest, and the Gulf of Mexico occupied the new gap between North and South America. At the beginning, the Gulf was a shallow sea, not well connected to the other oceans. It dried up often, creating vast salt pans about 166 to 162 million years ago that became the Louann Salt, mother lode of salt domes in the Gulf Coast. Limestones of the subsurface Smackover Formation were deposited along a carbonate shelf as the Gulf of Mexico deepened to the south about 160 million years ago.

In Late Jurassic time, rivers built deltas into the Gulf of Mexico, depositing the Cotton Valley Formation. Jurassic rocks are almost completely absent at the surface in Texas, though Jurassic limestone, sandstone, and shale beds can be seen along I-10 west of Sierra Blanca in westernmost Texas. By the end of the Jurassic Period, dinosaurs had diversified to become the dominant animals on Earth. During the same time, the earliest birds were also beginning to evolve alongside the dinosaurs.

At the beginning of Cretaceous time, the western edge of the North American plate was converging against other plates, which had all sorts of effects on the interior of the continent. The Rocky Mountains began developing about 140 million years ago and underwent major growth during the Laramide mountain building, which lasted from about 80 to 40 million years ago. The mountains in West Texas show folding and faulting associated with this mountain building. See the chapter about West Texas for more on the Laramide folding.

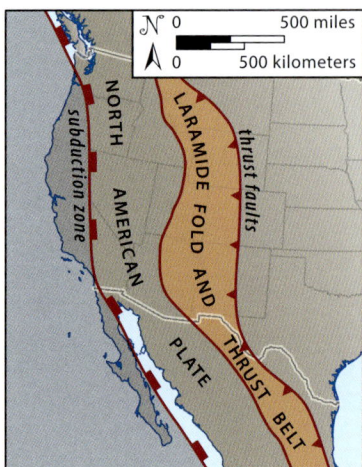

Tectonics in western North America during the Cretaceous.
—Modified from Gray and Page, 2008

Early Cretaceous rocks blanketed the center half of Texas and are exposed in limestone cliffs, caverns, and canyons. Abundant fossils and dinosaur tracks speak to the hospitality of the Cretaceous environment. A carbonate bank formed where shallow seawater of the expanding Gulf of Mexico spread over a low area on the continent known as the Comanche Shelf. The Ouachita highlands were mostly eroded at that time, so there was less sediment being shed into the Gulf of Mexico. The remains of the carbonate shelf are preserved as the 110-million-year-old Glen Rose Formation of the Trinity Group, the 105-million-year-old Edwards Limestone of the Fredericksburg Group, and the 104-million-year-old Segovia Formation following that. The Edwards Limestone, a porous, karstic rock, is a major water-bearing unit for Austin and San Antonio. The Edwards Plateau, where the limestone is at the surface today, is the recharge area for the aquifer.

Shallow seas advanced and retreated repeatedly over the continental margin. A trough developed to the east of the growing Rocky Mountains in Late Cretaceous time and filled with seawater. Known as the Western Interior Seaway, this body of water extended all the way from the Gulf of Mexico to the Arctic Ocean at times and existed over parts of Texas from about 95 to 79 million years ago. Units deposited on the sea bottom during this time include the Woodbine Formation, Eagle Ford Shale, and Austin Chalk.

The Edwards Limestone, seen here along I-10 east of Ozona, was a major component of the carbonate shelf that developed during Cretaceous time. (30.7108, -101.1683)

The Western Interior Seaway connected the Gulf of Mexico with the Arctic Ocean at various times during the Late Cretaceous.

The shallow seas were filled with calcareous-shelled organisms whose remains settled to become thick deposits of limestone. Dinosaurs roamed the sandy shorelines and mudflats, leaving evidence of their passing in fossilized footprints and trackways. Late Cretaceous rocks are found in a band from the Red River southward through Dallas–Ft. Worth to Austin and San Antonio, and westward to Del Rio and in Big Bend National Park. The Late Cretaceous rocks show a gradual shallowing of the seas and eventual shift from marine to river and floodplain deposition, represented in the Navarro and Taylor Groups.

In two pulses, about 80 and 72 million years ago, a swarm of magma from deep within the Earth intruded along a belt that trends eastward from Del Rio nearly to Waco. Most of these were submarine volcanic centers where eruptions created explosion craters on the Cretaceous seafloor, sending plumes of ash, rock, and steam upward. The lava came up along faults that followed the ancient line of the Ouachita Mountains and are in turn followed by the younger Miocene-aged fault zone of the Balcones Escarpment.

A major mass extinction marks the end of the Cretaceous. Dinosaurs, pterosaurs, mosasaurs, ammonites, and many marine plankton species all died out. For many years, the cause of this extinction remained a mystery debated by geologists. In 1980, Luis and Walter Alvarez proposed that a large meteorite impact had caused the extinction based on a widespread iridium layer at the Cretaceous-Paleogene boundary. While rare on Earth, iridium is much more abundant in space rocks. In the early 1990s, the discovery of a large impact crater in the Yucatán Peninsula confirmed that a 6-mile-wide meteorite struck the Earth at the end of the Cretaceous Period and played a significant role in the extinction event. Major volcanic eruptions of the Deccan Traps in India might also have played a role in the extinction. Scientists are still studying the details of these two events and their dramatic effect on life.

Geologists have scoured Texas looking for evidence of the meteorite impact in rocks of the right age. In southern Falls County, along the Brazos River southeast of Waco and east of Temple, are eroded riverbanks in rock that spans the Cretaceous–Paleogene boundary, also known as the K–Pg boundary. Here, the mudstones of the Corsicana Formation (Navarro Group) show erosion by the tsunami that immediately followed the meteor impact, as well as an iridium anomaly and nickel-rich spinel minerals.

The meteorite impact on the Yucatán Peninsula that formed the Chicxulub Crater about 66 million years ago had a profound effect on Texas and Earth in general.

This fossil ammonite, a cephalopod with a coiled or spiral shell, was preserved in the Comanche Peak Limestone. It went extinct at the end of the Cretaceous Period. The field of view is approximately 12 inches wide.

CENOZOIC ERA
66 million years ago to the present

The rocks east and south of the buried Ouachita Mountains represent the continuous sedimentary filling of the Gulf of Mexico since 66 million years ago. Rivers carried engorged sediment loads eroded from the high terrain of the newly uplifted Rocky Mountains and deposited layer upon layer of gravel, sand, silt, and clay into the Gulf. This wedge of sediment, estimated to be up to 40,000 feet thick, grew gulfward throughout the Cenozoic Era, creating the age bands you can see on the geologic map. Sediment continues to be added to the Gulf by the Colorado, Brazos, Sabine, Pecos, Rio Grande, and other Texas rivers. These inclined wedges of sediment are arranged like a tipped stack of books. The upper and lower Gulf Coast, coastal plain, and modern beaches, barrier islands, lagoons, and river deltas are all the product of this immense sediment-wedge-building process.

Volcanoes spurted lava and ash in West Texas 48 to 27 million years ago during the Eocene and Oligocene Epochs. These volcanic rocks form the Davis Mountains, Chisos Mountains, and many other peaks and ranges in the Big Bend country. As the volcanism progressed, caldera eruptions became more common. Similar eruptions occurred throughout the west from Montana south to Arizona. Many of the calderas are known only from the voluminous tuffs, rocks composed of ash and volcanic debris, that piled up around the eruptions. For more on these eruptions, see the section on Big Bend National Park in the West Texas chapter.

Volcanism continued until about 18 million years ago in Miocene time, but these later eruptions were related to the stretching of the crust that began in Texas about 25 million years ago. The crustal stretching created the series of fault-block mountains and linear valleys that form the topography in West Texas today. As the land pulled apart, valleys dropped down along normal faults, and ranges were uplifted across the vast area of the west called the Basin and Range Province. It extends from southern Oregon and Idaho south through Nevada and Utah to Arizona, New Mexico, and Mexico. From southern Colorado through New Mexico, the Rio Grande follows a

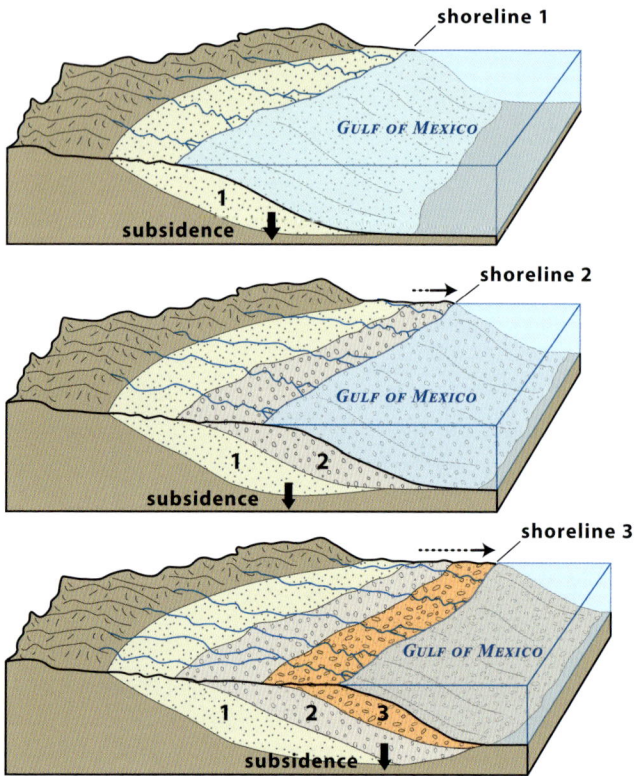

As rivers deposited sediments into the Gulf of Mexico, their weight downwarped the coastal area, providing more space for additional sediments to pile up. The shoreline moved progressively seaward.

Massive volcanic eruptions in the Eocene and Oligocene formed many of the peaks in West Texas, including the rugged Chisos Mountains in Big Bend National Park. (29.3322, -103.2415)

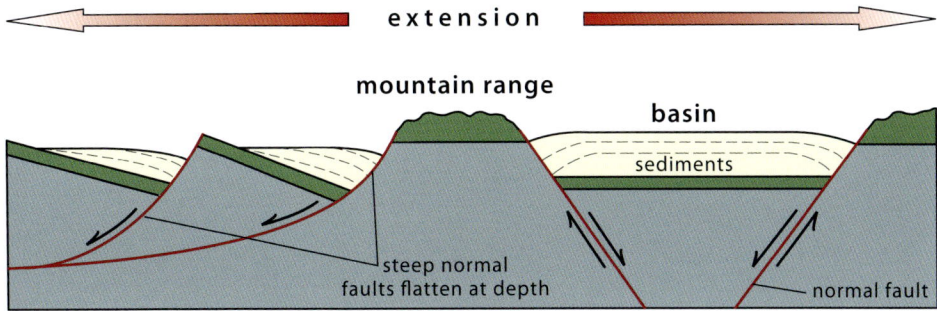

The stretching of the crust in West Texas uplifted mountain ranges and down-dropped valleys along normal faults.

rift valley that also formed during this extension. West Texas did not escape the pull. The Franklin, Hueco, and Guadalupe Mountains are uplifted blocks, and the Mesilla Bolson (which hosts the Rio Grande at El Paso), Hueco Bolson, and Salt Basin are linear, down-dropped basins.

At the same time that the Basin and Range was forming in the west, the Rockies went through a period of renewed uplift. The Earth heaved up as far east as central Texas, elevating the Cretaceous rocks nearly 2,000 feet above sea level. The near-vertical normal faults along which this last uplift occurred are known as the Balcones fault zone. Land to the southeast dropped down along these faults. The fault zone passes near San Antonio, Austin, and Waco and, not coincidentally, follows the trend of the Ouachita line. Weaknesses in the Earth's crust caused by earlier deformation are often reawakened when stresses are again applied to the crust. This regional uplift started a new cycle of intense erosion across Texas. The Edwards Limestone, which had been buried beneath 1,000 to 2,000 feet of younger sediments for millions of years, was exposed at the surface. This hard rock resisted erosion, so it became the elevated Edwards Plateau.

The resurgence of uplift in the Rocky Mountains in Miocene time sent huge volumes of sediment eastward. Rivers charged across the Panhandle of Texas, depositing a vast apron of gravel and sand that extended nearly to the point where Dallas and Ft. Worth now stand. This inclined wedge of sediment, the gangplank upon which nineteenth-century settlers climbed west toward the mountains, covered the colorful older Permian and Pennsylvanian-age rocks. What remains of the wedge today is called the Ogallala Formation. Groundwater contained within it is the main source of water on the High Plains.

Uplift of the Colorado Plateau in the Four Corners states at the end of Miocene time and into the Pliocene created a very gentle eastward tilt of the rocks in Texas. For the last 10 million years, the Red, Colorado, and Brazos Rivers have been eroding back the edge of the Panhandle's High Plains, until today, the steep Caprock Escarpment of the High Plains is about 200 miles west of its original edge near Dallas–Ft. Worth. The steep cliffs along this escarpment, such as at Palo Duro and Caprock Canyons, form some of the most enchanting and colorful scenery in Texas. The caprock is maintained in part by caliche, a calcium carbonate layer deposited during evaporation in the soil. Much of the caliche formed during the dry climate of the Pliocene Epoch.

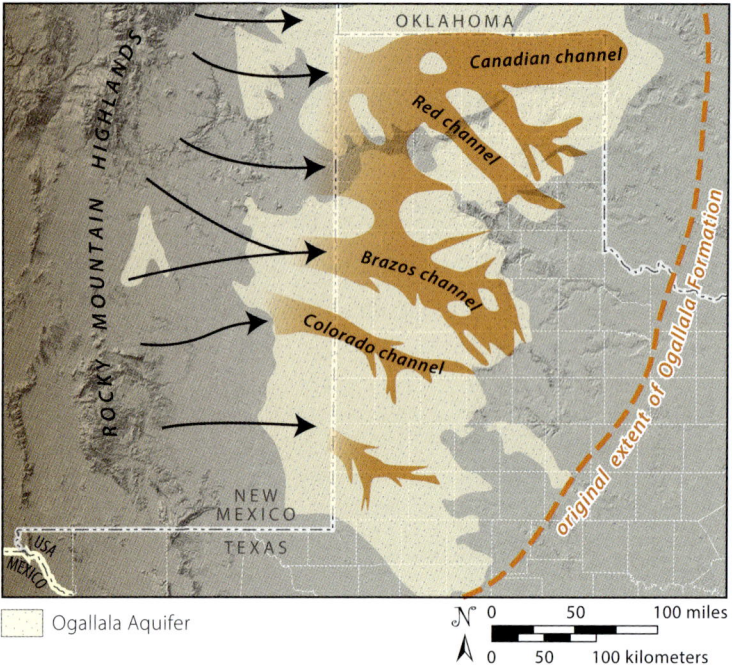

A sediment wedge built eastward across the Panhandle, deposited by rivers flowing east from the Rocky Mountains. The remains of this wedge are what forms the Ogallala Formation and its massive aquifer. —Modified from Ewing, 2016

An excellent interior view of the Ogallala Formation can be seen northwest of Post along US 84. (33.2138, -101.4201)

During the Pleistocene ice ages, from 2.6 million to 11,000 years ago, ice sheets reached as far south as Kansas. The cooler climate and abundant precipitation enhanced erosion in Texas. Streams dissected the High Plains, carving Palo Duro Canyon, the Caprock Escarpment, and the breaks of the Canadian River. Many other major streams were entrenched across Texas. As the ice sheets melted at the end of the Pleistocene, sea level rose, inundating Gulf Coast river mouths to form bays and estuaries. Extensive barrier islands grew along the Texas coast as sand, brought to the sea by rivers, was spread along the new shoreline by the action of waves and long-shore drift.

The Cenozoic Era is the age of mammals, with many of the common animals we know today arising during this time, such as primates, rodents, seals, elephants, horses, rhinos, pigs, hippos, camels, cattle, and deer. At the same time, the appearance of grasses profoundly affected the development of many of these mammal groups. Several million years ago, a spear-tossing, soon-to-be-auto-driving primate burst upon the African landscape, some say to change the scene forever.

Exactly when humans arrived in North America is not known. For many years, archeologists considered people of the Clovis Culture, arriving as long as 13,500 years ago, as the first people in North America; however, more recent discoveries of human footprints in New Mexico dated to more than 22,000 years ago have pushed back the arrival of humans in North America. Numerous sites of early human activity have been found throughout Texas. Lubbock Lake National Historic Landmark, northwest of Lubbock, preserves one such site where Clovis Culture people built a settlement near a spring-fed lake 11,500 years ago. Another location near Florence, the Gault Archeological Site, preserves a major tool manufacturing facility where more than 650,000 stone artifacts have been found, as well as numerous stone and bone art pieces dating to 13,400 years ago. An exciting recent discovery at Gault is stone arti-facts dated to as early as 20,000 years ago found in a layer below the Clovis Culture layer. Current research into who these people were is ongoing.

EROSION AND THE MODERN TOPOGRAPHY

The tale of Texas geology is not complete until we consider geologic processes oper-ating today. The wind constantly moves and shapes sand into marvelous dunes and sand sheets between Corpus Christi and Brownsville. West of Midland and Odessa, Monahans Sandhills State Park is a great place to study windblown deposits. Hurri-canes, with their high-velocity winds, mold and shape the barrier islands, lagoons, bays, and sand dunes along the 367 miles of Texas Gulf Coast. Adding to the shape of the coastal environments are the additional forces of waves, tides, and river flow. Geologists have studied these modern Texas environments closely in order to under-stand exactly how surface forces interact with sediments. It is only through this understanding of these geologic processes that ancient rocks, deposited in similar environments, can be recognized and interpreted.

The geology directly impacts the topography. The highest elevations in Texas are in the west, home to mountain ranges such as the Davis Mountains, the Chisos Mountains in Big Bend National Park, and the Guadalupe Mountains in Guada-lupe Mountains National Park where Guadalupe Peak (8,751 feet) is the tallest in the state. The High Plains rise topographically toward their Rocky Mountain source, reaching elevations of 4,000 feet in the northwest corner of the Panhandle, whereas the Edwards Plateau in the center of the state is about 2,000 feet above sea level. In general, the topography decreases eastward to about 300 feet above sea level at Texarkana, and the Gulf Coastal Plain slopes gently toward sea level.

The observed topography is the product of two factors, rocks and erosion. Guadalupe Peak, for example, stands high both because the rocks were uplifted and, equally important, because its reef limestones are hard and resist erosion better than the softer, surrounding rocks. The topography of the uplifted, north-south ranges in West Texas stands out clearly because of the faulted structure of these ranges, but again, they contain hard rocks that differentially resist erosion. It is also easy to spot the distinctive dome-like topography of the volcanic intrusions around Big Bend National Park. Erosion has had a tougher time attacking these hard intrusions than the less resistant shales surrounding them.

Weather fronts sweep across Texas mainly from west to east, though splashes of water-laden clouds frequently move landward from the Gulf of Mexico. Annual

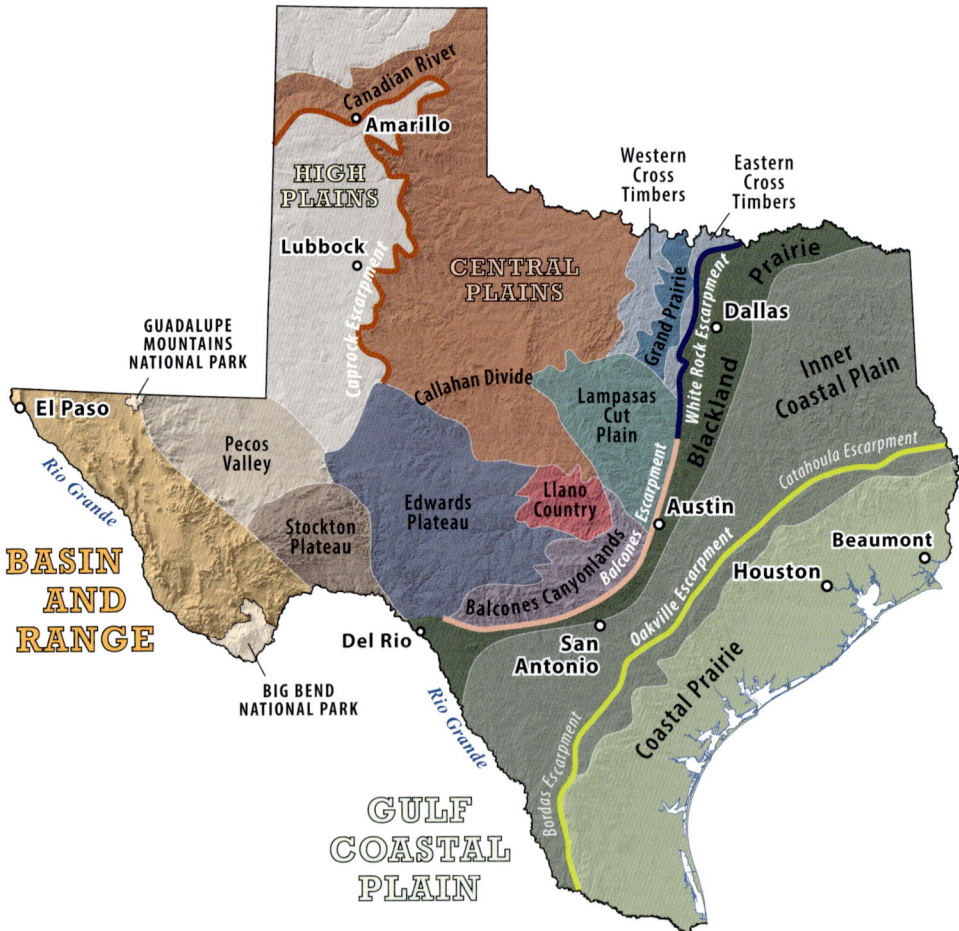

Physiographic provinces of Texas and major escarpments. —Modified from Ewing, 2016

rainfall increases west to east, from 8 inches per year in the dry western country around El Paso to more than 50 inches per year in the eastern semitropical forests near Beaumont. The lines of equal rainfall track almost directly north-south, or roughly perpendicular to the prevailing west to east weather fronts. However, higher rainfall does occur in West Texas, centered over the mountains. The westerly winds must climb over the peaks, and in the process, the flowing air is cooled, forcing its contained moisture to drop as rain on the flanks of the peaks.

Annual temperature and rainfall are dominant influences on the distribution of vegetation, but the underlying soil types also control plant distribution. A close look at a vegetation map of Texas reveals patterns that closely follow the patterns found on the geologic map of Texas. For example, the broad sweep of eastern pine forests follows the sandy ridges of Eocene- to Quaternary-aged rocks in the area because pines like well-drained sandy soils. The thin soils on top of broad expanses of rocky limestone terrain in the Edwards Plateau of central Texas support stands of live oaks, which prefer calcareous soils. The deciduous post oaks, on the other hand, prefer sandier soils, whereas mesquite grows preferentially on clay-rich soils where many other trees have a difficult time competing. Rich alluvial soils parallel to major rivers commonly host more lush vegetation than nearby hillsides.

In the dry west, where rainfall and plants are sparse, mechanical erosion by the wearing action of intense runoff after cloudbursts is the main erosive force. In West Texas, rocks are mechanically worn down, and hillslopes take on the form of mesas. Large rocks simply break and fall downslope under gravity. In the east, where rainfall is high, abundant plants create thick, humic soils that hold back water and prevent intense runoff. Water held in the ground, however, chemically erodes rocks, breaking them into small bits (soil). The chemical erosion causes softer, rounded hills because rocks disintegrate to small particles that move downslope via slow creep.

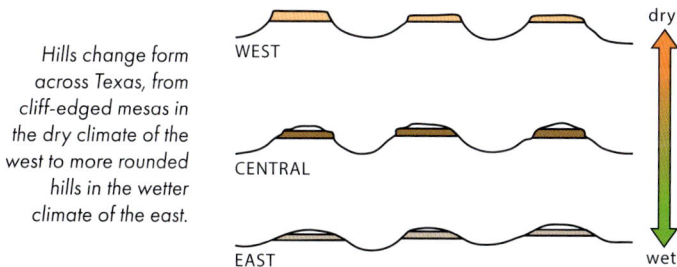

Hills change form across Texas, from cliff-edged mesas in the dry climate of the west to more rounded hills in the wetter climate of the east.

WEST

CENTRAL

EAST

dry

wet

WHY SO MUCH OIL IN TEXAS

Texas has a great amount of oil and gas because there has been not one, but two great periods of hydrocarbon generation. First, the oil in West Texas was generated and trapped in a number of basins that developed in Paleozoic time in response to the Ouachita mountain building. The second period of oil generation is the product of later tectonic forces in Mesozoic time that opened the Gulf of Mexico and allowed the deposition of thick, organic-rich sediments into the Gulf. The thick pile of Mesozoic and Cenozoic sedimentary deposits bore down heavily on the Jurassic-aged Louann Salt, and it responded by heaving upward and cutting through the overlying

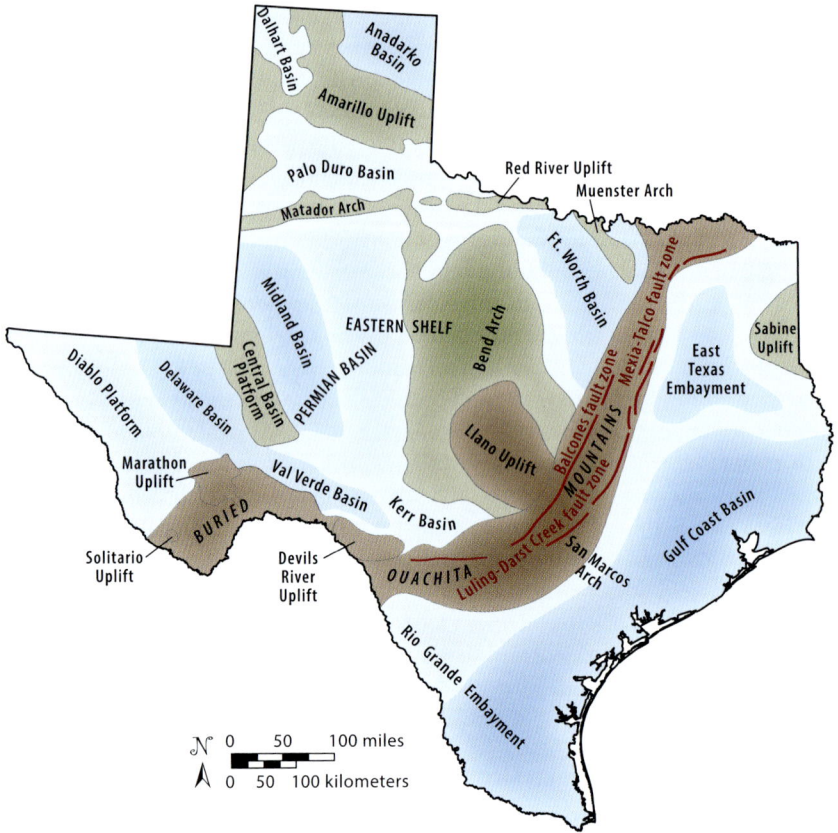

Simplified structural geology map of Texas. Most of these subsurface features are known from oil well drilling.

sediments to form tall spires, banks, and domes. Oil was trapped in profusion around these salt domes.

Most oil is generated from the remains of marine microorganisms in organic-rich shales and limestones. This organic material must pass through the oil window—the temperature conditions needed to turn the complex organic material into crude oil. Higher temperatures lead to the gas window that creates natural gas. If the temperature becomes too high, the organic material is destroyed. Once the oil and natural gas is created, it rises toward the Earth's surface until it is trapped by an impermeable layer of rock and held in a high-porosity reservoir rock. The thick accumulations of sediment and deep burial in Texas basins provided the right conditions for oil and gas formation. The various arches and uplifts provided structural traps to collect the hydrocarbons.

Native Americans living in Texas found and used oil from surface seeps long before Europeans arrived. Probably the earliest record of Europeans' use of oil is by the survivors of Spanish explorer Hernando De Soto's expedition, who caulked their boats with tarry oil in 1543 near Sabine Pass in the southeast corner of the state.

stratigraphic trap

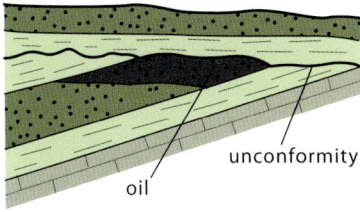

fault trap

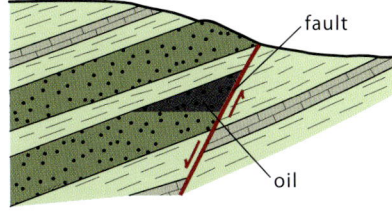

anticlinal trap

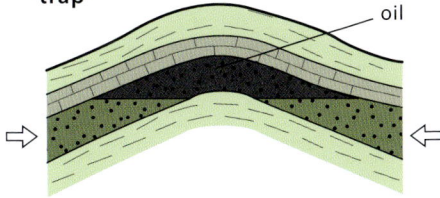

salt dome trap

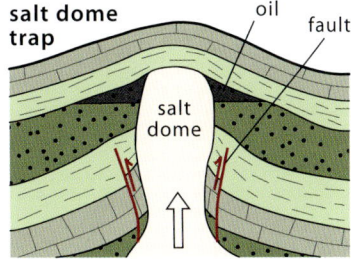

impermeable shale	porous sandstone	limestone	oil-saturated rock

Types of oil traps.

The first well to produce oil in Texas was drilled in 1866 by Lynn T. Barret near Melrose in Nacogdoches County. In 1867, Amory Starr and Peyton F. Edwards brought in a well at Oil Springs in the same area, giving Nacogdoches County the honor of the first commercial oil field, production, refinery, and pipeline activity in the state. However, the first major oil discovery came in 1894 when the city of Corsicana, south of Dallas–Ft. Worth, tried to drill a water well and discovered the Corsicana oil field instead. In 1901, the first great Texas gusher and giant field was brought in by Captain Anthony F. Lucas, who drilled Spindletop near Beaumont. In 1900, Texas produced a mere 836,000 barrels of oil and, in 1901, 4,400,000 barrels. By 1902, Spindletop alone poured out 17.5 million barrels of oil, which accounted for 94 percent of Texas's total production that year.

The biggest oil field in Texas, the East Texas field, was discovered in 1930 in Rusk County, near Kilgore, by the veteran wildcatter C. M. (Dad) Joiner. The East Texas field still produces oil today but has been surpassed in daily output by the Yates field, also a giant field, discovered in West Texas in 1926.

Sometimes when wells are drilled into rock, the oil and natural gas does not easily flow into the well, and the permeability of the rock must be artificially increased. Hydraulic fracturing, a process better known as "fracking," increases the permeability. This process was first used in the 1940s to increase the production of oil from typical reservoir rocks such as sandstone. By the 2000s, it was discovered that by combining horizontal drilling with fracking, oil and natural gas could be produced from unconventional reservoirs with very low permeability such as shale. In this process, a well is drilled into a hydrocarbon-bearing shale layer, then the drilling bends and follows the shale, sometimes for as far as 2 miles. Pressurized fluid cracks the rocks, and sand grains carried in the fluid prop the cracks open so the oil and natural gas can flow into the well. Fracking has greatly increased hydrocarbon production in Texas and other parts of the United States, but it comes with risks. Earthquake activity has significantly increased in areas where it is done, and the fluid injection raises concerns about groundwater contamination.

When the flow of oil or gas needs to be regulated from a well, a contraption known as a Christmas tree is installed.

THE GULF COASTAL PLAIN

The Texas Gulf Coast, a strip of low, flat terrain more than 300 miles long by 50 to 100 miles wide, lies adjacent to the Gulf of Mexico. The mostly flat surface is not quite horizontal because the coastal plain tips gently toward the Gulf at about 5 feet or less per mile. Although few rocks are seen along highways crossing the coastal plain, this mud and sand flat has a marvelous geologic story.

The western band of Cretaceous-aged limestones that arcs eastward near Dallas toward Texarkana is the approximate location and shape of the southern edge of North America about the time the Gulf began to open. The pull-apart, not coincidentally, follows the curve of the old Ouachita Mountain chain, pushed up by the collision of North and South America during the formation of the supercontinent Pangea about 300 million years ago. Though North and South America were glued together by the collision, the weld was not a permanent one. When it came time for Pangea to break apart in the latest episode of continental drift and seafloor spreading, North and South America split along the zone of crustal weakness defined by the Ouachita Mountains. The line where the two continents once came together also became the line where they eventually came apart.

The basin now occupied by the Gulf of Mexico began as a shallow salt pan when the crust sagged, and North and South America began to separate about 200 million years ago in the Early Jurassic. It wasn't connected to the other oceans at the time, and a thick section of salt, the Louann Salt, filled the closed depression as water in it continually evaporated. As pull-apart continued, Gulf waters opened to the other oceans, and by Early Cretaceous time, 140 million years ago, shallow marine seas lapped over the edge of the continental margin of North America. Limestones, built of the shells of marine organisms living in the shallow, Early Cretaceous seas, began to accumulate. Such rocks are seen around Ft. Worth. The basin continued to sag, and by Late Cretaceous time, 90 million years ago, the shoreline had built farther into the Gulf, and more rocks and limestones were deposited farther offshore. These rocks, of Late Cretaceous age, are found in the Dallas area today.

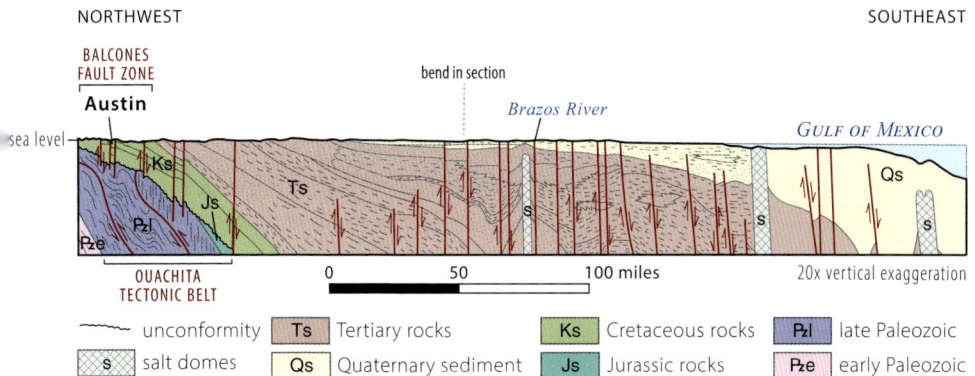

Cross section of East Texas from Austin to the Gulf Coast. —Modified from AAPG, 1973

GEOLOGIC AGE			NORTH GULF COAST		SOUTH GULF COAST	
ERA	PERIOD	EPOCH	ROCK NAME		ROCK NAME	
CENOZOIC	QUATERNARY	PLEISTOCENE	Willis Formation		Willis Formation	
	NEOGENE	PLIOCENE				
		MIOCENE			Goliad Sand	
					Fleming Group	Oakville Sandstone
	PALEOGENE	OLIGOCENE	Catahoula Formation		Anahuac Formation (in subsurface only)	
					Catahoula Formation	
					Frio Formation	
		EOCENE			Jackson Group	
			Claiborne Group	Sparta Sand	Claiborne Group	Yegua Formation
				Weches Formation		
				Queen City Formation		Laredo Formation
				Reklaw Formation		
			Wilcox Group	Carrizo Sand	Wilcox Group	Carrizo Sand
		PALEOCENE				
MESOZOIC	CRETACEOUS				Navarro Group	
					Taylor Group	
			Austin Group	Austin Chalk	Austin Chalk	
			Eagle Ford Group		Eagle Ford Group	
					Washita Group (Sixshooter Group)	Buda Formation
			Fredericksburg Group	Goodland Limestone	Fredericksburg Group	Edwards Limestone
				Edwards Limestone		
			Trinity Group	Antlers Formation	Trinity Group	Glen Rose Formation
				Glen Rose Formation		
	JURASSIC				Louann Salt (in subsurface only)	

Major formations and groups in the northern and southern regions of the Gulf Coast.

By 60 million years ago, in the beginning of the Cenozoic Era, highlands rose in the western interior of North America, and the sinking rate of the Gulf must have increased. Tremendously thick piles of sediment were dumped into the Gulf by ancestors of the Sabine, Trinity, Brazos, and Colorado Rivers. This process of coastal building in Texas extended the land more than 250 miles into the Gulf. Continuous mountain erosion, river transport of sediment, and deposition in the deep hole of the Gulf is the Earth's way of saying it doesn't like topography. Over geologic time, high mountains are reduced to flat plains, while deep oceans are filled up until they, too, are flat plains.

The coastal plain is the latest expression, then, of a series of sloping sediment wedges that have built gulfward since Cretaceous time. The recent geologic history of the coastal plain, however, is closely tied to sea level. For the last few hundred thousand years, sea level has risen and fallen in response to the growth and melting of the great ice sheets that once nearly covered North America. As ice melted, water was added to the oceans, causing sea level to rise. Conversely, as ice sheets grew, less water returned to the oceans via rivers, and sea level dropped.

About 135,000 years ago, during a warm interglacial period, sea level was higher by about 25 feet than it is today, and the coastline was then 20 miles inland from today's coastal position. Rivers built channels and floodplains at higher elevations to match the elevated coastline. Then sea level dropped as the last ice age reached its frigid apex about 18,000 years ago. The lowered coastline lay about 50 miles seaward with sea level about 300 to 450 feet lower than today. Rivers adjusted their gradients

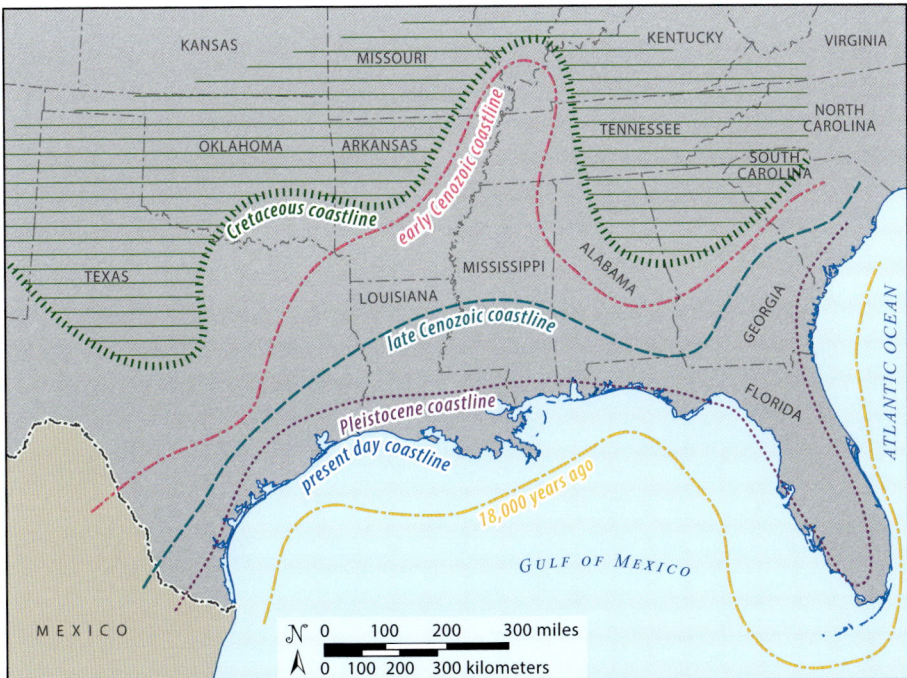

Map showing how the margin of North America has built progressively gulfward since Cretaceous time. The 18,000-year line shows where the shoreline was when glacial ice was at a maximum and sea level was lower than it is today.

accordingly by cutting downward to match the lower coastline level, and the coastal plain became a wider surface deeply dissected by entrenched rivers.

As the climate warmed, the great northern hemisphere ice sheets began to melt, and sea level rose, reaching present levels only about 5,000 years ago. The rising sea drowned river mouths and therein created bays and lagoons along the coastline, while rivers adjusted to the higher sea level by partly filling in their deep valleys. Present sea level is not as high as it was 135,000 years ago because today's climate is not as warm as it was then. A large volume of the Earth's water is still entrapped in glacial ice, mainly in Greenland and Antarctica, and if this remaining ice were to melt, part of the Texas coastal plain would be covered once again by shallow seawater.

Like an undulating snake or loose garden hose, rivers crossing the Texas coastal plain twist and turn in a sinuous pattern. Rivers tend to meander in this fashion because their banks aren't perfectly uniform, and the water flow is turbulent or highly irregular. Even if the river channel started out perfectly straight, it would soon develop bends and meanders because any small irregularity in the channel would deflect the stream flow. As soon as this happens, one bank is pounded harder by the force of the water, causing greater bank erosion on that side, which puts a slight bend in the channel. Over time, the current hits the outer bend with greater force than the inside bend, and the bend grows larger and becomes more accentuated. As the bank on the outer bend erodes sideways, the sediment carved from one bend is deposited in the slow-moving water on the next downstream inside bend. This sandy pile is called a point bar. Erosion of the outer bend and deposition on the inner bend cause the channel to migrate sideways through time. A loose garden hose simulates the motion, through time, of a meandering river.

The mechanics of river channels are only one part of the floodplain story, however. Much sediment, mainly mud, is carried by the river and deposited on the adjacent

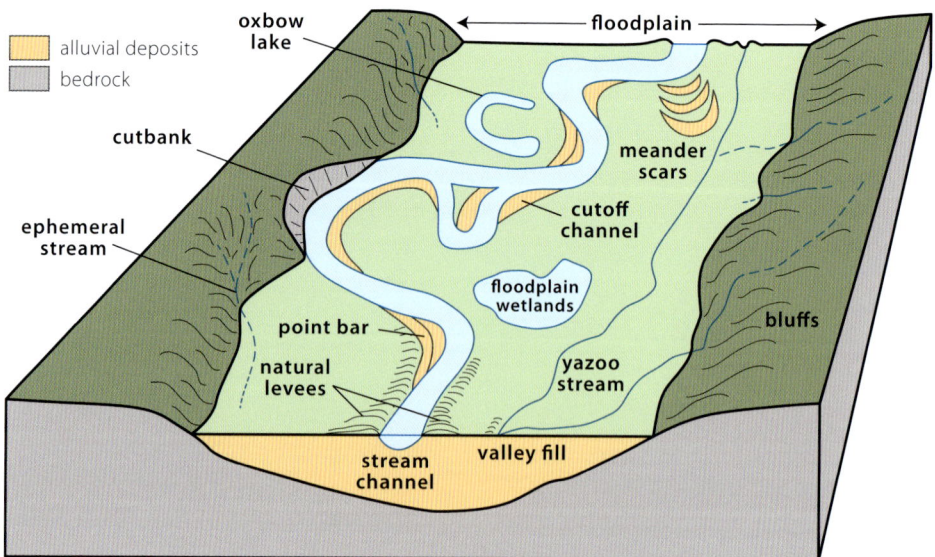

Anatomy of a meandering stream within a floodplain.

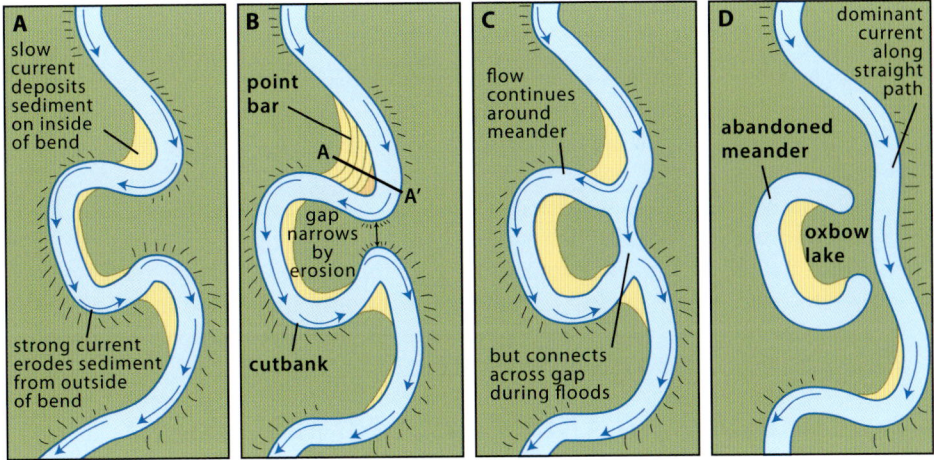

A
slow current deposits sediment on inside of bend

strong current erodes sediment from outside of bend

B
point bar

gap narrows by erosion

cutbank

C
flow continues around meander

but connects across gap during floods

D
dominant current along straight path

abandoned meander

oxbow lake

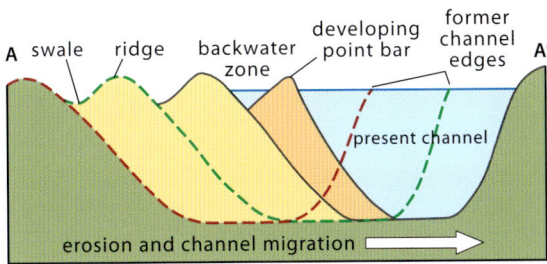

A swale   ridge   backwater zone   developing point bar   former channel edges   A'

present channel

erosion and channel migration →

Oxbow lakes form as meander loops grow and become more circular. As the river cuts into the outside bank of the curve, it eventually breaks through and creates a new path, leaving the meander bend "cut off" from the main stream.

Aerial view of a meander belt along the Brazos River.

plain during floods. The floodplain builds upward and seaward by the regular addition of these mud layers.

Lignite, which is compressed peat derived from plant material, is mined extensively along a wide band in northeast Texas. This Eocene deposit, 50 million years old, was laid down in the muddy areas between river channels, where plants grew abundantly.

With such rich organic sediments, as well as closely associated sands, it is no wonder northeast Texas is also rich in oil and gas deposits. Organic Cretaceous limestones, as well as the organic Eocene shales, are the sources for the hydrocarbons, while associated porous sandstones and limestones form the reservoirs into which the oil and gas migrated. These rocks were buried deeply where they could be cooked at the right temperatures to generate oil and gas.

BARRIER ISLANDS AND SAND MOVEMENT

The beautiful, long, curved, sandy rim of the Texas coastline is a string of elongated sandy beads, called barrier islands, that separate the mainland coast from Gulf waters. Padre Island, the longest single barrier island in the United States, stretches 113 miles under sunny Texas skies. Nearly one million visitors a year come to enjoy its beaches and natural beauty. The coastline and the adjacent barrier islands are a complex of ever-shifting environments driven by a dynamic interplay of wind, water, tides, waves, streams, and storms.

The great age of most geologic features is so often emphasized that it probably comes as a surprise to learn the barrier islands are only about 4,500 to 5,000 years old,

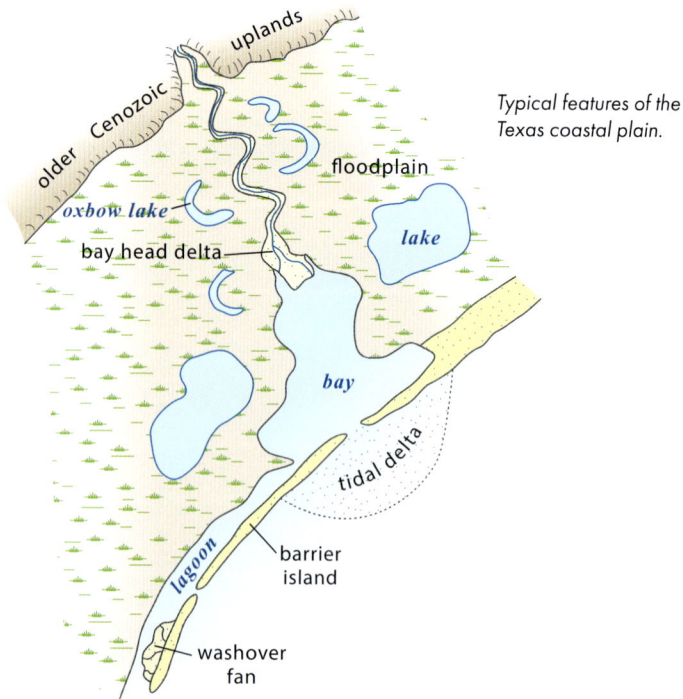

Typical features of the Texas coastal plain.

based on radiocarbon dating of shells. How the barrier islands came into being is intimately tied to far-away ice age glaciers and related changes in sea level. After sea level rose at the end of the last ice age, abundant sand lay on the submerged shelf, where it once was deposited in river channels, at river mouths, and along old shorelines. Now drowned and subjected to waves and storms, the sand was pushed toward shore as it was eroded and reworked. Sand built up, chiefly on the high, submerged drainage divides between the old ice age rivers. The barrier islands are constructed out of this sand, but exactly how is a matter of debate.

Geologists have settled on three main ideas for the process of barrier island construction. First, barrier islands may form from submerged offshore sandbars. On some coasts, as many as three ridges of offshore bars are found. These could migrate shoreward with the addition of sand to eventually emerge as barrier islands. Second,

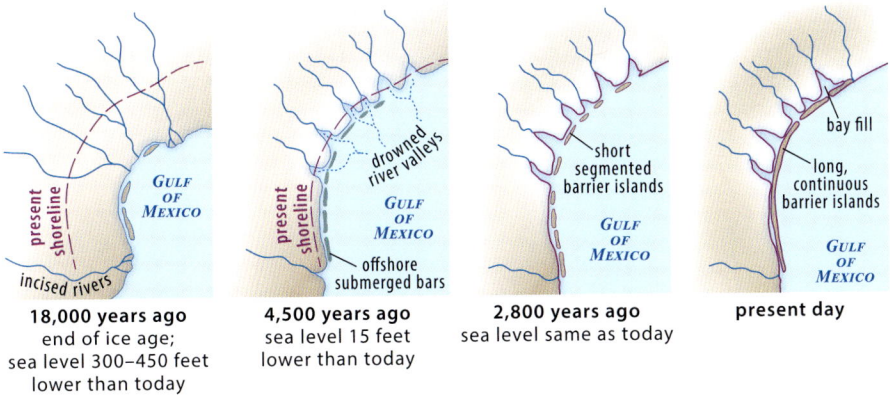

18,000 years ago
end of ice age;
sea level 300–450 feet
lower than today

4,500 years ago
sea level 15 feet
lower than today

2,800 years ago
sea level same as today

present day

Geologic history of Texas Barrier Islands.

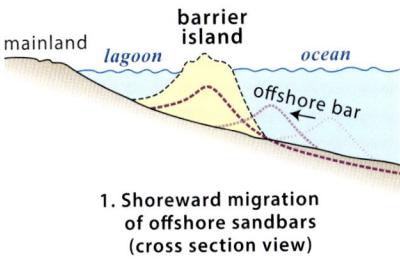

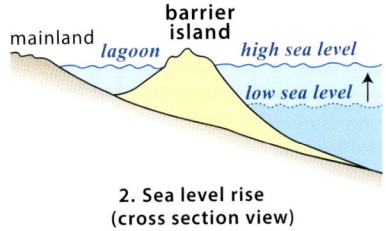

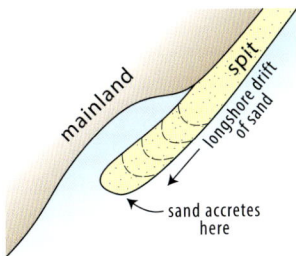

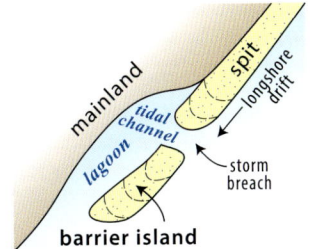

**1. Shoreward migration
of offshore sandbars
(cross section view)**

**2. Sea level rise
(cross section view)**

**3. Later accretion of spits
(map view)**

Three prevailing ideas about the formation of barrier islands.

barriers may be developed by drowning the area on the landward side of mainland beach sand ridges, as sea level rose. Third, barriers could form as sand is added to headlands as it moves parallel to the shore in a process called longshore drift that is driven by the consistent oblique approach of waves that strike the shoreline. All three processes operate on shorelines today, and a combination of them is the likely explanation for the development of the long barrier islands of the Texas coast.

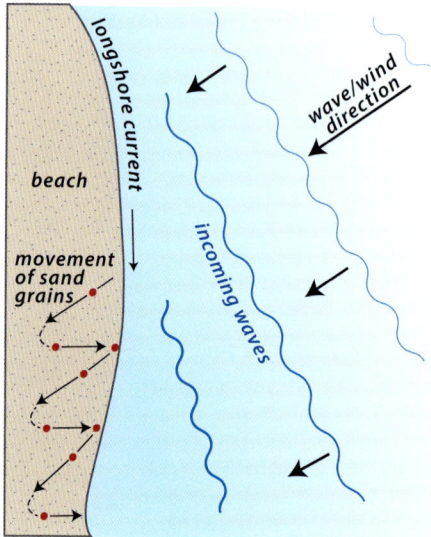

Waves approaching the shoreline at an angle drive sand up the beach at an angle, but the return flow of water and sand is straight down the beach. This arc of motion drives sand along the shore, the process called longshore drift.

As you stand on the beach and feel the energy of the waves breaking along the shore, it is readily apparent that sand trapped in this dynamic circulation is in constant motion. Rivers bring their sandy offerings to the sea, and in response, ocean waves return the sand to the shore. If waves always came straight into the coast, sand would merely pile up on the beach, and the beach would grow uniformly seaward. But waves more commonly attack the shore at an angle, which moves sand along the shore. Think about one sand grain as it is propelled up the beach at an angle by an oblique wave. As the swash of water then falls back down the beach, the sand grain falls back, not at an angle, but straight down the beach toward the water. This arc of motion has effectively moved the grain some distance down the beach. Now multiply the one grain and the one wave by billions, and you can see how the whole beach is moving. It is easy to see this process as you watch oblique waves strike any beach.

The long curve of the Texas barrier islands has a unique effect on longshore sand drift. Waves approach the Texas coast from different angles, of course, depending on the seasons and wind patterns, but the net wave approach is from the east-southeast. At the northern end of the coast, the net wave approach drives sand alongshore in a southwest direction, but because of the curve of the shoreline, sand at the southern end of the Texas coast is driven northward by the same wave approach. There is, therefore, a convergence of longshore drifting sand near the center of the arcing barrier island. This zone of convergence is located at the central part of Padre Island. Another piece of evidence supports the convergence idea: sandy beaches change gradually to shell beaches near central Padre Island. Shells on Little Shell Beach (small shells) come from the north, while the large shells on Big Shell Beach (just south of Little Shell Beach) come from the south. The converging longshore drift thus tends to pile both shells and sand toward the center of Padre Island. A second process concentrates the shells: as the wind blows sand off the beach into the dunes, the shells are left behind.

HURRICANES ALONG THE TEXAS COAST

The Texas Gulf Coast cannot be understood without considering the effect of hurricanes. Hurricanes are circular, counterclockwise motions of air that derive their energy from air rising off warm (greater than 80°F) ocean waters, attaining wind speeds of 75 miles per hour and stronger. The wind in hurricanes is asymmetrical, however, because the effect of the counterclockwise, circulatory winds is added to the forward motion of the storm system. Hence, the peak winds and storm surge is always a few tens of miles to the right of the point of landfall of the storm.

The Atlantic hurricane season runs from June 1 to November 30, with June through September being peak hurricane season in the Gulf of Mexico. Early summer hurricanes are more common along the Texas coast than elsewhere along the Gulf, due to tropical development in the warmer waters of the Caribbean at that time, as well as a tendency for early season Atlantic hurricanes to follow a straight westward course. As the summer progresses, paths of Atlantic hurricanes tend to move eastward and northward along the United States' eastern seaboard. By late August and into September, as the Gulf waters reach their warmest point, any storms that either develop in the Gulf, especially around the Bay of Campeche (northwest side of the Yucatán Peninsula), or develop in the Caribbean tend to affect Texas more frequently.

Since 1899, a significant hurricane has struck the Texas coast, on average, a little more than once every other year. For about a 20-year span in the 1930s and 1940s, the region saw warmer sea surface temperatures, and the frequency of Texas hurricanes was much higher than average. Because hurricanes derive their energy from ocean surface temperatures, warmer waters would be expected to increase the likelihood of hurricanes. Following this warm spell, the middle of the twentieth century brought cooler conditions and, thus, a reduction in the number of storms. By the end of the twentieth and into the twenty-first century, however, sea surface temperatures have risen at an alarming rate. This increase in temperature has not only increased the number of tropical storms and hurricanes in the Atlantic basin but has also made them stronger and intensify more rapidly.

In addition to being prolific wind machines, hurricanes can also drop intense rainfall and create deadly storm surge along the coasts. However, it does not take a hurricane to produce catastrophic damage to an area. Houston-area residents will vividly recall the 10-to-15-inch rainfall that dramatically flooded the east and northeast side of the city during Tropical Storm Allison in 1989, while only 3 to 4 inches fell west of the city. This discrepancy of rainfall occurred because the center of circulation with Allison passed along the west side of Houston.

Several other tropical storms and hurricanes have struck the Texas coast after Allison, including Hurricanes Bret, Claudette, Dolly, and Hanna and another Tropical Storm Allison in 2001 that dumped a whopping 37 inches of rain at the Port of Houston. While all major storm events are memorable, three hurricanes in particular that affected Texas after Allison will remain in residents' memories for years to come.

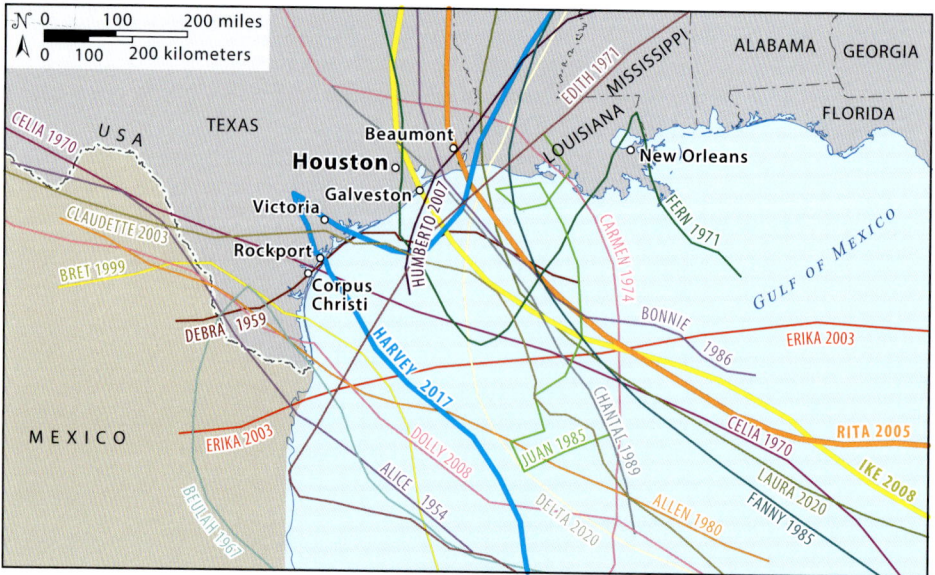

Tracks of major named hurricanes striking the Texas coast since 1950.
—Modified from Interactive Historical Hurricane Tracker

The first of these, Hurricane Rita, made landfall on September 23, 2005, as a Category 3 storm with 115 miles per hour winds near the border of Louisiana and Texas at Johnson's Bayou, Louisiana. At the time, Rita was the most intense tropical cyclone on record in the Gulf and the fourth-most intense Atlantic hurricane ever recorded. Before making landfall, Rita was a Category 5 storm with winds of 180 miles per hour and a central pressure of 895 millibars, the lowest pressure ever recorded in the Gulf of Mexico. While rainfall from the storm was not great (10.48 inches in the town of Center), major flooding occurred in Port Arthur and Beaumont, and many offshore oil platforms in the Gulf sustained significant damage. Hurricane Rita, however, will be most remembered for the massive evacuation of people from the Houston area, with more than three million traveling inland, making it the largest evacuation in US history.

Another memorable storm to affect Texas occurred on September 13, 2008, when Category 2 Hurricane Ike made landfall at Galveston, taking almost the exact path as the Great Galveston Hurricane of 1900. While *only* a Category 2 with sustained winds of 110 miles per hour at landfall, its large aerial size contributed to a storm surge as much as 20 feet high that washed completely over nearby barrier islands, including Bolivar Peninsula, where many structures were destroyed. In the town of Gilchrist, all but one structure was destroyed by the storm surge. Because of Ike's massive size, the storm affected an area along the coast from Corpus Christi to the Florida Panhandle. Ike also produced heavy rainfall in Galveston, where almost 19 inches of rain fell, causing flooding. At the time, Ike was one of the most destructive hurricanes to ever hit Texas; however, another storm that made landfall on the Texas coast in 2017 would replace Ike in the record books.

On August 25, 2017, Hurricane Harvey made landfall near Rockport as a powerful Category 4 storm with sustained winds of 130 miles per hour and peak wind gusts up to 145 miles per hour within its eyewall. In Rockport and Fulton, hundreds of buildings had their roofs blown off or their walls taken down. Storm surge up to 10 feet also inundated nearby bays, causing widespread major flooding. As with most hurricanes, once Harvey made landfall, it quickly weakened to a tropical storm. However, the worst of Harvey was still to come as weak atmospheric steering currents kept the storm nearly stationary for several days near Victoria. At the same time, the storm was also close enough to the Gulf to pick up more moisture. This setup caused the most prolific rainfall event in US history with at least eighteen reporting sites recording at least 48 inches of rain, and up to 60.58 inches fell in Needville, southwest of Houston. This scale of rainfall led to catastrophic flooding in the region and the displacement of almost 2 million residents, including more than 60,000 people requiring rescue. Parts of all twenty-two major freeways in the Houston region were also inundated with floodwaters, making travel impossible. In total, Harvey destroyed almost 17,000 homes and damaged another 290,000, making it the costliest hurricane in Texas and the second costliest in US history, when adjusted for inflation.

High winds in hurricanes are a major force on buildings, roofs, and windows, but water surge is probably the greater geologic force affecting the shoreline. As a hurricane moves across the ocean, the surface water is elevated because of the reduced atmospheric pressure. Wind stress also pushes water, creating a surge. Storm waves ahead of hurricanes can easily raise water level 3 feet before the storm arrives onshore. The elevated water level and waves at the shoreline do the most geologic work.

Property damage, flooding, and shoreline erosion are functions of wind intensity, rainfall, wave heights, and storm tide. The magnitudes of these effects are, in turn, dependent on the atmospheric pressure in the storm, its speed of travel across the ocean, the angle of the storm track to the coastline, the duration of the storm, the radius of the central funnel, the shoaling effects of the continental shelf offshore, and the height of the tide at the time of the storm.

The flat profile of Galveston Island, Bolivar Peninsula, and Louisiana's adjacent Grand Isle—the most hurricane prone areas along the Gulf Coast—are the direct result of hurricane hammering of dunes and beaches. Hurricanes also transport and deposit sediment in coastal bays, open channels through barrier islands, rip up biological communities both on and offshore, and lay down resultant shell hash layers. Water washing over the barrier islands deposit fans in lagoons. Hurricanes cause extensive inland, bay-head, and river flooding that, in turn, erodes and moves sediments.

Galveston Island is flat and mostly free of dunes. Farther south in the Padre Island area, beach profiles are steeper with higher dunes. This region receives fewer battering hurricanes that flatten islands, and the climate is drier. The lack of vegetation allows more sand to blow around freely and build up dunes.

SALT DOMES

Salt domes do not commonly reach the surface and therefore are not prominent features in the Texas landscape, but they do form an important part of the subsurface geology. About 200 million years ago in Late Triassic time, a vast, shallow sea lay to the east of the old Ouachita Mountains in what is now East Texas. Because of the hot, dry climate then, the waters continually evaporated, leaving behind salty deposits of evaporite minerals, such as salt, gypsum, and anhydrite. Thousands of feet of these

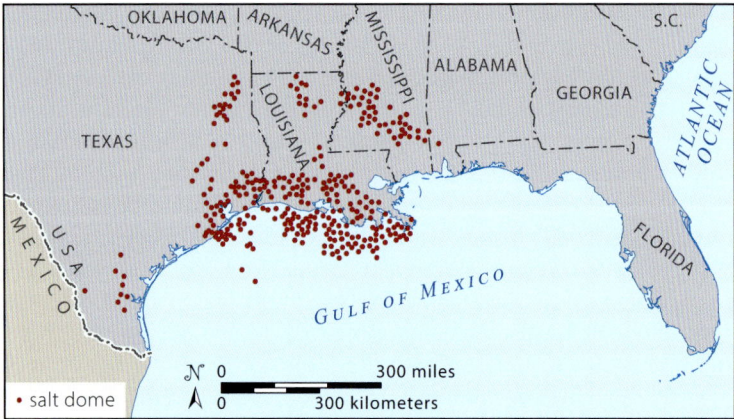

Salt domes in Texas and the US Gulf Coast.

salt deposits, known as the Louann Salt, continued to accumulate throughout Jurassic time, later to be buried beneath even greater thicknesses of younger sediments. Some geologists estimate the main layer of Louann Salt may be 30,000 to 50,000 feet deep beneath the surface of the Texas Gulf Coast.

Salt has curious properties. We normally think of it in terms of hard, brittle, little cubes, but when salt is subjected to heat and pressure, it becomes soft and plastic and can even creep and flow. Along the US Gulf Coast, as the pile of younger sediment thickened over the Jurassic salt layer, the salt was buried deeper and deeper and became hotter and softer and more pliable. The weight of the overlying sediment pile bore down on the warm salt, working much like your hand does when you push it down on a layer of wet mud. The salt oozed upward through the pressing sediment like the mud oozes up between your fingers, except the upward flowing salt created pillars, domes, ridges, and needles, some rising tens of thousands of feet above the mother salt lode.

Gulf Coast geologists who study the mechanics of salt dome formation see two dominant processes: In the first process, a dome or teardrop-shaped body of salt penetrates upward through the thick layers of overlying sediment. In a second major process, known as downbuilding, the overlying sediment continually collapses downward along arc-shaped faults to displace salt downward. Small, sediment-thickened basins are formed, and salt domes and ridges are left standing high next to these basins, while displaced salt below the basins moves laterally and downward toward the toe-end of the continental slope.

When the top of a salt dome is within a few thousand feet of the surface, groundwater leaches cavities in the salt mass and precipitates mineral residues such as gypsum, calcite, anhydrite, and sulfur. These minerals form a caprock over the top of the salt dome. Salt, sulfur, and gypsum are mined from the caprock of many East Texas salt domes. Salt dome caverns are also used as natural underground storage sites for oil, radioactive wastes, and even valuable files and records.

Rocks at the top and along the flanks of salt domes have yielded billions of barrels of crude oil in East Texas. Spindletop, one of the earliest, largest, and most famous oil fields in Texas, was discovered in 1901. The oil came from caprock reservoirs of a

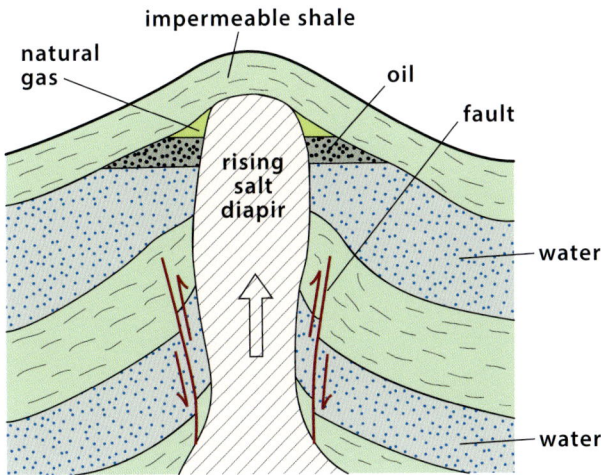

Diagram of a salt dome as it travels upward through layers of overlying sediments. Many oil wells in Texas, including Spindletop near Beaumont, are drilled in and around salt domes.

large salt dome near Beaumont, Texas. It took twenty-four more years before oil was discovered in the sedimentary rocks that pinch out against the flanks of salt domes, setting off a new wave of exploration.

Salt domes were the object of the first remote sensing methods applied in oil exploration. Because salt is less dense than other sedimentary rocks, salt domes can be located by carefully measuring gravity variations. Exploration wells were drilled where the gravity measured "low." Many salt dome oil fields were found by this technique.

In East Texas, more than five hundred salt domes have been mapped in the subsurface. Though surface expressions of salt domes are not common, two exceptional mounds stand up above the surrounding plains and can be visited easily by car. Damon Mound rises 80 feet above the flat coastal plain west of Houston near the small town of Damon, located on TX 36 between Rosenberg and West Columbia. High Island is a caprock mound located east of Galveston Bay at the coastline near the south end of TX 124. Three other surface mounds created by salt domes are located north of Houston, near Palestine, and a few miles from I-45, but they are not as obvious as the mounds at Damon or High Island.

NORTH GULF COASTAL PLAIN

AUSTIN

Austin, the Texas state capital city, straddles the line between two geologic provinces. To the west is an uplifted plateau of Cretaceous limestones surrounding the exposed Proterozoic rocks of the Llano Uplift, while the low Blackland Prairie stretches eastward for miles. Sharply dividing these two distinct topographic terrains is the Balcones fault zone, which forms a northeast-southwest line through the western half of Austin, creating the Balcones Escarpment. The escarpment has been the location for colonization and towns throughout Texas history because of the natural resources associated with the fault—plenty of natural building stone outcrops on the escarpment, water pours from springs, timber can be harvested from the wooded hills, and the adjacent

Geologic map of Austin

Texas Science and Natural History Museum (University of Texas)

BALCONES ESPARMENT

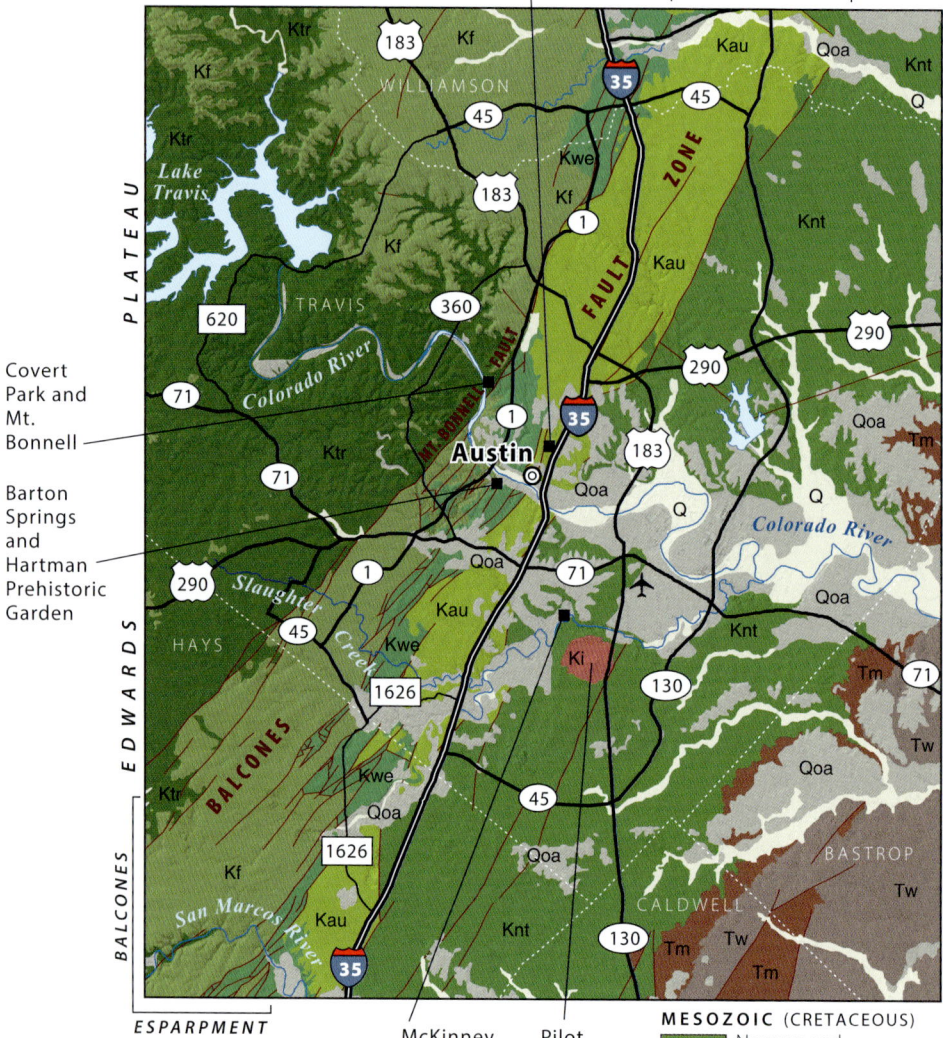

Covert Park and Mt. Bonnell

Barton Springs and Hartman Prehistoric Garden

McKinney Falls State Park

Pilot Knob volcano

CENOZOIC
QUATERNARY

Q — recent sediments; includes alluvium

Qoa — older surface deposits; includes terrace and high gravel deposits (Pleistocene)

NEOGENE and PALEOGENE

Tw — Wilcox Group (Eocene)

Tm — Midway Group (Paleocene–Eocene)

✈ airport

— normal fault

county boundaries shown as dotted white lines

MESOZOIC (CRETACEOUS)

Knt — Navarro and Taylor Groups

Kau — Austin Chalk

Kwe — Washita Group; includes Eagle Ford Group, Del Rio Clay, and Buda Formation

Kf — Fredericksburg Group; includes Edwards Limestone

Ktr — Trinity Group; includes Glen Rose Formation

VOLCANIC ROCKS

Ki — basalt of Pilot Knob volcano (Cretaceous)

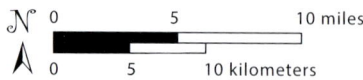

0   5   10 miles
0   5   10 kilometers

Geologic map of Austin

fertile prairies support a strong agricultural economy. In 1756, Bernardo de Miranda, a Spanish explorer, named the escarpment Los Balcones, meaning "balconies," which describes the stair-step, balcony-like topography rising above the plains.

At Austin, the total offset along the Balcones fault zone, a group of parallel faults, is about 1,200 feet, while 600 feet of displacement takes place along the Mt. Bonnell fault, the westernmost break of the fault zone. The offset can be seen on the cross section. Note how the Early Cretaceous Glen Rose Formation is at the surface west of the Mt. Bonnell fault, while east of the fault zone, younger rocks are at the surface. The younger rocks overlie the Glen Rose, which is buried 1,000 feet below.

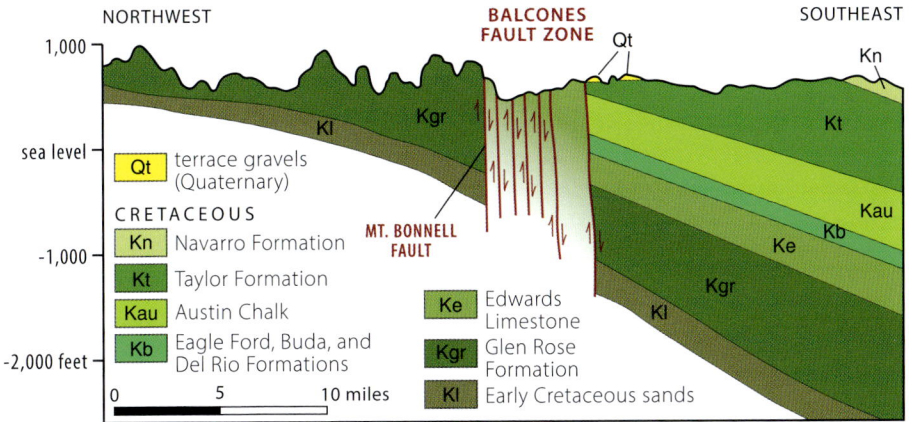

Section across the Balcones fault zone at Austin.

Austin is situated where the Colorado River cuts through the escarpment. The river begins its approximately 862-mile journey through Texas on the Llano Estacado near Lamesa before emptying into the Gulf at Matagorda Bay.

The rocks around Austin are mostly limestone, but dolomite, clay, basalt, tuff, sandstones, and river gravel are also present. Limestone ledges are commonly seen around the edges of Lake Travis, a reservoir on the Colorado River. They are part of the Glen Rose Formation, famous for its dinosaur tracks in many places in Texas. The Edwards Limestone is another prominent, thick unit seen in roadcuts and outcrops northwest of town along US 183. A band of white limestone, called the Austin Chalk, cuts a swath down the center of its namesake, with many exposures. East of Austin, in the prairie country, are softer shales, sandstones, and marl of the Late Cretaceous Taylor and Navarro Groups, but they aren't hard enough to resist erosion, so outcrops are not so common.

Springs emanate from limestone aquifers up and down the length of the Balcones Escarpment. One good example in Austin is Barton Springs, a favorite swimming and picnicking site on the south side of the Colorado River. The water comes to the surface at Barton Springs where a fault has offset a porous unit within the Edwards Limestone, which is honeycombed with caves and solution cavities.

Another nearby site of interest to geology fans is the Hartman Prehistoric Garden, part of the larger Zilker Botanical Gardens, located along Barton Springs Road near the MoPac Expressway (TX 1). It offers a rare look into the botany of the area approximately 70 million years ago when dinosaurs roamed the land. Founded in 1992 after several dinosaur tracks were discovered in an old limestone quarry, the garden was developed to include plant species that would have lived in this part of Texas during the end of the Cretaceous Period, including ferns, horsetails, liverworts, cycads, conifers, ginkgos, magnolias, and other plants. In all, more than one hundred species of prehistoric plants thrive at Zilker Gardens.

Pilot Knob, southeast of the city center along US 183 near Austin-Bergstrom International Airport, is a cluster of small, rounded hills that expose the core of a Cretaceous explosion crater. The assemblage of volcanic rocks includes reddish tuff or volcanic ash, pyroclastic debris blown from the volcano, and black, fine-grained nephelinite—a volcanic rock related to basalt. About 80 million years ago, this area of Texas was a shallow marine shelf where limestone rock was deposited by lime-secreting organisms that lived in the sea. Fractures broke the Earth's crust in a line following the Balcones fault zone, and hot, molten lava rose through the fractures from deep in the Earth's mantle. As the lava pushed upward through the sediments, explosions driven by steam rocked the seafloor. Craters formed around these explosion vents, one of which is Pilot Knob. Ash, debris, and lava filled the crater, and eventually, a dome extended above the seafloor. When volcanic activity ceased and the lava had cooled, reefs developed around the mound and, where the mound broke the sea surface, beach deposits formed as waves lapped the margins of the mound. At McKinney Falls State Park southeast of Austin, deposits of gray limestone (Austin Chalk) compose the ledge at Upper McKinney Falls, and ash from Pilot Knob can be seen under the ledge.

Limestone of the Austin Chalk forms the thick ledge at Upper Mckinney Falls. Underneath is a layer of ash and pyroclastics from nearby Pilot Knob and other small volcanoes in the area. (30.1847, -97.7255)

Finally, the mound was covered as the sea bottom subsided throughout the Cretaceous Period, and clay and marl of the Taylor Group were deposited over the top of the mound. About 10 million years ago, in Miocene time, uplift occurred along the Balcones fault zone, and the Cretaceous rocks were elevated, exposed, and eroded. The harder volcanic rock resisted erosion a little better than the limestone, so Pilot Knob, at 710 feet above sea level, today stands 180 feet above the surrounding terrain. Pilot Knob, is perhaps the best known of dozens of volcanic craters and mounds that string out along the Balcones fault zone. The largest concentration of Cretaceous volcanic mounds is found around Uvalde to the southwest of Austin.

A pleasant view of the city and surrounding countryside can be had from the top of Mt. Bonnell, which rises 775 feet above sea level. The Glen Rose Formation, elevated by uplift on the west side of the Mt. Bonnell fault, is the limestone bedrock around Mt. Bonnell. At Covert Park, steps made of limestone take you to the top.

Pilot Knob, one of many Cretaceous volcanic hills in the area southeast of Austin, can be seen looking east from McKinney Falls Parkway. (30.1675, -97.7231)

View looking north and upstream along the Colorado River from the top of Mt. Bonnell at Covert Park. Note the plateau of the Hill Country in the distance. (30.3212, -97.7736)

The Texas State Capitol in Austin was constructed of locally quarried Edwards Limestone and Town Mountain Granite. —Courtesy of Dadarot, Creative Commons CC0

The state capitol building at 103 West 11th Street is a geologic monument in itself. The Edwards Limestone was quarried in Travis County for part of the stonework, whereas the Texas pink granite came from Granite Mountain near Marble Falls, where beautiful granite building stone is still actively quarried. Construction on the capitol began in 1882 under the direction of architect E. E. Meyers of Detroit, Michigan. The edifice was completed and dedicated in 1888.

Geology displays are featured in the Texas Science and Natural History Museum at 2400 Trinity Street on the University of Texas campus, while a treasure-trove of Texas geological literature can be purchased from the Bureau of Economic Geology, located in the J. J. Pickle Research Campus at 10611 Exploration Way in north Austin.

I-20
DALLAS (I-45)—MARSHALL—LOUISIANA
165 miles

The original settlement of Dallas was situated along the Trinity River where it crosses the outcrop band of Austin Chalk. The bedrock created a ford across an otherwise muddy river bottom. I-20 crosses the Trinity River about 2 miles east of the I-45 interchange. Just east of the river, a small outcrop of Austin Chalk can be seen by exiting onto Dowdy Ferry Road at exit 476, then traveling north a very short distance to Great Trinity Forest Gateway on the left. Look for a small ledge of chalk across the pond. East of Dowdy Ferry Road, I-20 crosses the mainly flat, agricultural Blackland Prairie region east of Dallas. Rocks are nowhere to be seen, except for white limestone rubble here and there in fields, where Cretaceous limestones form the bedrock.

Blocky layers of Austin Chalk are visible at the edge of the pond at Great Trinity Forest Gateway. (32.6797, -96.6789)

The pine-covered, rolling hills of northeast Texas follow the outcrop band of Eocene Claiborne sandstones. For about 70 miles west of Marshall, I-20 rolls along on the hilly topography, where river drainages have created low hills and valleys as they carve into the soft sandstone bedrock. Pine trees grow in profusion on these sandy soils.

A great example of one of these old-growth pine stands is at Tyler State Park, located to the north of exit 562 along FM 14. While few rocks are exposed here, geology still plays a major role in shaping the land. The large pine trees grow in and around the park for two important reasons. First, the underlying rock type is the Eocene-aged Queen City Formation (Claiborne Group), which produces the well-drained, sandy soil required by pines. Second, pines need abundant moisture to survive. At Tyler State Park, and East Texas in general, rainfall is greater than to the west near Dallas.

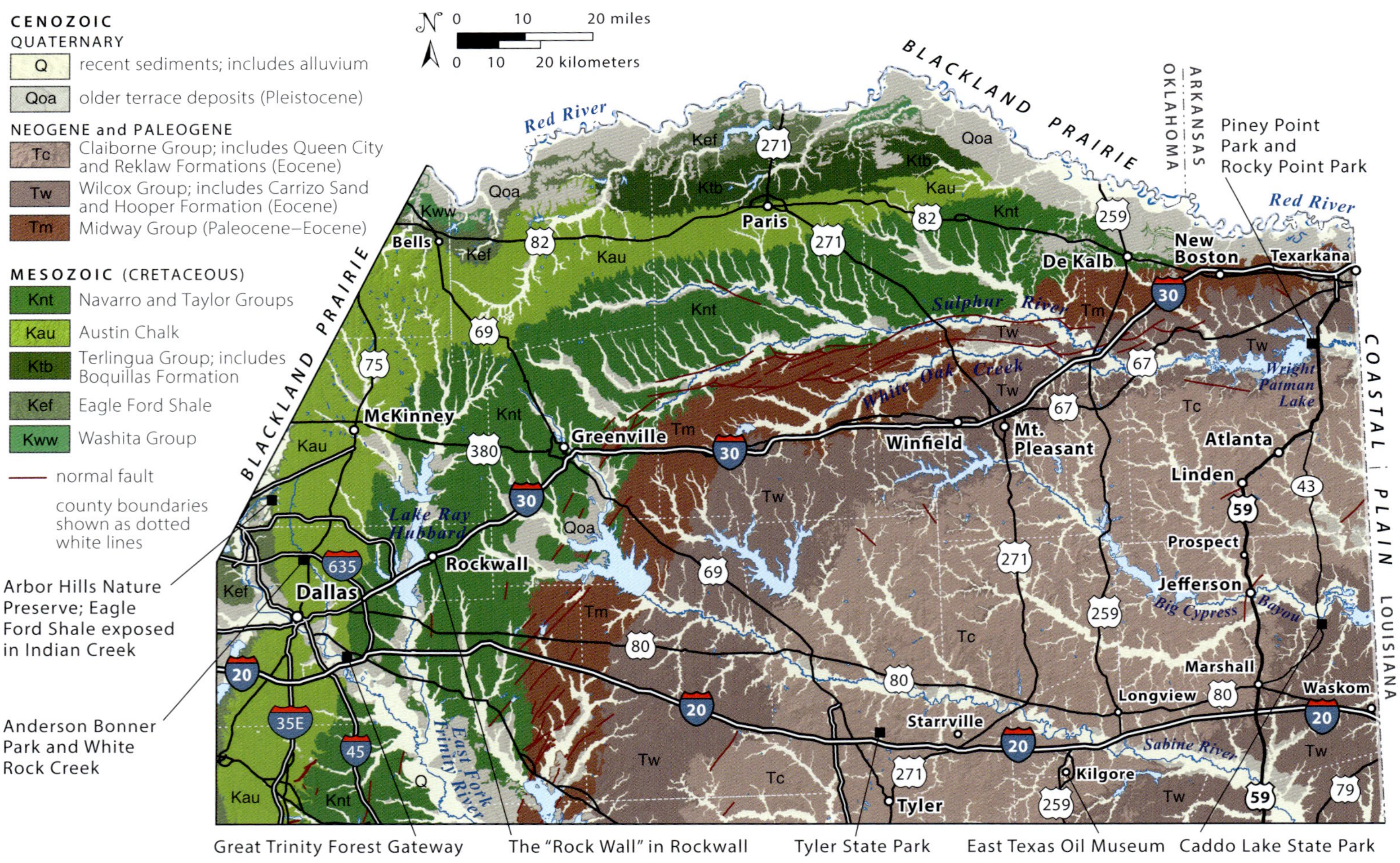
CENOZOIC
QUATERNARY
Q recent sediments; includes alluvium
Qoa older terrace deposits (Pleistocene)
NEOGENE and PALEOGENE
Tc Claiborne Group; includes Queen City and Reklaw Formations (Eocene)
Tw Wilcox Group; includes Carrizo Sand and Hooper Formation (Eocene)
Tm Midway Group (Paleocene–Eocene)
MESOZOIC (CRETACEOUS)
Knt Navarro and Taylor Groups
Kau Austin Chalk
Ktb Terlingua Group; includes Boquillas Formation
Kef Eagle Ford Shale
Kww Washita Group
normal fault
county boundaries shown as dotted white lines
0 10 20 miles
0 10 20 kilometers
BLACKLAND PRAIRIE
BLACKLAND PRAIRIE
COASTAL PLAIN
OKLAHOMA
ARKANSAS
LOUISIANA
Red River
Red River
Sulphur River
White Oak Creek
Wright Patman Lake
Lake Ray Hubbard
East Fork Trinity River
Big Cypress Bayou
Sabine River
Bells
Paris
De Kalb
New Boston
Texarkana
McKinney
Greenville
Winfield
Mt. Pleasant
Atlanta
Linden
Prospect
Jefferson
Marshall
Longview
Waskom
Rockwall
Dallas
Starrville
Tyler
Kilgore
Piney Point Park and Rocky Point Park
Arbor Hills Nature Preserve; Eagle Ford Shale exposed in Indian Creek
Anderson Bonner Park and White Rock Creek
Great Trinity Forest Gateway
The "Rock Wall" in Rockwall
Tyler State Park
East Texas Oil Museum
Caddo Lake State Park

Geology along I-20 and I-30 between Dallas–Ft. Worth, Marshall, and Texarkana.

Layers of gray sandstone and claystone, and a small layer of lignite coal, compose the Queen City Formation that can be seen along I-20 east of Tyler State Park. (32.4474, -95.1781)

In the central segment of I-20, small roadside exposures are more common. Northeast of Tyler, just east of exit 567 (TX 155), gray sandstone, claystone, and a black lignite bed appear on the south side of the highway. Watch for additional, smaller roadcuts in this area—they are small windows into the bedrock of Eocene rocks that form the substrate in northeast Texas.

Don't miss the oil derrick picnic tables at the rest area east of Starrville between mileposts 573 and 574. The importance of oil and gas to this region is demonstrated in the excellent East Texas Oil Museum in Kilgore. See the US 59/US 259 roadlog for more information about the museum and Kilgore.

Watch for the Sabine River crossing about 4 miles east of the US 259 interchange. Gray sandstone and shale of the Reklaw Formation are exposed in riverbanks, though they are not spectacular by any means; better exposures in the Sabine River can be seen just north of exit 591 where FM 2087 crosses the river. The best roadcut of Eocene-aged Carrizo sandstone and Reklaw Formation claystone occurs just west of the Sabine River on the south side of eastbound lanes on I-20. This is the best roadcut along the entire stretch between Dallas and the Louisiana state line. Here, sandstone channels display cross-bedding and ripples, gray claystones are exposed, and beds of bright-red to black ironstone concretions can be seen. The outcrop is stained yellow and red from limonite, a hydrous iron oxide mineral that forms when the iron weathers. The rocks here were laid down on a floodplain during the Eocene.

Around Longview, large surface (strip) coal mines recovered lignite from the Eocene-aged Wilcox Group in the 1970s and 80s but have since been closed and the land reclaimed. The lignite mined here was burned in coal-fired power plants

Bedded gray to white claystones of the Reklaw Formation sit on sandstones of the Carrizo Sand in a roadcut just west of the Sabine River crossing and Longview. Scattered within the Carrizo are iron-rich concretions. (32.4326, -94.7624)

to provide electricity. Lignite, a hard peat or soft coal, is not the highest grade of coal—nor does it have the heating efficiency of the harder, higher grade bituminous or anthracite coals with their high carbon percentages—but its abundance here and proximity to the power plants made it economical. It was also easy to mine. The plant material that was compressed to form lignite was deposited in floodplains between river channels during Eocene time around 50 million years ago.

No rocks crop out along the roadway between Longview and the Louisiana state line. Look for patches of red sands of the Queen City and Reklaw Formations and gray mudstones of the Hooper Formation, all indicative of the underlying geology in this part of Texas. The Hooper Formation is part of the Wilcox Group.

I-30
DALLAS (I-45)—TEXARKANA

176 miles

See map on page 44.

The western third of I-30 between Dallas and Texarkana traverses a band of Creta-ceous limy rocks in the Blackland Prairie area, while the eastern two-thirds crosses soft, sandy to clayey sediments of early Paleocene and Eocene age. You will see more on the geologic map than along the highway because the region has few roadcuts or outcrops. Light-colored limestone rubble in fields near Dallas and gray clay in stream banks farther east are about the only roadside indicators of the geology. Some parks in and around Dallas, where streams have cut into the bedrock, provide an opportunity to see what lies under your feet.

Notice on the geologic map on page 44 how the bands of Paleocene and Eocene rocks swing eastward toward Texarkana. This curve represents the position of the edge of the early Gulf Coast shortly after the Gulf opened as North America split from South America in Triassic time. After the Western Interior Seaway receded at the end of Cretaceous time, wedges of sand and clay sediment have been progressively added to fill in the Gulf, resulting in today's shoreline at Galveston, about 225 miles away, and a current pile of sediment about 40,000 feet thick.

Six miles east of the I-30 and I-635E junction (exit 56C), I-30 crosses Lake Ray Hubbard, an artificial lake created by the damming of the East Fork of the Trinity River. The Trinity heads in the Rolling Plains and Western Cross Timbers country

ANDERSON BONNER PARK

A few miles north of I-30 near the I-635/US 75 interchange is Anderson Bonner Park. Flowing through the park is the aptly named White Rock Creek, which has carved its way downward through a massive bed of stark-white Austin Chalk. This soft, limy rock was deposited between 89 and 85 million years ago in the Late Cretaceous when the Western Interior Seaway covered this part of Texas. It is thought that the water depth in this area of the sea was as much as 800 feet, too far from shore to deposit sand and mud. Microscopic plankton called coccolithophores thrived in the warm sea, and when they died, their remains settled to form layer upon layer of seafloor sediment that is now the Austin Chalk. During the Cretaceous, Earth's climate was warmer than now, a veritable greenhouse, and coccolithophores lived in all the world's oceans. Thick layers of chalk formed across Earth. Today, we can see immense white chalk banks in places such as the White Cliffs of Dover in England and the German island of Rügen.

Anderson Bonner Park, sometimes referred to as the White Cliffs of Texas, provides an excellent exposure of Austin Chalk along White Rock Creek in Dallas. (32.9203, -96.7780)

ARBOR HILLS NATURE PRESERVE

One place to see a rare outcrop of a rock that has had a profound effect on the economy of Texas is in the Arbor Hills Nature Preserve on the north side of Dallas. Here, in the streambed and walls along Indian Creek, are dark-gray layers of the Eagle Ford Shale, a series of organic-rich shale and muddy limestone marls. The Eagle Ford was deposited during the Cretaceous Period when the Western Interior Seaway covered this part of Texas. These sediments were deposited in deeper waters off the coast where there was little to no oxygen, which prevented algae, plankton, and organic-rich material from decaying. The organic matter gives the Eagle Ford its dark color.

In much of Texas, the Eagle Ford Shale is buried thousands of feet underground. Because it is buried so deep, it has been affected by Earth's geothermal gradient. As the shale is heated, the unit passes through what is called the "oil window," the temperature and time required to turn organic matter into crude oil. Because of the vast amounts of oil discovered, as well as new crude oil recovery methods such as horizontal drilling and hydraulic fracturing (fracking, for short), a new oil boom is taking place in Texas.

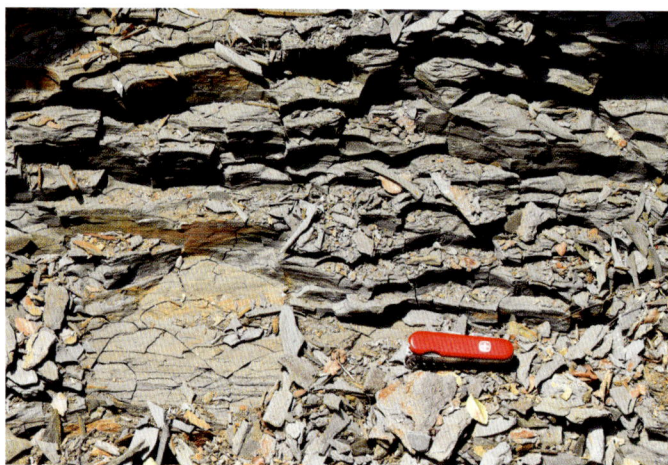

This close-up shows the shaly bedding of the Eagle Ford Shale exposed along Indian Creek in the Arbor Hills Nature Preserve. Swiss Army Knife for scale. (33.0478, -96.8504)

northwest of Dallas–Ft. Worth and becomes one of the major streams of East Texas, ultimately flowing into Galveston Bay east of Houston.

After climbing out of the valley that contains Lake Ray Hubbard and the Trinity River, I-30 passes by the town of Rockwall, so named for the mysterious feature that is found nearby. In 1852, local farmers digging a well discovered an unusual rock wall that later was thought to cross all of Rockwall County. The original wall was approximately 23 feet tall and extended for 20 miles. Later work determined that there were many "walls" throughout the county of varying sizes. By 1900, geologists who examined the wall hypothesized that the walls were clastic sand dikes formed naturally when a slurry of sand was rapidly injected into an open fracture that cut across sedimentary rock layers and other rock types. These dikes are commonly vertical or near vertical and can vary in width from a few inches to several feet across. Shortly after 1900, rumors began to swirl that some prehistoric people had created the wall. Even to this day, the debate rages on whether the wall is natural or manmade. The wall still exists but is buried and on private property, so public access is not allowed. A portion of the wall, recreated with rocks from the first site of discovery in 1852, can be seen at

Sections of rock from the original "rock wall" were used to create this replica located at Rockwall County Historical Foundation Museum in Rockwall. (32.9302, -96.4494)

the Rockwall County Historical Foundation Museum located at 901 E. Washington St. in Rockwall.

A mile east of Greenville, the freeway crosses the Sabine River, barely a creek here. The Sabine flows southeastward from its headwaters east of Dallas in the Blackland Prairie. It becomes a sizeable river by the time it marks the state border with Louisiana before emptying into the Gulf of Mexico at Port Arthur.

Near Winfield, just west of Mt. Pleasant, 50-million-year-old lignite beds contained in the Wilcox Group were mined in open pits. Lignite is a black, highly compressed plant material on its way to being coal. The plant material was deposited around the early edge of the Gulf Coast in swampy areas between river drainages. Similar environments exist today landward of the Texas barrier islands, and the drive from Houston to Galveston emulates a trip through the environment of deposition of this 50-million-year-old lignite.

Coal mining around Winfield began in 1974 at the Monticello-Winfield Mine and continued until 2015, when the mines were shut down. During that period, more than 15,500 acres were mined and subsequently reclaimed. Today, one would never know that this was a major coal mining area of Texas.

In riverbanks where the freeway crosses White Oak Creek (between mileposts 174 and 175) and the Sulphur River (milepost 181), gray clays of Eocene age are exposed. The outcrops are small along I-30. A better view of these clays can be seen in riverbanks where US 259 crosses the Sulphur just north of I-30.

Between Mt. Pleasant and Texarkana, notice how pine trees become more common eastward. Pines prefer well-drained, sandy soils, and the soils are sandier because sandstone is more common in this part of the Eocene Wilcox Group. Rainfall also increases from 36 inches per year measured at Dallas to 48 inches per year at Texarkana, and pines require this added moisture. The countryside becomes a little hillier near Texarkana, as the road crosses the topographic relief created by lateral drainages of the Sulphur and Red Rivers.

Just north of Texarkana, the Red River (which is a major tributary to the Mississippi River) forms the boundary between Texas and Arkansas, and farther to the west,

Gray, Eocene-aged clays appear along the riverbanks of the Sulphur River. (33.3082,-94.7282)

the boundary between Texas and Oklahoma. The river has been used as a boundary since at least the late 1700s when it was the dividing line between territories of Spain and France. In 1805, a royal document proclaimed the river to be the northern and eastern boundary of the Spanish province of Texas. When later surveyors mapped out the borders between the states, using a major river such as the Red made perfect sense. Geology, however, had other ideas. Rivers move and meander periodically, in some cases shifting their channel many miles. Look at any map or Google Earth satellite image today, and you will see a state border that does not follow the modern river but instead follows where the river used to flow when the state border was laid out hundreds of years ago.

This satellite image of northeast Texas shows the Red River and its convoluted path that has changed over the years. Oxbow lakes occupy many former channels. —Image courtesy Google Earth, November 2023

The Red River, as viewed from TX 8 north of New Boston, is slow and meandering, which allows sandbars to form. (33.5694, -94.4118)

I-820/I-35W
FT. WORTH — GAINESVILLE — OKLAHOMA
78 miles

We begin this road guide on I-820 on the west side of Ft. Worth at the junction with I-30. North of the junction, I-820 travels along the flat terrain of the Grand Prairie. Subdivisions and urban development have encroached on what was once rich soils and farmland. Although few outcrops or roadcuts occur along the way, the Grand Prairie sits on a belt of flat-lying Fredericksburg Group limestone, sandstone, and shale beds that were deposited near the shoreline of a warm, shallow sea in the Early Cretaceous that would become the continent-spanning Western Interior Seaway by the Late Cretaceous. As this sea advanced across middle America, shorelines pushed outward into the clear marine water, laying sandstone and mudstone over limestone. When sea level rose, marine limestone was laid back over the retreating sandy and muddy shoreline. This pattern repeated many times; alternating layers of sandstone, limestone, and mudstone (shale) are common in this interval of the Cretaceous.

As I-820 leaves the I-30 intersection, note that the highway is built on a high spur, which is held up by Cretaceous Goodland Limestone, though no rocks are visible. As I-820 crosses the bridge over Lake Worth, look east to see an exposed bank of Early Cretaceous Paluxy Sand and Goodland Limestone at water level. Note the beds are flat lying. To see this rock up close, head to Mosque Point Park along Cahoba Drive, accessed from exit 8.

About 9 miles north of the I-30 junction, watch for a huge former quarry on both sides of I-820 between exits 12 and 13. Flat-lying beds of Cretaceous gray shale, limestone, and marl of the Kiamichi Formation, a unit in the Fredericksburg Group, were once beautifully exposed.

Sandstone ledges of the Late Cretaceous Woodbine Formation are exposed along the water's edge on the northern shore of Grapevine Lake at Rockledge Park. These sands were deposited along the shore of the Western Interior Seaway. To reach the

GRAND PRAIRIE

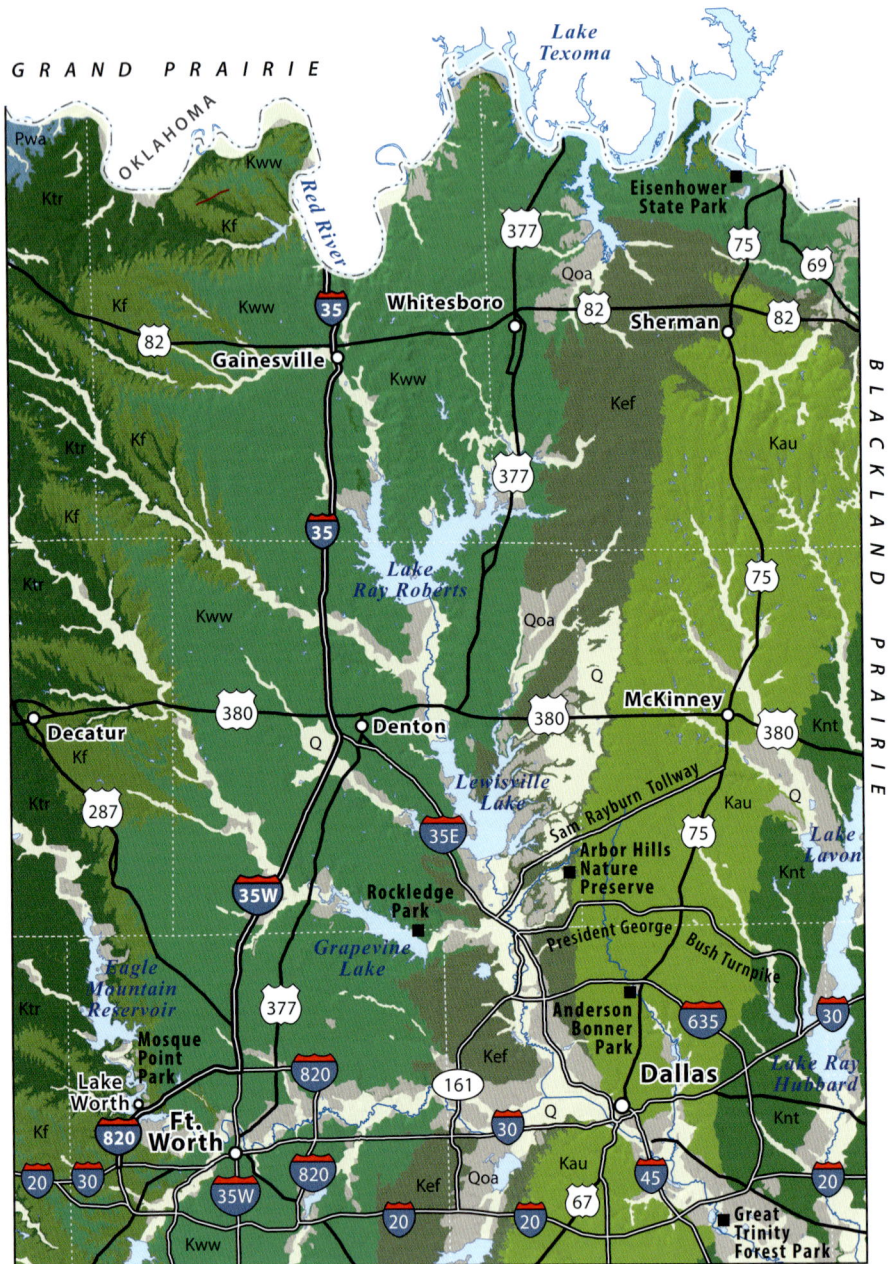

OKLAHOMA

Lake Texoma

Red River

Pwa

Ktr

Kww

Kf

Eisenhower State Park

377

75

69

82

Qoa

Whitesboro

82

Sherman

82

Kef

Kau

Kf

Kww

35

82

Gainesville

Kww

Ktr

Kf

377

Kf

Lake Ray Roberts

B L A C K L A N D

Ktr

35

Kww

Qoa

75

Kf

380

Decatur

Q

Denton

380

McKinney

380

Knt

P R A I R I E

Kf

287

Q

Lewisville Lake

Sam Rayburn Tollway

Kau

Lake Lavon

Ktr

35E

75

Arbor Hills Nature Preserve

Knt

35W

Rockledge Park

Q

Eagle Mountain Reservoir

377

Grapevine Lake

President George Bush Turnpike

Ktr

Mosque Point Park

Anderson Bonner Park

635

30

Lake Worth

820

Kef

Dallas

Lake Ray Hubbard

Kf

Ft. Worth

820

161

30

Kau

Knt

820

20

30

Kef

Qoa

45

Great Trinity Forest Park

20

35W

20

20

67

Kww

CENOZOIC

QUATERNARY

| Q | recent sediments; includes alluvium |

| Qoa | older terrace deposits (Pleistocene) |

—— normal fault

county boundaries shown as dotted white lines

MESOZOIC (CRETACEOUS)

Knt	Navarro and Taylor Groups
Kau	Austin Chalk
Kef	Eagle Ford Shale
Kww	Woodbine Formation and Washita Group; includes Caddo Formation
Kf	Fredericksburg Group; includes Goodland Limestone and Kiamichi Formation
Ktr	Trinity Group; includes Antlers Formation and Paluxy Sand

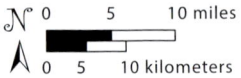

0   5   10 miles

0   5   10 kilometers

PALEOZOIC
(PERMIAN)

| Pwa | Wichita Albany Group |

Geology along I-35W between Ft. Worth and Gainesville.

EISENHOWER STATE PARK

Located on Lake Texoma, a reservoir of the Red River, Eisenhower State Park captures a moment in time where seas advanced across the continent during the Cretaceous. As sea level rose, the old Paleozoic land surface sank below the water. At the same time, up to 300 feet of sandstone was deposited as rivers drained the continental landmass to the northwest. This sandstone, the Antlers Formation of the Trinity Group, contains numerous cross-bedding features, evidence of these major river systems. As sea level rose further, clearer and calmer waters deposited Caddo Formation marlstone, as well as the Goodland Limestone, a resistant caprock in the Fredericksburg Group. Exploring the many trails within the park, visitors can also find younger deposits away from the lakeshore cliffs, such as a dark shale with nodular limestone cobbles associated with the Caddo Formation, as well as the Kiamichi Formation, another lighter-colored limestone from the Early Cretaceous. In addition, small fossils of seashells and clams can be found within the limestone outcrops.

Small ledges of Goodland Limestone along the shores of Lake Texoma within Eisenhower State Park. (33.8214, -96.6059)

Cretaceous shell and clam fossils can be found in outcrops of Goodland Limestone around Eisenhower State Park. Swiss Army Knife for scale. (33.8191, -96.6046)

park, take Cross Timbers Road (exit 74 from I-35W) east to Long Prairie Road and turn south (right) to Lakeside Village Boulevard and turn right. At the roundabout, turn onto Fairway Drive.

North of Ft. Worth, I-35W/I-35 crosses flat terrain all the way to Gainesville. North of Gainesville, hills are encountered as the expressway nears the Red River, which forms the Oklahoma border. Side drainages have cut into the bedrock here, and a few exposures of light-colored limestone occur at the Gainesville rest area (exit 502 for southbound travelers). The steep river bluffs are cut into Cretaceous limestone, but none can be readily seen from the bridge.

The Red River is aptly named, for it often runs bright red because it carries red sediment. The headwaters of the river are on the High Plains of the Texas Panhandle, where the Red River's tributaries have cut canyons through bright-red Permian shales and sandstones. The river continues to run eastward, forming the Texas-Oklahoma border where more red Permian rocks are exposed. Downstream from the I-35 crossing, Lake Texoma allows this red sediment to drop out and settle to the bottom in the upper reaches of the impoundment.

I-35W AND I-35
Ft. Worth — Waco — Austin
183 miles

From the junction of I-20, I-35W heads south along the flat terrain of the Grand Prairie. The underlying bedrock is Early Cretaceous limestone, though none shows up along the highway. In the distance to the east is a prominent, tree-covered topographic ridge, known as the Eastern Cross Timbers, that is held up by the Late Cretaceous Austin Chalk. The ridge swings closer to the highway near Alvarado. South of Alvarado, the highway traverses a low, flat area eroded into the soft Eagle Ford Shale, also of Late Cretaceous age. Unfortunately, few rocks appear along this stretch of road between Alvarado and Hillsboro, where I-35E joins to reform I-35.

Lake Whitney State Park, 17 miles west on TX 22 from exit 368A of I-35 in Hillsboro, is a popular place to fish and cool off when the weather turns warm. Scant outcrops of the Early Cretaceous Washita Group clays and limestones occur along the lakeshore. Across the lake are bluffs of Fredericksburg Group claystone and limestone, also Early Cretaceous in age. The Rock at Walling Bend Park, on the west side of the lake, is a small knob of Fredericksburg Group that protrudes above the lake surface.

Near Waco, the hills to the west are on the edge of the Comanche Plateau, an uplifted area where older Cretaceous limestones form the bedrock. The highway crosses the Brazos River in the center of Waco. The Brazos ("arms" in Spanish, originally called Brazos de Dios, or Arms of God) nearly bisects Texas from northwest to southeast. Tributaries of the Brazos have their headwaters on the High Plains in the Texas Panhandle. The Brazos frequently flows reddish brown from the red-colored sediment it picks up as it winds across Permian red beds in the country north of Abilene.

From Waco, the Brazos River continues flowing southeastward, paralleling TX 6 past Bryan, College Station, Washington (the first capital of Texas), Rosenberg, and finally Freeport, where it empties into the Gulf of Mexico. The Brazos River was considered by Stephen F. Austin to be the future major waterway for trade across Texas when

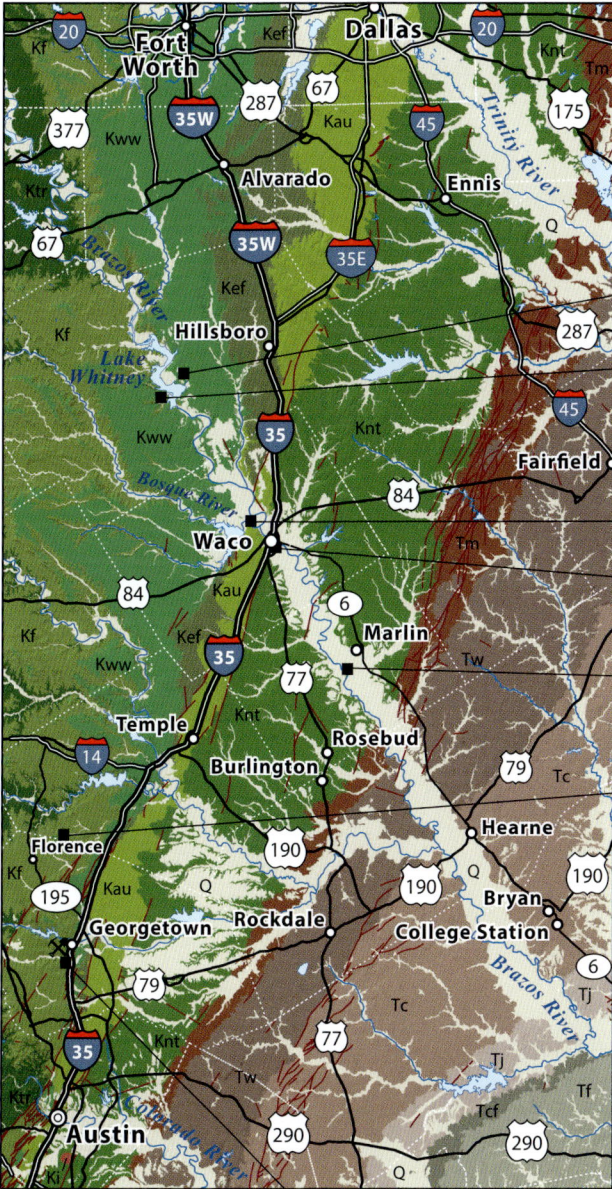

EASTERN
CROSS TIMBERS

COMANCHE PLATEAU

LAMPASAS CUT PLAIN

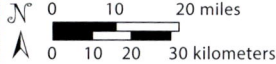

0   10   20 miles
0 10 20   30 kilometers

Fort Worth
Dallas

Alvarado
Ennis

Hillsboro

Lake Whitney

Waco

Fairfield

Marlin

Temple

Rosebud

Burlington

Hearne

Florence
Bryan

Georgetown
Rockdale
College Station

Austin

Inner Space Cavern

BALCONES
ESPARPMENT

Lake Whitney State Park

Walling Bend Park

Waco Mammoth
National Monument
and Brazos Park

Mayborn Museum
at Baylor University

Falls on the Brazos Park

Gault Archeological Site

Trinity River
Brazos River
Bosque River
Colorado River
Brazos River

⚒ quarry

normal fault;
movement is down
to the southeast

MESOZOIC
CRETACEOUS

Knt	Navarro and Taylor Groups; includes Pecan Gap Chalk of the Taylor Group
Kau	Austin Chalk
Kef	Eagle Ford Group
Kww	Woodbine Formation and Washita Group; includes Georgetown Limestone of the Washita Group
Kf	Fredericksburg Group; includes Edwards Limestone
Ktr	Trinity Group

NEOGENE and PALEOGENE

Tf	Fleming Group (Miocene)
Tcf	Catahoula Formation (Oligocene to Miocene)
Tj	Jackson Group (Eocene and Oligocene)
Tc	Claiborne Group (Eocene)
Tw	Wilcox Group (Eocene)
Tm	Midway Group (Paleocene–Eocene)

CENOZOIC

Q	recent sediments, undivided (Holocene and Pleistocene)

VOLCANIC ROCKS

Ki	basalt of Pilot Knob (Cretaceous)

Geology along I-35 between Ft. Worth and Austin.

The Pecan Gap Chalk forms a ledge in the Brazos riverbed at the Falls on the Brazos Park, southwest of Marlin. (31.2461, -96.9199)

he established a colony at Ft. Bend (modern day Richmond, southwest of Houston) in 1821. Paddle-wheel boats moved up and down the Brazos carrying cotton and trade goods for many years until the advent of railroads and dams.

Southwest of Marlin, a town on TX 6 southeast of Waco, the Brazos flows over a small sliver of Pecan Gap Chalk, a Late Cretaceous carbonate of the Taylor Group that is exposed in the riverbed. The Falls on the Brazos Park, on FM 712, provides an opportunity to observe the Pecan Gap, as well as a small "falls" that the river flows over.

Between Waco and Temple, the eastern edge of the uplifted Lampasas Cut Plain is visible to the west as a low line of skyline hills held up by resistant Cretaceous limestone. I-35 lies on the Austin Chalk, though none is seen except in fleeting glimpses of small exposures beneath highway bridges and stream banks.

East of Temple around 1907 (the exact date is not known), a 125-pound meteorite fell and was later plowed up by a field hand on a local plantation near Burlington. The meteorite was a well-known curiosity to people in the area and was later moved to nearby Rosebud, serving as a hitching post in front of the drug store. In 1915, the meteorite was presented to the University of Texas in Austin for study and resides there today.

Between Temple, the wildflower capital of Texas, and Austin, the state capital, I-35 runs northeast-southwest, paralleling the edge of the Hill Country, which lies to the west of the highway. The road is mostly on the flat terrain that lies at the foot of the Balcones (the Balcony) Escarpment. South of Temple, on the skyline to the west, the elevated hills marking the edge of this escarpment are readily apparent. Between Temple and Georgetown, I-35 follows a band of Early Cretaceous Georgetown Limestone of the Washita Group. The freeway swings eastward between Georgetown and Austin, encountering the younger Late Cretaceous Austin Chalk. Here and there, a glint of white chalk can be seen in stream cuts beneath bridges, or in small roadcuts, but you must look carefully to find these few, rather poor exposures.

Located 6.5 miles northeast of Florence on FM 2843, the Gault Archeological Site is an ongoing research project that started in the early 1900s when Henry Gault found artifacts on his farm while tilling. A short-lived scientific investigation of the farm began in 1929, but local landowners prevented any additional excavations after that point until the 1990s when Clovis points and mammoth bones were discovered. Recent excavations show that Clovis people inhibited the site around 13,000 years ago. The area was favorable for settlement, having a spring-fed creek and a diversity of game and resources, including an abundant supply of chert from the nearby Edwards Limestone. However, the most surprising discovery to date is a layer of

WACO MAMMOTH NATIONAL MONUMENT

Along the Bosque River northwest of Waco is Waco Mammoth National Monument, which preserves what is believed to be the only known nursery herd of Columbian mammoths in the fossil record. The story begins about 1.5 million years ago in the Pleistocene Epoch when mammoths migrated to North America across a temporary land bridge in the Bering Strait. Some of these evolved into the Columbian mammoth and truly lived up to their name; these pachyderms stood more than 14 feet tall and weighed more than 11 tons! Around 65,000 years ago, a small group of female and juvenile mammoths (and one camel) were in this area when it is hypothesized they were caught in a catastrophic event and buried in river sediments. The most likely explanation is that a sudden flood caught the mammoths by surprise and trapped them in the rising waters because they were unable to climb the slippery riverbanks to safety. Another hypothesis is that this was the last watering hole during a severe drought, and the animals died of dehydration. A second group of mammoths, which were fossilized around 51,000 years ago, were also found in the same stream channel, again the likely victims of a flash flooding event.

To date, twenty-six mammoths and the lone camel have been discovered at the site since it was discovered in 1978. In addition, fossils of a dwarf antelope, alligator, giant tortoise, a juvenile saber-toothed cat, and other animals have been found. Baylor University's Mayborn Museum serves as the repository for all fossils discovered at the monument, allowing researchers to study the fossils and recreate the conditions of this area during the Pleistocene. Located on Steinbeck Bend Drive, the monument is accessed via exit 339 (Lake Shore Drive) for southbound travelers and exit 335C (Martin Luther King Jr. Boulevard) for northbound travelers.

Beautifully preserved fossils of mammoths can be seen inside the protective enclosure at Waco Mammoth National Monument. (31.6058, -97.1743)

sediment below that of the Clovis Culture that contains stone projectile points with a completely different design than the Clovis points above it. These artifacts have been dated to between 16,000 and 20,000 years old, challenging the belief that Clovis people were the first to inhabit North America. The Gault site is private, and entry is by appointment only.

Inner Space Cavern, a commercial enterprise at Georgetown, offers tours. The cave formed in limestone uplifted along the Balcones fault zone as subterranean groundwater etched its way through cracks in the limestone.

South of Georgetown is a huge limestone quarry west of I-35 between mileposts 259 and 257. The Texas Crushed Stone Company mines Cretaceous Edwards Limestone here, mainly for road aggregate. Blocks of limestone are lined up along the feeder road to form a stone fence, where you can get a close look at the rocks. The Edwards Limestone is the main rock unit that forms the Edwards Plateau west of Austin and north of San Antonio. Springs, caves, and sinkholes are common in the Edwards Limestone, and the water in the subterranean labyrinth of the Edwards supplies San Antonio's drinking water.

A limestone block fence along I-35 south of Georgetown near milepost 258 provides a closer look at the Edwards Limestone being mined in a quarry to the west of here. (30.5910, -97.6922)

I-45
DALLAS—HOUSTON
237 miles

I-45 begins its southward path from near downtown Dallas in an appropriately named interchange with I-30 and I-35E called the "mix-master" by local residents. For the first couple of miles, the roadway traverses a typical cityscape. About 2.5 miles south of the I-30 interchange, you enter into the broad, shallow floodplain of the Trinity River, the longest river with its entire watershed within the state of Texas. The river has cut its way into the Austin Chalk, the predominant bedrock type in this part of Texas.

To see a wonderful roadcut in the Austin Chalk, take exit 277 and turn left (east) on Simpson Stuart Road, then right (south) on S. Central Expressway (TX 310) for 1.5 miles. This roadcut (on both sides of the road just north of I-20) is in white, marly

Geology along I-45 between Dallas and Houston.

Map labels:

exit 277/TX310

Corsicana oil field

INNER COASTAL PLAIN

East Texas Oil Museum

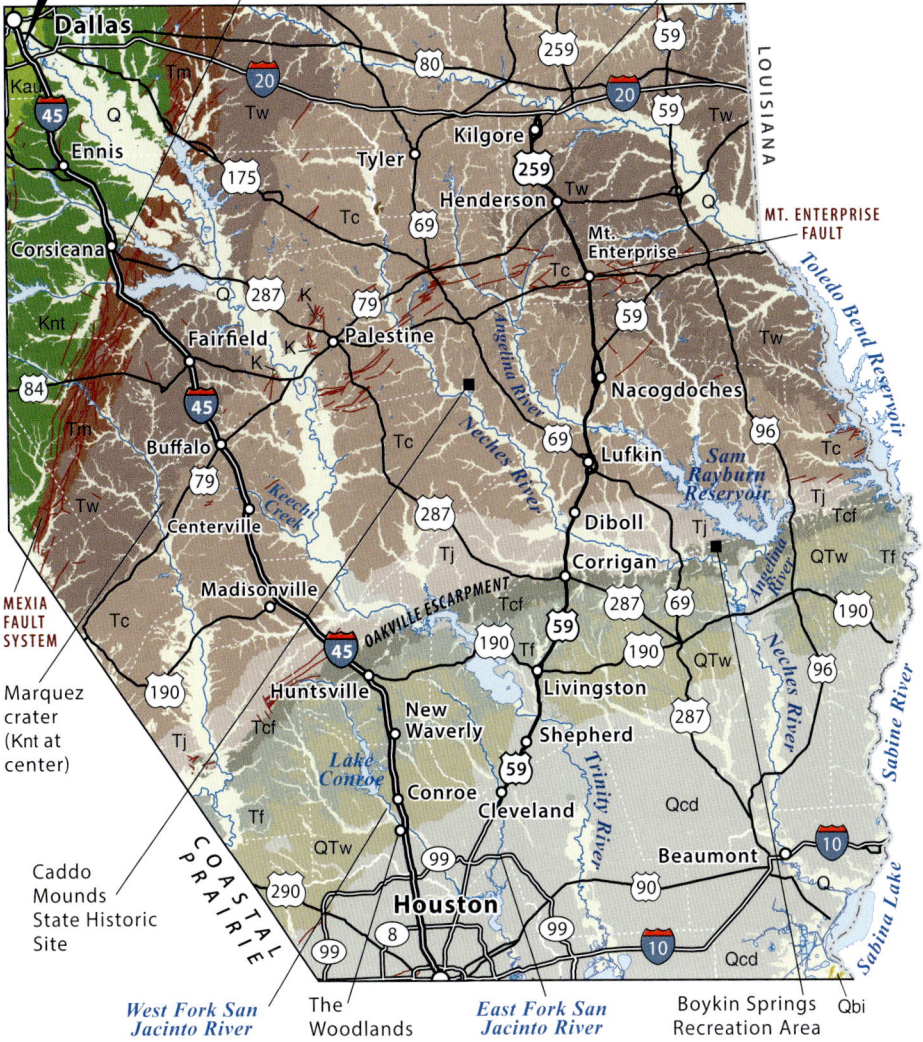

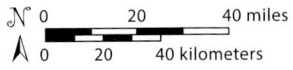

Dallas

Tm

Kau

Q

Ennis

45

259

80

259

59

20

20

59

LOUISIANA

Tw

Kilgore

Tyler

175

Henderson

259

Tw

Q

MT. ENTERPRISE FAULT

Corsicana

Tc

69

Mt. Enterprise

Toledo Bend Reservoir

Knt

Q

287

K

79

Palestine

Tc

59

Fairfield

K

K

Ingelina River

Nacogdoches

Tw

84

45

Tc

Tc

96

Buffalo

79

Tc

Neches River

69

Lufkin

Sam Rayburn Reservoir

Tj

Tcf

Keechi Creek

287

Diboll

Tj

QTw

Tf

Tw

Centerville

Tj

Corrigan

MEXIA FAULT SYSTEM

Madisonville

OAKVILLE ESCARPMENT

Tcf

287

69

190

Angelina River

Marquez crater (Knt at center)

45

190

Tf

59

190

QTw

96

Huntsville

Tcf

Livingston

287

Neches River

Sabine River

190

Tj

New Waverly

Shepherd

Caddo Mounds State Historic Site

Tf

Lake Conroe

59

Trinity River

COASTAL PRAIRIE

QTw

Conroe

Cleveland

Qcd

Beaumont

Sabine Lake

290

99

10

Houston

8

90

Q

99

10

99

Qcd

Qbi

West Fork San Jacinto River

The Woodlands

East Fork San Jacinto River

Boykin Springs Recreation Area

Legend:

N 0   20   40 miles
0   20   40 kilometers

—— normal fault

county boundaries shown as dotted white lines

CENOZOIC

QUATERNARY

Q recent sediments; includes alluvium and older terrace and high gravel deposits (Holocene and Pleistocene)

Qbi barrier island deposits

Qcd coastal deposits; includes Lissie, Beaumont, and Deweyville Formations (Pleistocene)

QUATERNARY–NEOGENE

QTw Willis Formation (Miocene to Pleistocene)

NEOGENE and PALEOGENE

Tf Fleming Group (Miocene)

Tcf Catahoula Formation (Oligocene to Miocene)

Tj Jackson Group (Eocene and Oligocene)

Tc Claiborne Group; includes Weches and Reklaw Formations and Sparta Sand (Eocene)

Tw Wilcox Group; includes Carrizo Sand (Eocene)

Tm Midway Group (Paleocene–Eocene)

MESOZOIC (CRETACEOUS)

K sedimentary rocks, undivided

Knt Navarro and Taylor Groups

Kau Austin Chalk

Geology along I-45 between Dallas and Houston.

A large roadcut of Austin Chalk occurs on both sides of TX 310 near the I-45/I-20 interchange south of Dallas. In the close-up, the wavy pattern along the left side of the rock is a shell fragment of the clam Inoceramus, *which could be several feet across. (32.6667, -96.7237)*

limestone of Austin Chalk and is full of fossils. Clams and oysters are most common, but pieces of huge clams up to several feet across are also common and characteristic for this time period in the Late Cretaceous. Named *Inoceramus*, they appear as flat dinner-plates, about 0.25 inch thick. The clam built its shell with calcite crystals arranged like stand-up toothpicks, perpendicular to the shell surface. Shells of this clam are so big, you probably won't find a whole one!

Between Dallas and Corsicana, I-45 crosses bands of Late Cretaceous marlstone, limestone, and shale, though the landscape is fairly flat, and rocks are not common along this stretch. At Ennis, the road swings east, then south again in a few miles, toward Corsicana. The Corsicana oil field, located near the southward bend in the interstate, is where the first major oil in Texas was discovered in 1894. Corsicana marks the edge of the Late Cretaceous belt and approximates the position of the original edge of the Gulf Coastal Plain that began to receive huge amounts of sand, silt, and clay sediments about 60 million years ago, a process that continues today.

South of Corsicana, the freeway traverses a series of sand, silt, and clay bands that represent successive episodes of Gulf-filling. Sandy ridges south of Corsicana are Paleocene, the oldest in the sequence, becoming younger and younger southward toward Houston. To the west are low skyline ridges held up by Cretaceous rocks. Cretaceous rocks are separated from the younger Eocene rocks by a series of linear, north-south faults that are related to the timing and trend of the Balcones fault zone farther west. Here, the Mexia fault system was also activated in Miocene time, 10 million years ago, and like the Balcones fault near Austin, this one follows the edge of the old, buried Ouachita Mountain range trend.

Low, rolling hills of sand and shale beds characterize the countryside between Corsicana and Fairfield. Near Fairfield are three major salt domes that have pushed older Cretaceous caprocks to the surface to nestle among the surrounding Paleogene terrain. Upper Gulf Coast salt domes are numerous in the subsurface around this area, and the surface domes are merely an expression of this extensive array of subterranean salt upheavals.

South of Fairfield between mileposts 186 and 184, the interstate crosses a rather prominent ridge of Carrizo Sand of Eocene age. Look for redness of the soil and a few sandy exposures. Between Buffalo and Huntsville, the road rolls up and down over successive ridges of Eocene-aged Claiborne Group sandstones. These ridges are the result of different beds of resistant sands that dip toward the Gulf Coast. A prominent ridge of Sparta Sand is located about 10 miles south of Buffalo between mileposts 167 and 166. Here, the freeway climbs out of Keechi Creek and over the sand ridge, though there are no obvious outcrops or roadcuts of this Eocene unit.

About 58 million years ago, a meteorite struck the coastal plain about 15 miles west of Centerville near Marquez. The resulting 8-mile-diameter Marquez Crater is not visible at the surface, having been buried by the pile of Paleocene and Eocene sediments deposited in the growing Gulf of Mexico. Its presence is known from the disruption of subsurface Cretaceous-aged rock.

The hilliness south of Centerville is mainly due to a ridge of Sparta Sand, which is a quartz sand that weathers yellow and brown. All the Pleistocene to Eocene sand and clay deposits encountered along this section of highway are part of the thick wedge of sediment that Texas rivers dumped into the Gulf of Mexico beginning about 66 million years ago as sea levels fell from their high stand during the Cretaceous. The wedges of sediment tip southward; units exposed in this area lie deep under the surface below Galveston. Many of these sedimentary rocks were exposed in freeway roadcuts when the highway was first built but are now hidden behind the grassy slopes bordering the road.

Between Madisonville and Huntsville, younger Eocene sandstone units cross the highway at nearly right angles, creating an even more noticeable up-and-down topography. Roadcuts are rare, but you can see a sandy exposure here and there along the way. The Oakville Escarpment, the boundary between an older Jackson Group sandstone and the overlying Catahoula Formation, is a slight rise near milepost 125. South of Huntsville, low-rolling hills mark the presence of Miocene sandstones and shales.

I-45 crosses the northern edge of the Coastal Prairie south of New Waverly, about 50 miles north of Houston. The freeway rolls along on the flat floodplain of ice age (Pleistocene) sand and clay deposits, crossing the West Fork of the San Jacinto River a few miles south of Conroe. Lake Conroe, west of I-45 north of Conroe, is a reservoir on this river. It may seem almost impossible to construct a dam across a river where no hard bedrock exists to support it. However, engineers have come up with an interesting way around this. Instead of using concrete to hold back the water, dams in this part of the world are constructed using earth-fill embankments, which are large mounds of compacted soil and rock built across a river valley. This method is preferred due to the generally flat terrain and readily available material to build such structures, as well as for its cost of construction. In this type of design, spillways, constructed as separate channels, are a crucial component to maintain the strength of the dam during excessive water events.

South of The Woodlands, urbanization obscures all geology along I-45. In downtown Houston, the Houston Museum of Natural Science is a must visit for anyone interested in prehistoric life, Texas geology, and the Gulf Coast petroleum industry.

US 59 AND US 259
HOUSTON — NACOGDOCHES — KILGORE (I-20)

205 miles

See map on page 59.

This south-to-north trek is another transect across the Coastal Prairie and onto the low hills of the adjacent Inner Coastal Plain. The highway crosses sandy bands of Pleistocene to Eocene sedimentary rocks. The 40 miles north of Houston to Cleveland is on the flat coastal plain, which is broken only by crossings of the West Fork and East Fork of the San Jacinto River, streams that begin at the edge of the Oakville Escarpment. Tan, sandy soils that predominate here are covered by extensive stands of pines that love such soils.

About 50 miles north of Houston, look for orange-tan, sandy soils and poor road-cuts around Shepherd, where the Trinity River creates a bit of topography in the ice age deposits of the Willis Formation. The more rolling topography between the Trinity River and Livingston marks the southern edge of the Inner Coastal Plain, where Miocene orange-tan sands and soils are found.

The highway crosses the Neches River between Corrigan and Diboll, about 100 miles north of Houston. The abundant pines growing on the sandy soils atop Miocene sandstones in this area support an active timbering industry.

Along the southern shoreline of Sam Rayburn Reservoir are weakly consolidated sandstones of the Oligocene-aged Catahoula Formation. (31.0627, -94.1054)

Between Diboll and Nacogdoches, the roadway crosses the orange to tan Sparta Sandstone of the Claiborne Group. The Eocene-aged Sparta is generally a ridge former, and hilly topography can be traced across the countryside where the Sparta comes to the surface. A few poorly exposed roadcuts of these sandstones can be seen along this section of highway. Rolling hills around Lufkin are on Sparta sandstones as well. Note how the topography increases to the north as the underlying formations get older. The sandstones harden with age, becoming a little more resistant to erosion, so hills stand higher.

Between Lufkin and Nacogdoches, the highway crosses the Angelina River, a tributary of the Neches River, which ultimately feeds into the Sabine River at Port Arthur. East of Lufkin is the large Sam Rayburn Reservoir, which is a dammed segment of the Angelina River. The Angelina National Forest surrounds the reservoir and offers recreational opportunities in the piney woods.

BOYKIN SPRINGS RECREATION AREA

East of Lufkin near Sam Rayburn Reservoir is Boykin Springs Recreation Area, where a spring with clear, cool, flowing water was discovered by Sterling Boykin around 1850. A dam created the 9-acre lake. The rock in this area is the Oligocene-aged Catahoula Formation consisting of sandstone, siltstone, and mudstone. This package of sediments was deposited on a gently sloping plain along the shoreline of the Gulf Coast, where large stands of palm trees flourished. Today, the tree wood is preserved as *Palmoxylon*, a petrified wood that is the state stone of Texas.

During the Oligocene, supervolcanoes were erupting hundreds of miles to the west, sending their ash high into the atmosphere. Most of this ash settled in West Texas, forming ash layers more than 65 feet in the Catahoula, but some of this ash was carried eastward and fell onto the coastal plain. The siltstone and mudstone found at Boykin Springs contains shards of volcanic glass associated with the ash. Silica from volcanic ash helped preserve the wood and makes the Catahoula one of the more resistant rock layers found in the Coastal Bend region of Texas.

Water from Boykin Springs emerges from the blocky Catahoula Formation. (31.0591, -94.2791)

CADDO MOUNDS STATE HISTORIC SITE

Northwest of Lufkin, the Caddo Mounds State Historic Site preserves a small village where, around 800 CE, a group of Caddo people settled on a terrace of the Neches River. This area offered fertile soils for growing crops, forests teeming with game, and a spring that provided a reliable source of water. The Caddo people built grass-thatch homes in which to live and three large ceremonial mounds: one for burial and two others as platform mounds for a temple. The site was also part of a major trade network, with turquoise from New Mexico, copper from Michigan, and seashells from Florida discovered at the settlement. Around 1250, the site was abandoned for unknown reasons. By the mid-1800s, the Caddo people were forced to relocate to Indian Territory (now Oklahoma) and are now centered around Binger, Oklahoma.

In Nacogdoches, watch for a good roadside cut on Business US 59 (South Street) at Smith Street, where the Weches Formation, another Eocene unit of the Claiborne Group, can be viewed up close. Its red, weathered siltstones and sandstones overlie green, unweathered siltstones and sandstones. Careful examination shows extensive churning by marine organisms. Marine fossils are abundant, not only in this roadcut but also in many other rock exposures in and around Nacogdoches. The green color is from the iron-rich mineral glauconite, which forms in shallow marine shelf waters. As the iron in these rocks weathers, it oxidizes, or rusts, and turns red. Nacogdoches is home to Stephen F. Austin State University and claims to be the oldest town in Texas. The old stone fort near the college was built in 1779.

The width of this band of Sparta Sand, which lies directly above the Weches Formation, extends north from Nacogdoches for a few miles, and a few roadcuts of the orange-tan sand are seen near the junction of US 59 and US 259. From Nacogdoches, this guide follows US 259 to Kilgore.

Weches Formation siltstone and sandstone in a roadcut in central Nacogdoches at the Business 59/Smith Street intersection. (31.5964, -94.6613)

Iron staining brings out the individual layers of the cross-bedding in the Reklaw Formation. (31.7911, -94.6782)

Seven miles north of the US 259/US 59 junction, watch for an excellent roadcut in the Eocene-aged Reklaw sandstone, the oldest unit in the Claiborne Group. Beautifully preserved cross-bedding is accentuated with iron-staining. The straightness of the cross-beds indicates they were deposited in slow-flowing water. The red color from weathered glauconite further indicates deposition in shallow marine water.

Around the town of Mt. Enterprise, notice the hilliness, bright red soils, and occasional roadcuts as the highway crosses the colorful outcrop band of the Carrizo Sand, a unit in the Wilcox Group. A red Carrizo band lies south of Mt. Enterprise, and another one can be found north of town. The Carrizo is repeated here because of the east-west-oriented Mt. Enterprise fault, which US 259 crosses just south of town. The north side of the fault is the down-dropped side.

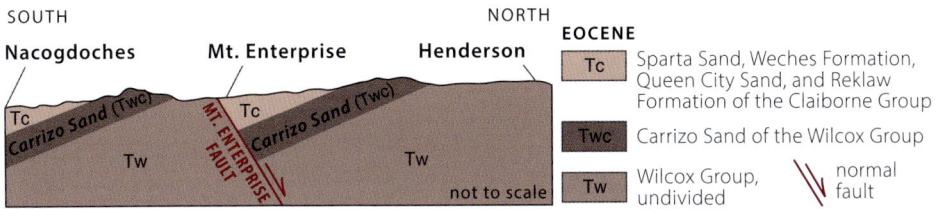

This cross section shows how two Carrizo sand bands, one on either side of Mt. Enterprise, are caused by faulting.

At the junction of US 79 and US 259 in Henderson, bright-red, orange, and tan Eocene Carrizo Sand exposures form a backdrop to the buildings along the roadway. The darker, reddish portions are sand channels, whereas the lighter, nearly white beds are clay layers deposited by floods next to the original Eocene channels. Between Henderson and Kilgore, the topography continues to roll along with more sand and shale alternations. The Carrizo Sand extends northward about half the distance to Kilgore.

Bright-red Carrizo sand-stone in a large outcrop behind a hotel near the US 79/US 259 junction in Henderson. (32.1483, -94.7900)

In Kilgore, home to Kilgore College, is the marvelous East Texas Oil Museum, located on Business US 259. The museum offers a historical look at a full-size street scene of a 1930s oil boom town, complete with a muddy street, shops, movie theatre, post office, and barbershop where you can hear all the news of the day. Murals, displays, and original equipment and tools tell the exciting story of discovery of the gigantic East Texas oil field by C. M. "Dad" Joiner in 1930. The field virtually engulfs the town of Kilgore and is the largest oil field in the contiguous United States, having produced 5.2 billion barrels of oil as of 2024; this is second only to Prudhoe Bay in Alaska, which has produced more than 12.5 billion barrels since 1977. The field still produces oil trapped in sandstone in the pinched out edge of the Late Cretaceous Woodbine Formation. The museum charges a small entry fee, but it is one of the best small museums in Texas.

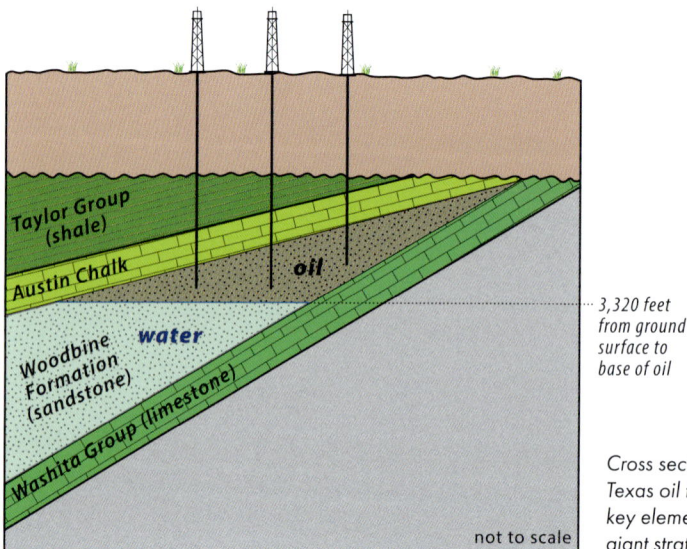

Taylor Group (shale)

Austin Chalk

oil

Woodbine Formation (sandstone)

water

Washita Group (limestone)

3,320 feet from ground surface to base of oil

not to scale

Cross section of the East Texas oil field, showing key elements of this giant stratigraphic trap.

US 59
TEXARKANA (I-30) — MARSHALL (I-20)

68 miles

See map on page 44.

The road between Texarkana and Marshall, a lovely drive through dense pine forests and small towns, traverses noticeably hilly topography in this part of Texas. The reason for the hilly relief is that bedrock is cut deeply by side drainages that run perpendicular to the road as they make their way toward the Red River only a few miles to the east. The road goes up and down across one drainage and drainage divide into another and then another. Between Texarkana and Wright Patman Lake (a reservoir on the Sulphur River), the bedrock is Eocene Wilcox Group sandstone and shale. Poor exposures of Wilcox and Reklaw sandstones can be seen along the beaches of Wright Patman Lake at both Rocky Point Park and Piney Point Park just west of US 59 south of the Sulphur River.

A red soil (top) caps the Queen City sandstone in an outcrop behind the Pizza Hut in Atlanta. (33.1203, -94.1791)

From Wright Patman Lake southward, the towns of Atlanta, Linden, Jefferson, and Marshall reside on younger Eocene Claiborne sandstones and shales. Watch for an excavation wall behind the Pizza Hut in Atlanta, where you can see a section of red soil over Eocene Queen City sandstone and claystone.

Halfway between Linden and Jefferson is a bright-red to tan Queen City sandstone and claystone roadcut on the east side of the highway at the tiny community of Prospect. The claystones were deposited in flooded areas between river channels in Eocene time, while the sandstones were laid down as sandy beds in the river channels.

Bright-red sandstone and pale-gray clay-stones, both from the Queen City Formation, are found in a low roadside outcrop north of Jefferson. (32.8421, -94.3662)

CADDO LAKE STATE PARK

Approximately 13 miles east of Jefferson off TX 43 is Caddo Lake State Park. Billed as one of the only natural lakes in Texas, Caddo has a unique history of formation. About 2,000 years ago, the Mississippi River claimed the Red River as one of its tributaries, causing the Red River's current to slow dramatically. Unable to efficiently move vegetation and sediment down the river, the debris piled up and created a dam. As the dam grew larger, water began spilling out of the Red River's channel, flooding low-lying areas. Over time, the debris pile grew to more than 100 miles long. Called the Great Raft by locals, this debris jam and subsequent backup of water spilled into the Big Cypress Creek drainage basin around 1770, creating Caddo Lake. In the mid-1830s, the US Army Corps of Engineers removed the Great Raft, only for it to begin rebuilding itself. Finally, in the 1870s, the raft was removed for good.

In the early 1900s, freshwater pearls were discovered in mussels that lived in the lake. However, when an earthen dam was completed in 1914, it effectively raised the water level of the lake and made reaching the mussels impossible. Another dam was built in 1971 that holds the water level of Caddo Lake at the level the Great Raft did naturally. Today, the shoreline is home to one of the largest inland bald cypress forests in the United States.

Bald cypress trees with Spanish moss line Big Cypress Bayou, where Big Cypress Creek is backed up in the upper reaches of Caddo Lake. (32.6918, -94.1794)

SOUTH GULF COASTAL PLAIN

HOUSTON AND THE GALVESTON AREA

Contrary to popular belief, the Houston-Galveston region has as much geology as anywhere in the state. If you have just returned from a trip to Big Bend National Park, however, this statement is undoubtedly stretching your credibility quotient as you peer across terrain so flat that the highest topography is found on freeway overpasses! But the comparison between the geology of Houston-Galveston and the geology of Big Bend National Park is a matter of geologic time. The rocks in Big Bend tell a story of events that happened in distant past millennia, while the events are

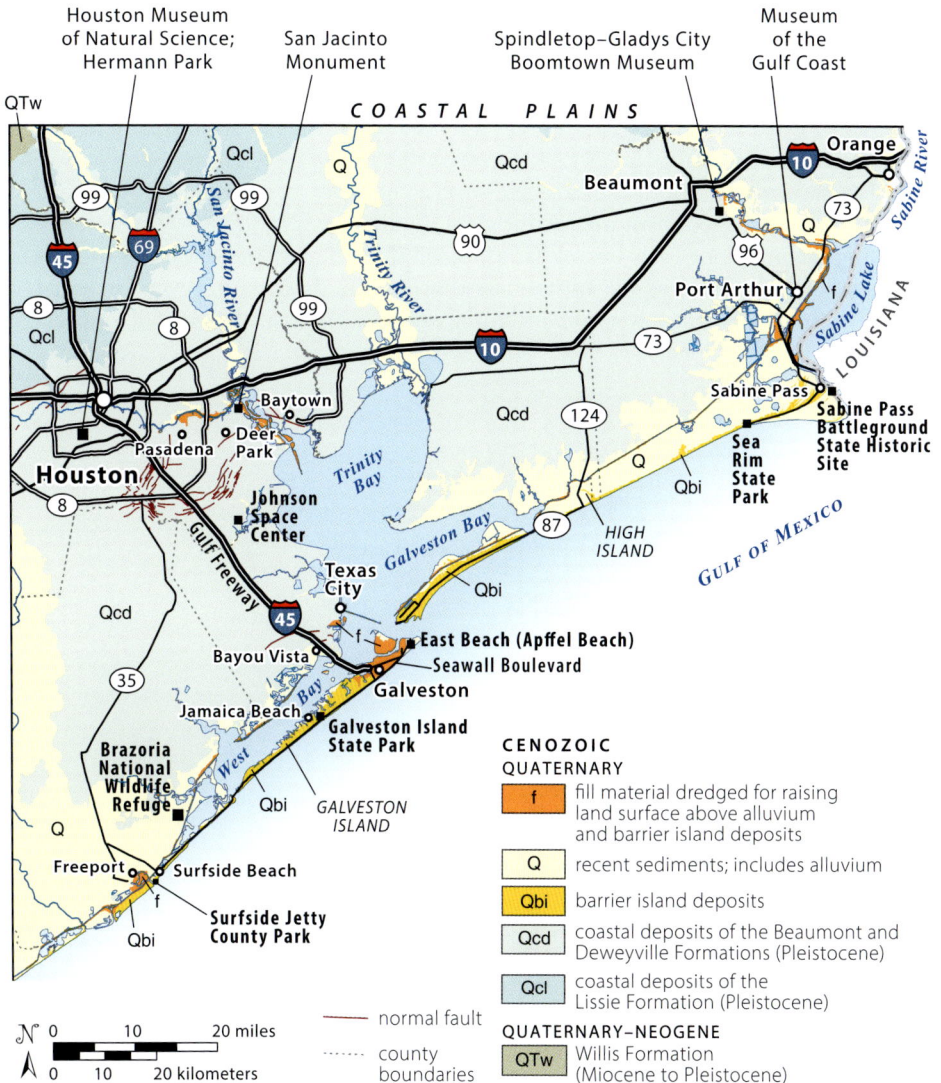

CENOZOIC
QUATERNARY

f	fill material dredged for raising land surface above alluvium and barrier island deposits
Q	recent sediments; includes alluvium
Qbi	barrier island deposits
Qcd	coastal deposits of the Beaumont and Deweyville Formations (Pleistocene)
Qcl	coastal deposits of the Lissie Formation (Pleistocene)

QUATERNARY–NEOGENE

QTw	Willis Formation (Miocene to Pleistocene)

normal fault

county boundaries

Geology of the Houston and Galveston area and along I-10 between Houston and Beaumont.

taking place right now around Houston-Galveston. Here, to be observed in real time, are active faults, subsiding land, rivers that deposit and erode sediment, hurricane-blasted barrier islands, tidally washed bays, beaches where sand movement is driven by pounding surf and waves, and swamps and marshes where organic-rich muds belch bubbles of fermented gas. This gas foretells of hydrocarbons to come in some distant future millennium when all the sediments in today's Houston-Galveston environments will be buried, in their turn, under thousands of feet of sediment.

Houston is the nerve center for the worldwide petroleum industry, with every major and many smaller world petroleum corporations either headquartered or have offices here. More petroleum geologists reside and ply their trade in Houston than any other city in the world. The worldwide search for petroleum, as well as the financing, production, refining, shipping, pipelining, supplying, and marketing of the world's principal energy material in the twenty-first century largely emanates from Houston.

The Houston metropolis rests on the flat surface of the coastal plain, which is at the top of a giant wedge of mud and sand deposited into the Gulf Coast during the Pleistocene ice ages by ancestral Texas rivers. These floodplain, river, and delta muds support a dense growth of semitropical vegetation that thrives in the mild climate and 50-inch annual rainfall.

Houstonians do not live on a stable surface, however, because numerous active faults traverse the sediment wedge. The giant mud wedge is naturally sliding into the Gulf of Mexico along long, curved faults, called slump faults. These features form at the bottom of a block as the entire mass slowly slides downward. Several external factors can cause a block to move, including earthquakes, excessive rainfall, or overloading of the block itself.

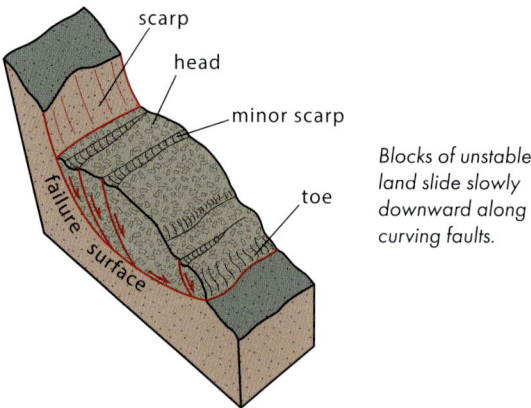

Blocks of unstable land slide slowly downward along curving faults.

At the same time, the sediment wedge sinks as the mud compacts under its own weight, and water is squeezed out. Houston rests on the nation's third largest aquifer, from which the city merrily pumped groundwater from two hundred wells until the 1960s, when the realization came that the entire metropolitan area was sinking at an accelerated rate. For example, the area around the San Jacinto Monument sank 6 feet between 1900 and 1964.

Subsidence was also noted at the Brownwood subdivision near Baytown, where the entire subdivision sank by more than 10 feet throughout the 1960s and 1970s. Removal of oil and gas early in the century also contributed to local sinking problems.

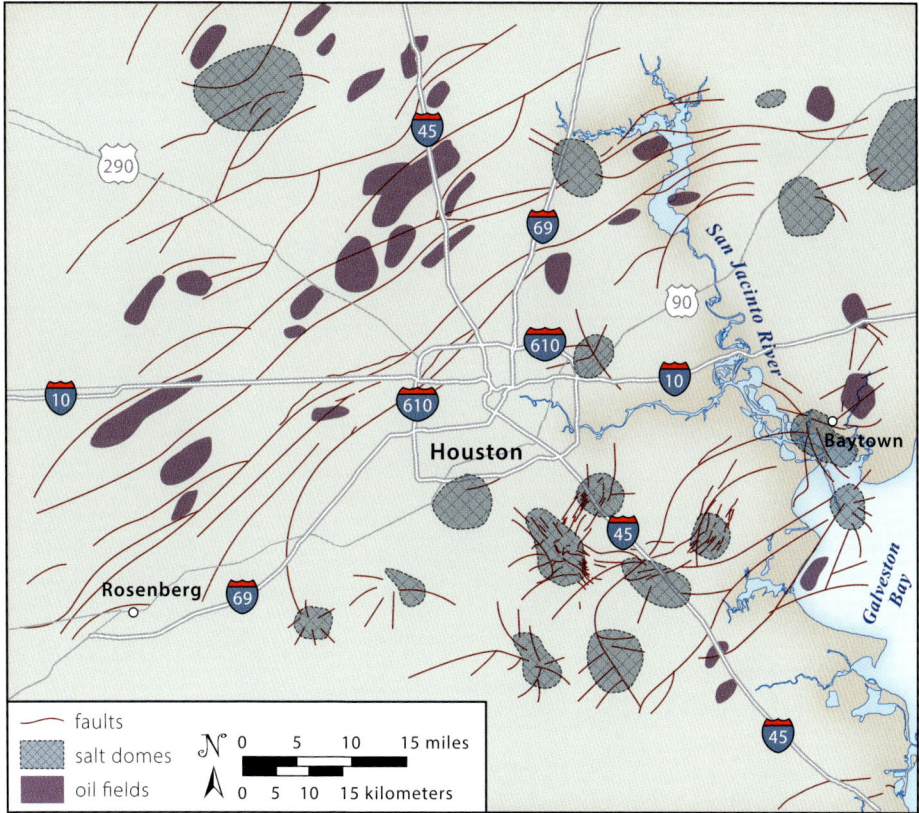

Subsurface faults, oil fields, and salt domes in the Houston area

Concrete bulkheads are all that remain from the Brownwood subdivision after much of the land subsided because of excessive groundwater usage. (29.7586, -95.0526)

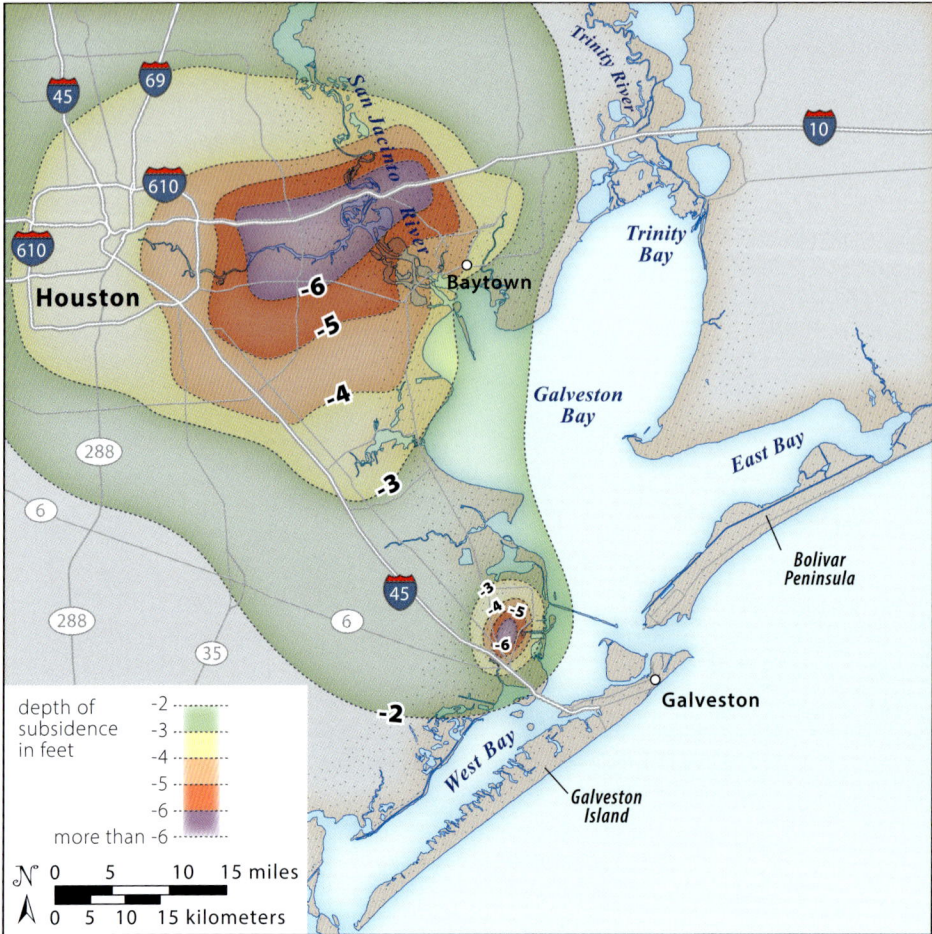

Subsidence in the Houston-Galveston area due to removal of petroleum and water from the subsurface.

The water pumping was stopped, and drinking water now comes from Lake Livingston and Lake Houston. Also, less and less oil and gas are removed from depleting reservoirs, so the rate of subsidence in Houston has decreased significantly. The natural subsidence from compaction, however, will continue, as will natural sliding along faults. Hurricane and thunderstorm flooding can be expected to become more severe in subsiding areas.

The Houston Museum of Natural Science in the northern part of Hermann Park is one of the premier museums of the world. The museum houses the Morian Hall of Paleontology featuring dinosaurs and prehistoric life, the Cullen Hall of Gems and Minerals with its world-class displays, and the Weiss Energy Hall, which explains the petroleum industry from exploration and recovery to processing and its end products.

THE GULF FREEWAY (I-45)

Between Houston and Galveston, the Gulf Freeway (I-45) rides on the flat surface of Pleistocene river sediment and delta muds. Note the intermittent swamps, incised bayous (creeks), and rich flora along the way. When this landscape is buried under a few thousand feet of sediment in a few million years, which it will be some day, it will be a source for new oil and gas.

The freeway follows the west shore of Galveston Bay, though you can't see the bay from the road. The bay occupies the drowned mouths of the San Jacinto and Trinity Rivers. The drowning occurred when sea level rose nearly 400 feet as climate warmed and the world's vast glaciers melted after the last ice age. Bay waters are only about 12 feet deep, so to accommodate shipping, a channel was dug early this century. As of 2023, the Port of Houston is the nation's busiest port in terms of overall tonnage moved.

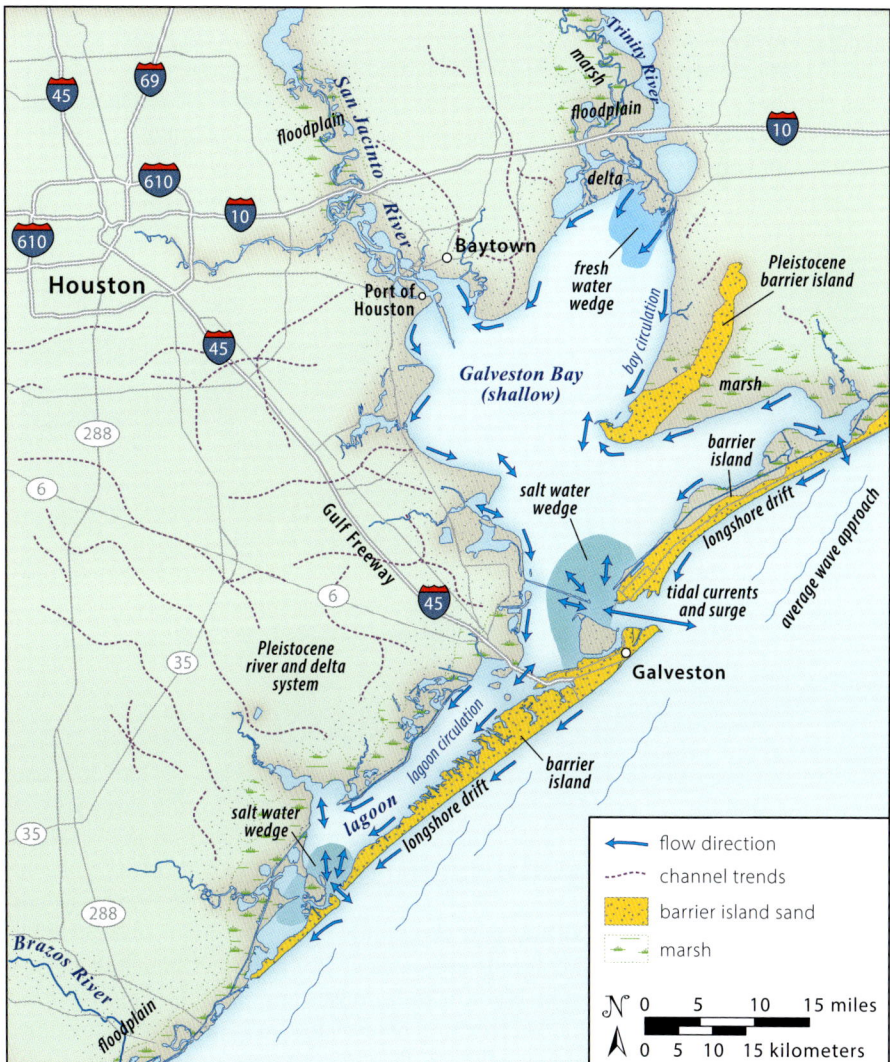

Surface environments and water circulation in the Galveston Bay area.

The water in Galveston Bay circulates mainly in a counterclockwise pattern, driven by dominant winds blowing out of the southeast and by low tidal surges that enter at the passes on either end of Galveston Island. Oyster shell banks are common in Galveston Bay, an indicator of the rich biological productivity that characterizes such a shallow bay.

Watch for NASA Road-1 (exit 24), about 13 miles south of Houston, that accesses the Johnson Space Center. In addition to Mission Control, the museum, and outdoor rocket displays, be sure to visit the Lunar Sample Laboratory in Building 31A, where the largest collection of moon rocks is stored and displayed and where analytical equipment for studying these important geologic samples is located.

Farther south, tall refinery columns rise like a metallic forest north of the freeway near Texas City. Some of the largest refineries in the country, which supply gasoline, diesel, and other petrochemicals, are located here, as well as farther north in nearby Pasadena and Deer Park. The shallow lagoonal waters of West Bay come into view as the freeway nears the causeway bridge to Galveston. Look for circular patches of salt marsh grass, caused by dense plant roots that hold and retain sand and organic material, thus building mounds, west of the road. The grass flats and salt marshes border both the mainland and barrier island sides of the lagoon, forming important habitats for birds and underwater dwellers.

Salt marshes and grassy mounds protect West Bay near Bayou Vista from coastal storms, as well as provide habitat for birds and other animals. (29.3265, -94.9304)

GALVESTON ISLAND

The city of Galveston, on Galveston Island, is located on a Karankawa site. It is also where the Spanish explorer Cabeza de Vaca was storm-tossed from a shipwreck in 1528. The pirate Jean Lafitte started the first European settlement on the island in 1817, and since then, Galveston has attracted sun worshipers with its 32 miles of sandy beaches.

The island is not always all fun and sun, however. On the night of September 8, 1900, a hurricane with wind speeds estimated at 140 miles per hour (equal to a Category 4 storm) hit Galveston Island with little warning. A 15-foot storm surge swept over the island, which had an elevation of just 9 feet above sea level at the time. The enormous storm surge, along with relentless storm waves and powerful winds,

The seawall at Galveston was built after the 1900 hurricane, then rebuilt higher after another storm in 1915. Large rocks in front of the wall help break up waves before they reach the wall. (29.2464, -94.8611)

destroyed more than 3,600 buildings in Galveston and claimed an estimated 8,000 to 12,000 people. The Great Storm of 1900 remains to this day the deadliest natural disaster in United States history. Shortly thereafter in 1902, the Galveston seawall was built, houses were jacked-up, and the city was elevated 11 feet with fill. For the most part, the 10.3-mile-long seawall has protected Galveston from the ravages of Gulf storms. The wall was no match in 2008, however, when large waves from Hurricane Ike overtopped it, damaging large portions of the wall and the city of Galveston.

A series of long stone jetties are part of the Galveston beach protection system. The large blocks in the jetties are native Texas granite, mined from the Proterozoic-age Town Mountain Granite in quarries northwest of Austin at the town of Marble Falls. This same granite is the principal building stone used in the Texas state capitol in Austin.

To see the natural environments of Galveston Island, drive southwest on Seawall Boulevard, beyond the end of the seawall. Note the rather flat, sandy profile to the island. Much sand is tied up by plants, so not a lot of free sand is available to build dunes, though low dunes are found behind the beach in many places. Hurricanes also play a vital role in flattening the profile of the island. A drive north from Seawall Boulevard on any number of back streets will take you across the beach ridges and give a view of the muddy, shelly banks, salt marshes, and grass flats that characterize the lagoon-side of the island.

Galveston Island State Park, southwest of Galveston in Jamaica Beach, preserves natural beaches, beach ridges, salt marshes, and grass flats. The park is what the rest of Galveston looked like before twentieth-century urbanization.

A highway with various names follows the barrier island chain southwestward and connects Galveston to Surfside Beach and Freeport. Brazoria National Wildlife Refuge, north of Freeport, hosts thousands of birds that enjoy the wetlands. At Freeport, the intercoastal waterway separates the coastal beach from the mainland.

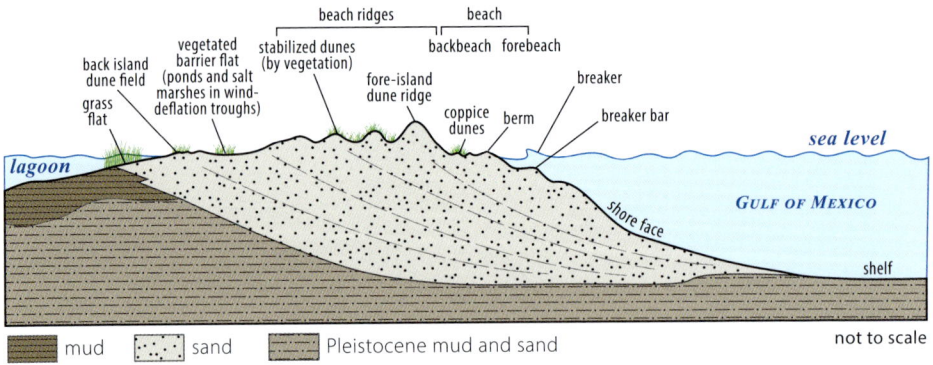

Cross section of Galveston Island, a barrier island.

At the northeast end of Galveston Island at East Beach (Apffel Beach), wind-blown dunes display wind ripples and cross-bedded internal structures, while a wide expanse of vegetated back-dune flats host swampy pools where waterfowl abound.

I-10
HOUSTON—BEAUMONT—LOUISIANA

112 miles

See map on page 69.

This trip encompasses the complete story of petroleum in Texas. At one end is Houston, where the modern-day petroleum industry is headquartered, while at the other end is Beaumont, where it all started, and in between, the road traverses depositional environments that tell the tale of how petroleum is formed.

Between Houston and Beaumont, I-10 rolls along the flat terrain so typical of the Gulf Coastal Plain. East of Houston, the highway crosses the San Jacinto River (between exits 786 and 787), and a few miles farther, it crosses the wide, swampy, and lake-studded valley of the Trinity River. Both rivers empty into Trinity Bay, which is at the upper end of Galveston Bay. The bays occupy the drowned mouths of both rivers. When sea level was lower during the ice age, these rivers extended many miles to the south, even beyond the present-day coastline at Galveston. In the last few thousand years, sea level has risen and literally drowned the river mouths to form bays. A rich mix of freshwater and saltwater organisms occupy the waters of the shallow bays, lakes, swamps, and channels. An equally organic-rich, near-tropical plant assemblage grows in the wet climate and contributes to the soup of organic debris.

The Trinity River valley is a modern-day model for environmental conditions that have existed along the Gulf for many millions of years. These environments, now buried deeply in the continually sinking sedimentary wedge of the Gulf, are responsible for the extensive petroleum deposits in this area. Many thousands of feet beneath the land surface, temperatures become hot enough to cook organic material to form oil and gas, while the sands in the buried stream channels and beaches become the reservoirs as the newly cooked oil occupies the tiny, myriad spaces between the sand grains.

Drowned mouth and delta of the Trinity River along I-10 east of Houston. —Image courtesy Google Earth, November 2023

High Island is the surface expression of a large underground salt dome. (29.5435, -94.4059)

Salt domes pushing up the thick layers of organic mud form the structure for the indigenous hydrocarbons to migrate into and be entrapped. A great place to see the surface expression of one such salt dome is near the coastal community of High Island, named because it is the highest point along the Gulf Coast between Mobile, Alabama, and the Yucatán Peninsula. The "island" protects local residents from hurricanes during times of high storm surge. Because salt domes make excellent oil traps, numerous oil wells can be seen around the edges of the dome.

While Houston is the modern-day Goliath of petroleum, the town of Beaumont, 90 miles away, is the earthmother, where the petroleum age dawned at 10:30 a.m. on the cold morning of January 10, 1901. As reported in *Spindletop* by J. A. Clark and M. T. Halbouty, the first great American gusher "roared in like a shot from a heavy cannon and spouted oil a hundred feet over the top of the wooden derrick out on the hummock that the world would soon know as Spindletop. This oil discovery changed the world. Before Spindletop, oil was used for lamps and lubrication. The famous gusher of Captain Anthony F. Lucas changed that. It started the liquid fuel age, which brought forth the automobile, airplane, the network of American highways, improved railroads and marine transportation, the era of mass production, and untold comforts and conveniences."

According to the Clark and Halbouty book, Spindletop is "a little knob of land rising out of a swampy prairie in the southeast corner of Texas," south of the town of Beaumont, at what was Gladys City in 1901. The hill is the surface expression of a large, subterranean salt dome, which trapped enough oil to produce 100,000 barrels per day in the initial flow from the Spindletop discovery well. Beaumont became an immediate boom town, and large and famous oil companies, including Gulf and Texaco, got their start here. All the elements of a burgeoning petroleum industry existed in the Beaumont-Orange area in 1901. Clipper ships called at the Port of Beaumont in the 1800s, so a shipping network was well established. Timber and shipbuilding were major industries at Orange, which had ready access to the Gulf Coast via the Sabine River. Wooden derricks and other oil field equipment could thereby either be locally built or easily shipped in from elsewhere.

The Spindletop–Gladys City Boomtown Museum at Lamar University in Beaumont features artifacts and documents from the early oil days, a film explaining the Spindletop discovery, and a reconstructed turn-of-the-century oil town complete with wooden derricks, early oil field equipment, and wooden front stores, including a blacksmith shop, saloon, and post office.

A replica of the Lucas Gusher oil derrick at the Spindletop–Gladys City Boomtown Museum at Lamar University. Daily reenactments use water to simulate the gushing oil on the day it was discovered. (30.0330, -94.0791)

South of Beaumont in Port Arthur is the Museum of the Gulf Coast featuring geological, historical, and cultural exhibits related to the region. Farther south is Sabine Pass, where the Sabine River cuts through the barrier island and coastal marshes to reach the Gulf. Access to the natural environment can be had at Sabine Pass Battleground State Historic Site, where a Civil War battle was fought in 1863, and Sea Rim State Park, where 5 miles of public beach are available for birding, nature walks, and wildlife viewing.

I-10
HOUSTON — SAN ANTONIO
186 miles

Between Houston and Columbus, I-10 traverses the distinctly flat terrain of the Gulf Coastal Plain. You won't see any rocks along this stretch of highway, but to a geologist's way of thinking, the flatness of the coastal plain is as interesting a phenomenon as rugged mountains. The plain tells a story—one of eons of streams and meandering rivers carrying infinite loads of sediment from the interior of the North American continent and depositing the sediment in swamps and low areas between rivers to form the flat coastal plain. The North American continent has been building the Gulf Coast continuously southward in this manner for at least 60 million years.

Along this stretch between Houston and Columbus, I-10 crosses the Brazos and Colorado, major Texas rivers responsible for much of the coastal plain deposition, and the smaller San Bernard that lies between them. At bridge crossings, you will get a quick look into the incised channels cut about 50 feet down into the coastal plain. This downcutting occurred during the Pleistocene ice ages, when sea level was much lower, climate was wetter, and rivers carried more water and sediment load, on average, than they do today.

Look for meander bends, sand and gravel bars, point bars, steep outer banks, and low inner banks—all features of meandering streams—where I-10 crosses the Brazos, San Bernard, and Colorado Rivers. The San Bernard in particular displays gravel bars when the water level is low. Attwater Prairie Chicken National Wildlife Refuge, south of I-10 along the San Bernard River, is recognized as a national natural landmark because it preserves original Gulf Coastal Prairie. Construction gravel is mined in

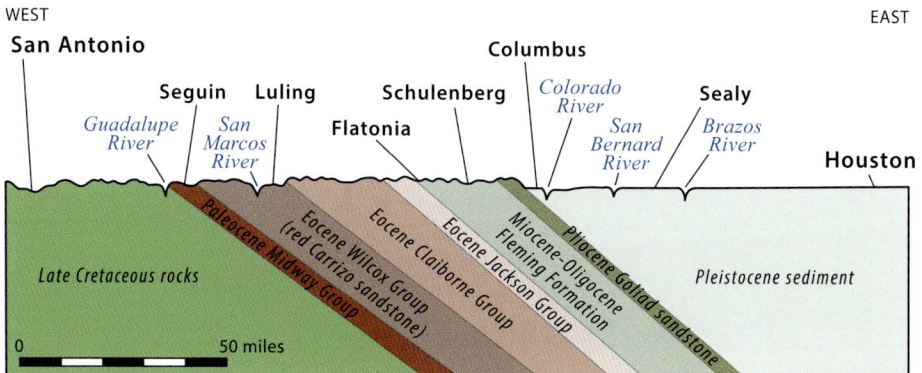

Cross section along I-10 between San Antonio and Houston.

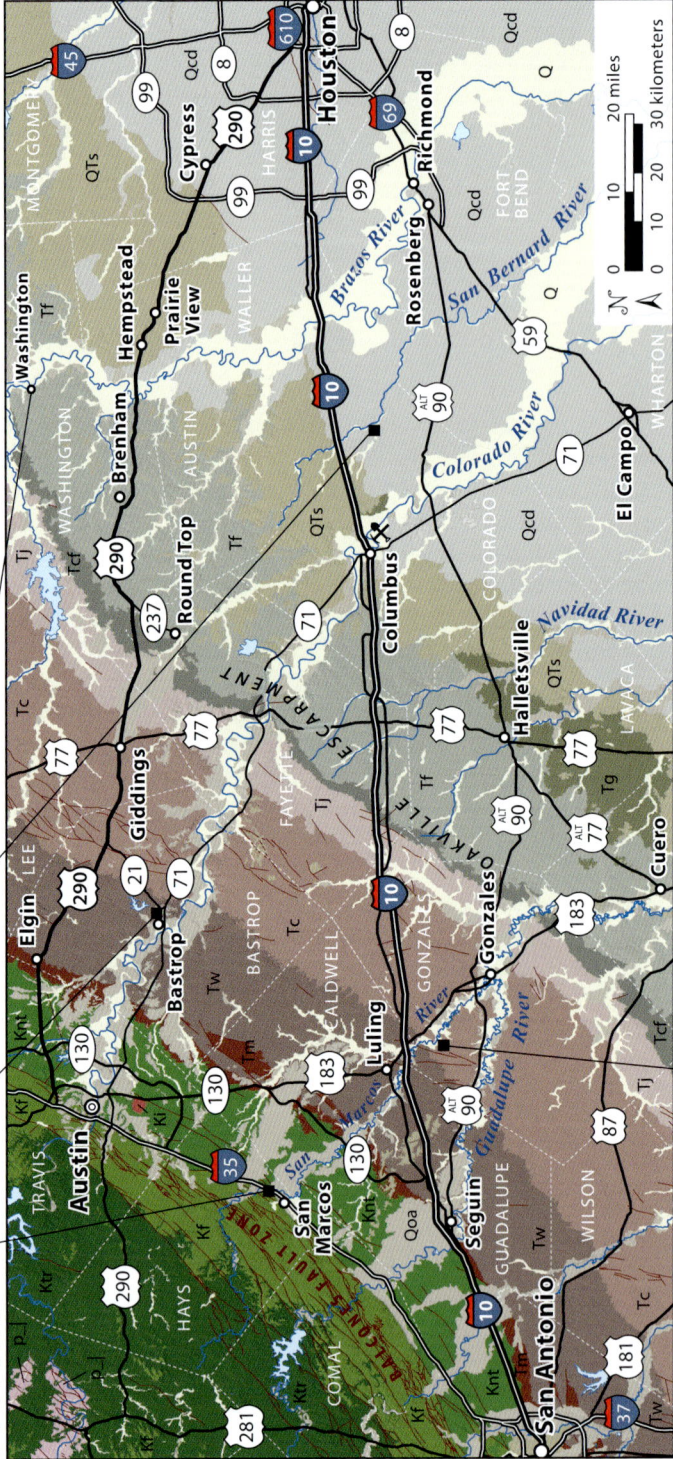

Geology along I-10 between Houston and San Antonio.

pits on the south side of the freeway near milepost 698, just east of the bridge over the Colorado River.

Between Columbus (Colorado River) and San Antonio, the topography is hilly and rolling in sharp contrast to the flatness of the coastal plain eastward toward Houston. This marked change begins just west of the Colorado River crossing. With one exception, no outcrops of rock are visible along this stretch of freeway, but the topography is a key to the underlying geology. I-10 runs perpendicular to a series of northeast-southwest-oriented sand ridges and intervening clay swales, or depressions, produced by the differential erosion of the inclined, mostly Paleocene- to Eocene-aged rocks.

Look carefully for the one outcrop of red (Eocene-aged) sandstone on a ridge on the south side of I-10, 2 miles east of exit 632 to Luling at the Caldwell-Gonzales county line sign between mileposts 635 and 634. The red soil color is common in high rainfall areas where the chemical weathering of groundwater leaches out all the calcium and potassium, leaving behind oxidized iron (rust). Moreover, the Carrizo Sand here also contained abundant iron to begin with.

Blocks of iron-rich Carrizo Sand can be seen in Palmetto State Park, just south of I-10 along the San Marcos River. The park was opened in 1936, and with the help of the Civilian Conservation Corps, many stone structures were built using locally quarried sandstone. The park is accessed via exit 632 (US 183).

Between Seguin and San Antonio, I-10 rolls across mudstones and siltstones of the Late Cretaceous Taylor and Navarro Groups and the Paleocene Midway Group. These rocks are quite soft and easily eroded, so outcrops along this stretch are virtually nonexistent. Seguin, along the Guadalupe River, rests on a high alluvial terrace deposited by an ancestral version of the river—one of many terraces that parallel the courses of modern-day Texas rivers.

This hilltop exposure east of Luling shows the red soil that is common in this area due to the underlying Carrizo Sand. (29.6527, -97.5563)

Red sandstone blocks of locally quarried Carrizo Sand are used in Palmetto State Park for traffic control and building material. (29.5894, -97.5841)

I-35
AUSTIN—SAN ANTONIO
75 miles

I-35 crosses the Colorado River at the south edge of downtown Austin, though outcrops are not seen in the riverbanks near I-35. Between Austin and San Marcos, I-35 rides on flat Cretaceous terrain. South of Austin, watch for an exposure of white Austin Chalk on the east side of I-35 at Slaughter Creek, about 6 miles south of the TX 71 crossover. South of Austin near the Onion Creek neighborhood, the road surface is on Late Cretaceous Taylor and Navarro Group mudstones, which are soft and do not stand up as natural outcrops. North of San Marcos, the Balcones Escarpment is obvious across the flat fields west of the freeway.

Wonder World Cave, a commercial park in San Marcos, formed when an earthquake along the Balcones fault zone created a fissure in the Cretaceous Georgetown Limestone. The cave is billed as the only commercial "dry-formed cave" in the United States because the open cavity was not dissolved by water. Also in San Marcos, Aquarena Springs (now part of the Meadows Center for Water and the Environment) is a group of more than 200 springs that discharge water from the Edwards Aquifer at a rate of over 123 million gallons daily. Artifacts discovered near the spring date back more than 12,000 years, suggesting that this site was inhabited by Clovis People.

The interstate between San Marcos and San Antonio is built on Late Cretaceous Pecan Gap Chalk (Taylor Group), which does not form good outcrops because of its easily erodible nature. Just to the west is a line of skyline ridges that are upheld by hard, resistant Edwards Limestone, about 25 million years older than the chalk on which I-35 is constructed. The older limestone is found higher than the younger

CENOZOIC
QUATERNARY
Q   recent sediments; includes alluvium

Qoa   older surface deposits; includes terrace and high gravel deposits (Pleistocene)

QUATERNARY–NEOGENE
QTs   sedimentary deposits; includes Uvalde Gravel and Willis Formation (Miocene to Pleistocene)

NEOGENE and PALEOGENE
Tj   Jackson Group (Eocene and Oligocene)

Tc   Claiborne Group (Eocene)

Tw   Wilcox Group (Eocene)

Tm   Midway Group (Paleocene–Eocene)

MESOZOIC (CRETACEOUS)
Knt   Navarro and Taylor Groups; includes Pecan Gap Chalk of the Taylor Group

Kau   Austin Chalk

Kwe   Washita and Eagle Ford Groups, undivided; includes Georgetown Limestone of the Washita Group

Kf   Fredericksburg Group; includes Edwards Limestone

Ktr   Trinity Group; includes Glen Rose Formation

VOLCANIC ROCKS
Ki   basalt of Pilot Knob volcano (Cretaceous)

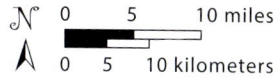

0   5   10 miles

0   5   10 kilometers

quarry

airport

normal fault

county boundaries shown as dotted white lines

Pilot Knob volcano

San Marcos Springs; Wonder World Cave

Canyon Lake Gorge; Hidden Valley fault

Comal Springs on Comal River

Slaughter Creek

Colorado River

TRAVIS

Austin

HAYS

San Marcos River

Canyon Lake

San Marcos

COMAL

BALCONES FAULT ZONE

BALCONES ESCARPMENT

New Braunfels

BEXAR

Guadalupe River

GUADALUPE

GONZALES

WILSON

San Antonio

EDWARDS PLATEAU

BLACK PRAIRIE

CALDWELL

COASTAL PLAIN

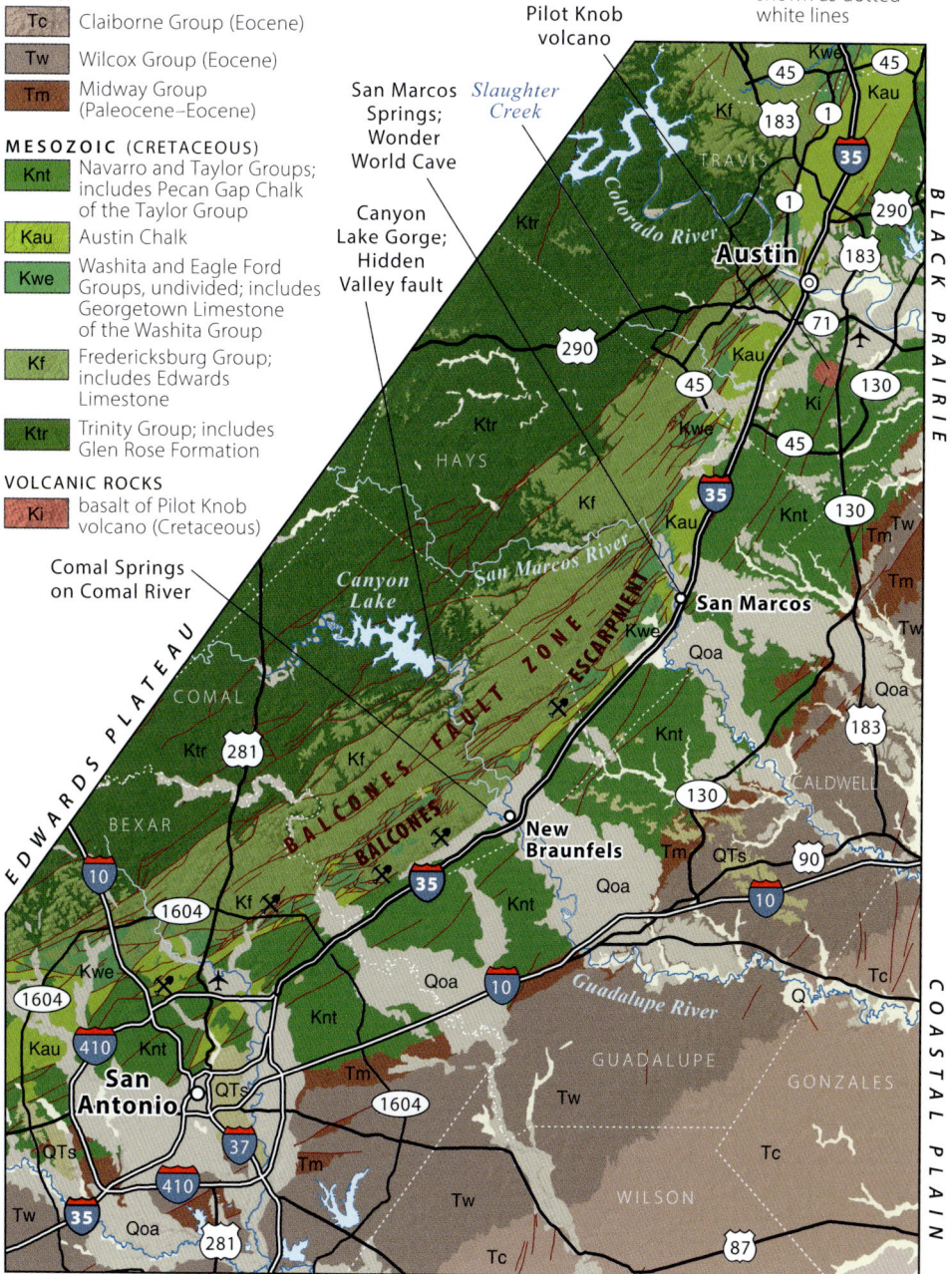

Geology along I-35 between Austin and San Antonio.

claystone marls because one of the main faults of the Balcones fault zone lies between the highway and the skyline ridge. The limestone was uplifted along this fault in Miocene time, about 20 to 10 million years ago. This uplift raised the Hill Country and Edwards Plateau northwest of San Antonio to about 2,000 feet above sea level. This movement on the Balcones fault zone occurred about the same time that the Colorado Plateau in the southwest United States was elevated and the Rio Grande rift in New Mexico opened. A large part of western North America underwent simultaneous uplift and faulting during Miocene time.

CANYON LAKE GORGE

In 1964, Canyon Lake Dam was completed on the Guadalupe River, effectively creating Canyon Lake. When the dam was built, an emergency spillway was also completed to protect the dam in case of torrential rainfall. The spillway was put to the test in July 2002 when as much as 35 inches of rain fell in the area, and on July 4, water overtopped the spillway for the first time. This water raced down a small creek leading to the Guadalupe River and, in the process, gouged a canyon approximately 1 mile long, up to 150 feet wide, and 80 feet deep. The newly formed gorge in limestone bedrock became a laboratory for geologists who study catastrophic flooding and canyon formation. The flood also revealed a part of the Glen Rose Formation, which was deposited along the shoreline of a warm, shallow sea about 110 million years ago. The Glen Rose of the Trinity Group is well known for the abundance of fossils, including snails, sea urchins, and other sea life, but perhaps the most impressive fossils found as a result of the flooding were dinosaur tracks. In addition, ancient ripple marks, created when waves wash back and forth across a beach, can also be seen. Structural geologists also got to see previously hidden portions of the Hidden Valley fault, part of the Balcones fault zone. Several springs were also exposed along the riverbed.

Overflow from Canyon Lake in July 2002 eroded this gorge in the Glen Rose Formation. (29.8615, -98.1893)

Numerous springs emanate from the limestone along the Balcones fault zone. Comal Springs in New Braunfels and San Marcos Springs in San Marcos are the largest springs in the southwestern United States and the sources for the Comal and San Marcos Rivers, respectively. These springs are partial outlets for the water in the immense Edwards Aquifer, which supplies San Antonio's drinking water.

Huge limestone quarries lie west of the road between New Braunfels and San Antonio, one at the Comal County line a few miles south of San Marcos, another south of New Braunfels at Rueckle Road (exit 184), and yet another at exit 177 on the Guadalupe County line nearer San Antonio. The large amounts of cement, road gravel, and aggregate from these quarries have fueled the economy of this area. The cement and aggregate are mainly shipped to coastal cities such as Corpus Christi and Houston, where limestone and rock are in short supply. The Edwards Limestone is the main stone quarried in these and similar quarries at many places along the Balcones Escarpment north of San Antonio.

SAN ANTONIO

San Antonio is built on Texas geology. Not only is it a center for geologists who make their living in the state looking for geologic resources such as oil, gas, water, and minerals, it also sits astride a fundamental juncture in the state's geology. The Balcones fault zone, nearly 30 miles wide, slices northeast-southwest right through the city, separating the high, upthrown Cretaceous limestone terrain of the Edwards Plateau and Hill Country north of town from the flat, lowland sandstone and mudstone terrain of the Gulf Coastal Plain south of town.

The Balcones fault zone popped up about 10 million years ago in the Miocene Epoch, elevating the Edwards Plateau nearly 2,000 feet above sea level. The fault follows the edge of the old, buried Ouachita Mountains, which means both the fault and the range lie along a deep-seated suture in the Earth's crust. The fault is not a single break in the rocks around San Antonio, nor a single line on the geologic map, but rather a zone of stair-stepping faults. As you drive north through San Antonio— and particularly north on I-10—the up and down topography and knolls of limestone result from this pattern of fault slices and intervening fault blocks. Along US 281, this block pattern also shows up quite well, especially around Encino Park.

Groundwater from the high Edwards Plateau to the north and from sinks in rivers to the west flows toward San Antonio. Fractures, cracks, and cavities in the limestone blocks form a tortuous pathway for the water, which eventually surfaces along faults in the Balcones fault zone, creating springs in San Antonio such as those in San Pedro Springs Park.

Cretaceous limestone, particularly the Late Cretaceous Austin Chalk, has been quarried in San Antonio for centuries—mainly for building stone. The Austin Chalk is not as hard as the Edwards Limestone, nor as riddled with solution holes, so it is easier to mine and produces a more uniform building stone. One of the best places to see limestone close-up in a quarry is at Brackenridge Park, where the Japanese Tea Garden is built against the backdrop of limestone quarry walls. The outer walkway of the Japanese Tea Garden winds along the quarry wall, where fossils of Cretaceous oyster shells and giant clams (*Inoceramus*) are abundant. Remnants of the nineteenth-century Alamo Cement Company mill are also part of Brackenridge Park, including kilns and a smokestack.

Geologic map of San Antonio.

EDWARDS PLATEAU

San Pedro Springs Park

Brackenridge Park

COMAL

BALCONES FAULT ZONE

BEXAR

San Antonio

Encino Park

Calaveras Lake

Victor Braunig Lake

Medina River

San Antonio River

WILSON

ATASCOSA

COASTAL PLAIN

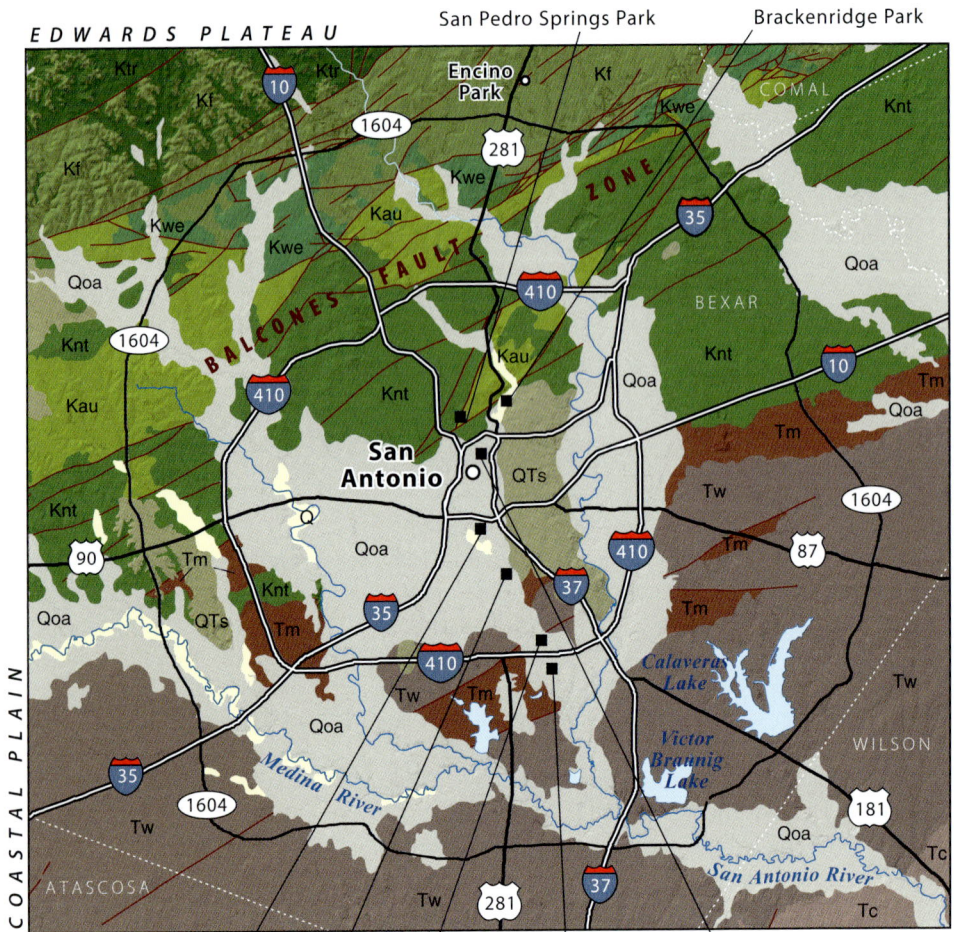

Mission Concepción   Mission San Jose   Mission San Juan   Mission San Francisco de la Espada Catholic Church   Alamo (Mission San Antonio de Valero) and River Walk

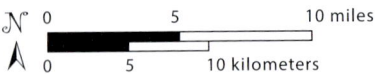

0   5   10 miles
0   5   10 kilometers

CENOZOIC

QUATERNARY

Q   recent sediments; includes alluvium

Qoa   older surface deposits; includes terrace and high gravel deposits (Pleistocene)

QUATERNARY–NEOGENE

QTs   Uvalde Gravel (Pliocene or Pleistocene)

NEOGENE and PALEOGENE

Tc   Claiborne Group (Eocene)

Tw   Wilcox Group; includes Carrizo Sand (Eocene)

Tm   Midway Group (Paleocene–Eocene)

—— normal fault

county boundaries shown as dotted white lines

MESOZOIC (CRETACEOUS)

Knt   Navarro and Taylor Groups

Kau   Austin Chalk

Kwe   Washita Group; includes Eagle Ford Group, Del Rio Clay, and Buda Formation

Kf   Fredericksburg Group; includes Edwards Limestone

Ktr   Trinity Group; includes Glen Rose Formation

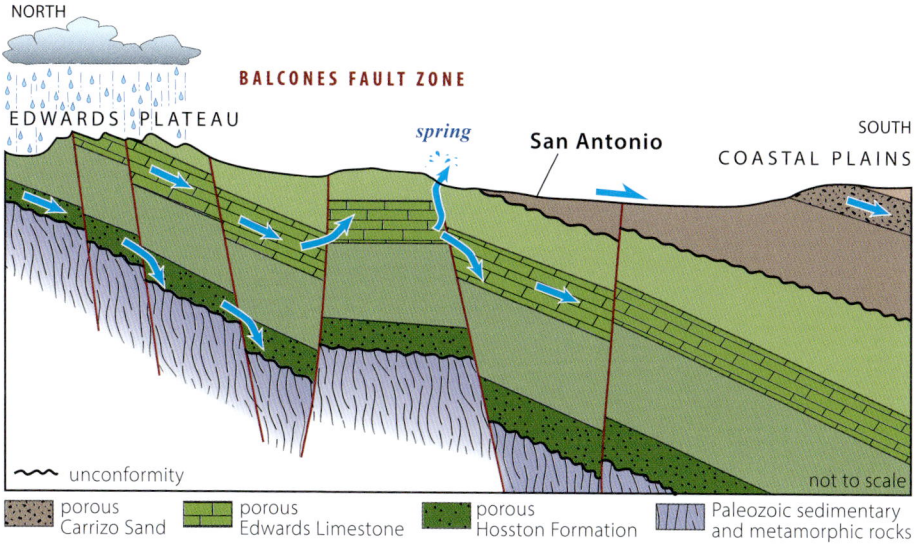

NORTH

BALCONES FAULT ZONE

EDWARDS PLATEAU

spring San Antonio

SOUTH
COASTAL PLAINS

| porous Carrizo Sand | porous Edwards Limestone | porous Hosston Formation | Paleozoic sedimentary and metamorphic rocks |

~~ unconformity

not to scale

Cross section showing recharge of porous rocks at the surface and flow of groundwater beneath the surface at San Antonio as it intersects the Balcones fault zone.

Buildings at Brackenridge Park were constructed from rock quarried within the former Alamo Stone Quarry. (29.4608, -98.4768)

Another historic geologic site in San Antonio is San Pedro Springs Park. Just off I-35 at San Pedro (exit 157A) and north on San Pedro Avenue, San Pedro Springs Park hosts a natural spring that bubbles freshwater from a fault cutting the Austin Chalk. Exposures of Austin Chalk are found in San Pedro Springs Park and also near Trinity University, 1.5 miles northeast of the springs. The Payaya people relished the springs prior to the arrival of the Spanish, who first visited the site in 1709 and named

Water from San Pedro Springs is diverted and flows into a pond used for swimming during the summer months. (29.4467, -98.5016)

Mission San Antonio de Valero, better known as the Alamo, was built with locally obtained Austin Chalk. (29.4258, -98.4865)

Close-up view of rocks cemented in place in the Alamo. The brownish-tan rock at center is a chert nodule. The view is approximately 12 inches across. (29.4257, -98.4863)

it El Agua de San Pedro. The original San Antonio village and presidio (fort) were located near San Pedro Springs. The springs were set aside as a public park in 1734, the oldest in Texas, and perhaps the second oldest in the United States. Notice the large cypress trees in the park—they like to have their feet in water—evidence for the continuous activity of the springs and drainageway here. Cypress trees grow along the San Antonio's River Walk for the same wet-footed reason. During the winter months, the surface discharge from the actual springs does not flow.

Most historic San Antonio buildings are built of native stone. The most famous, the Alamo—originally a Spanish mission (Mission San Antonio de Valero) constructed in 1718 and moved to its present site in 1724—is walled with Cretaceous limestone of the Austin Chalk, which probably came from former quarries located at the present-day San Antonio Zoo. Mission Concepción and Mission San Jose have walls of calcareous tufa, a soft, limy rock chemically deposited in nearby springs. Missions Espada and San Juan, farther south in the youngest sandstone and shale units, have walls of Eocene Wilcox Group sandstone and pebbly gravel cemented by caliche. Historic buildings commonly tell the story of the local geology in their walls because builders used the nearest stone.

I-35
SAN ANTONIO — LAREDO
154 miles

I-35 between San Antonio and Laredo is a long, dry stretch of highway. The underlying bedrock along the entire length of road is Eocene sandstone and mudstone deposited during sedimentation into the Gulf of Mexico about 50 million years ago.

Near the I-410 and I-35 crossover in San Antonio, watch for sand and gravel pits, where construction gravels are mined from the coarse sediment deposited by ancient versions of the San Antonio River. Low ridges composed of Eocene-aged Wilcox Group sandstones occur south of the I-410 crossover. Farther south near Lytle, lignite coal from the Wilcox Group was mined beginning in the 1880s until the advent of oil-burning locomotives drastically reduced the demand for coal. The mines later closed in the 1930s. Low, rolling topography forms where the alternating sandstone and mudstone beds of the Wilcox Group are eroded at different rates. The sands stand high; the mudstones are in the bottomlands.

About 15 miles south of the I-410 crossover, I-35 climbs a ridge of reddish Eocene Carrizo Sand, a member of the Wilcox Group. This unit characteristically forms red sandy hills and ridges in a distinctive band across a large part of the state. To the south, the Carrizo Sand extends downward into the subsurface to form an aquifer that waters the Winter Garden Region, the agricultural area between San Antonio and Laredo that produces year-round. This exposed sand ridge crossed by I-35 is an important recharge area where rainwater enters the Carrizo unit. Iron in the sandstone comes from an iron-rich mineral called glauconite that forms as tiny pellets, sometimes as animal fecal pellets on shallow marine shelf areas. The iron is so rich in rocks of this type in northeast Texas that the iron was once commercially mined.

Sandy red roadsides and fields continue for a few miles to Moore, where a younger set of Eocene sands is encountered, as evidenced by the color change to tan. Some white caliche can be seen on hilltop ledges in this area.

The band of Eocene deposits swings southward between Devine and Pearsall; for the remainder of the distance to Laredo, the road runs along the middle Eocene Claiborne band of sandstone and mudstone, which outcrop sporadically. Oil fields and their nodding pumps dot the landscape, and drill rigs indicate the area is still being explored for new oil. North of Cotulla, the fields produce mostly oil, while south of Cotulla, gas is the main product. The Pearsall oil field, named for the nearby town of Pearsall, was discovered in 1936. Early production was strong, with some wells producing over 20,000 barrels of oil per day from the Late Cretaceous Austin Chalk.

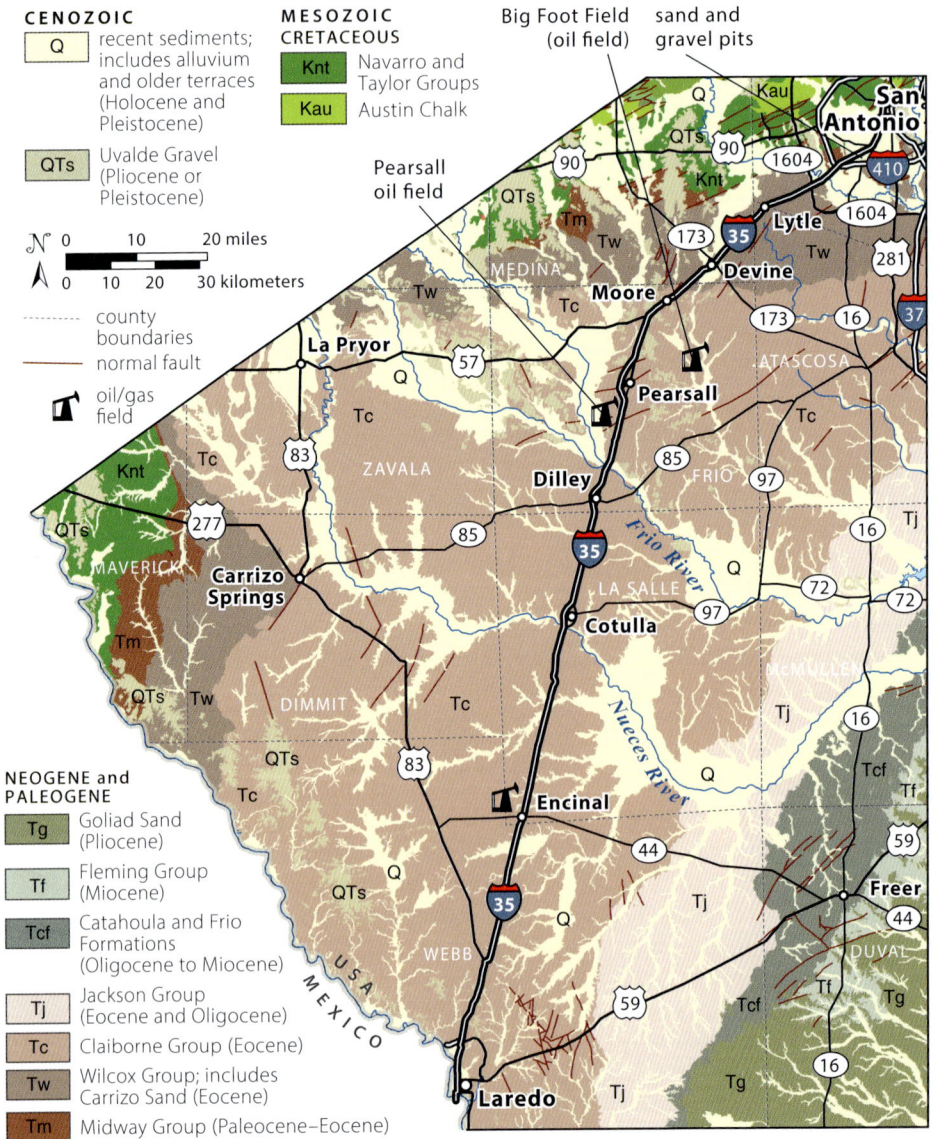

Geology along I-35 between San Antonio and Laredo.

The 1980s saw another expansion of drilling with the wide spread use of hydraulic fracturing in the Pearsall Shale. Nearby Big Foot Field, north of Pearsall, produces oil from delta sandstones of Paleogene age. Stuart City and Encinal Fields, near the town of Encinal, produce gas from reservoirs in Cretaceous limestone and Paleogene sandstone.

South of Cotulla, the highway crosses the Nueces River, which originates in the limestone uplands of the western Hill Country and empties into the Gulf of Mexico at Nueces Bay near Corpus Christi. Note the chert-rich limestone gravels here, which streams have eroded from the Edwards Plateau.

Thornbrush and cactus become prominent in the countryside approaching Laredo. These hardy survivors are well adapted to a climate where the searing desert sun results in high evaporation rates, and where rainfall averages only 20 inches per year.

Southward from the US 83 junction (about 20 miles north of Laredo), note the gravels on high hills and drainage divides among the generally low, undulating topography. The gravels were laid down by ancient streams that flowed on top of those high surfaces, when stream level and sea level were higher than they are today. Since the interglacial warm period of 135,000 years ago, streams cut downward to adjust to the lowered sea level during the final ice age of the Pleistocene Epoch.

I-37
CORPUS CHRISTI—SAN ANTONIO
140 miles

I-37 heads north out of Corpus Christi, gradually climbing the river bluffs on the west bank of the Nueces River. The road follows Nueces Bay for a few miles, still running along the top of the terraces adjacent to the river. The road continues to ride the drainage divide on the edge of the Nueces River valley, then it descends into the valley and crosses the river at Labonte Park (formerly Nueces River Park).

North of the Nueces River crossing, I-37 traverses the low, flat river floodplain before climbing out of the valley and onto the Pleistocene coastal plain surface. Rich, black soil and farms dominate the nearly treeless landscape for many miles. Occasional oil field pumps tell of additional black riches beneath the surface. Northeast-southwest bands of subsurface reservoirs have been tapped in this area. Oligocene-aged barrier sandstones of the Frio Formation provide the main producing reservoirs around Corpus Christi. Northwest of Corpus Christi, river and deltaic sandstones of the Frio Formation are the principal oil and gas reservoirs, while the Eocene Jackson Group and Yegua Formation sands form the reservoirs about halfway between Corpus Christi and San Antonio. Closer to San Antonio, the reservoirs are found in older Paleogene sandstones and Late Cretaceous chalks and limestones.

Near Mathis is Lake Corpus Christi, a dammed part of the Nueces River that is accessed at Lake Corpus Christi State Park. Note the topography around the lake is more rolling than the Pleistocene coastal plain between Corpus Christi and Mathis. Northwest of Mathis, the road is on older Pliocene-aged sandstones of the Goliad Sand. This unit is a bit more resistant than clay beds, so it stands up to erosion, forming topographic relief. Look for sand quarries and white sand on hillcuts around here, in combination with views to the west of Lake Corpus Christi. The Goliad is a noted ridge-former that parallels the coastal plain and holds up ridges in a long band across much of southern Texas.

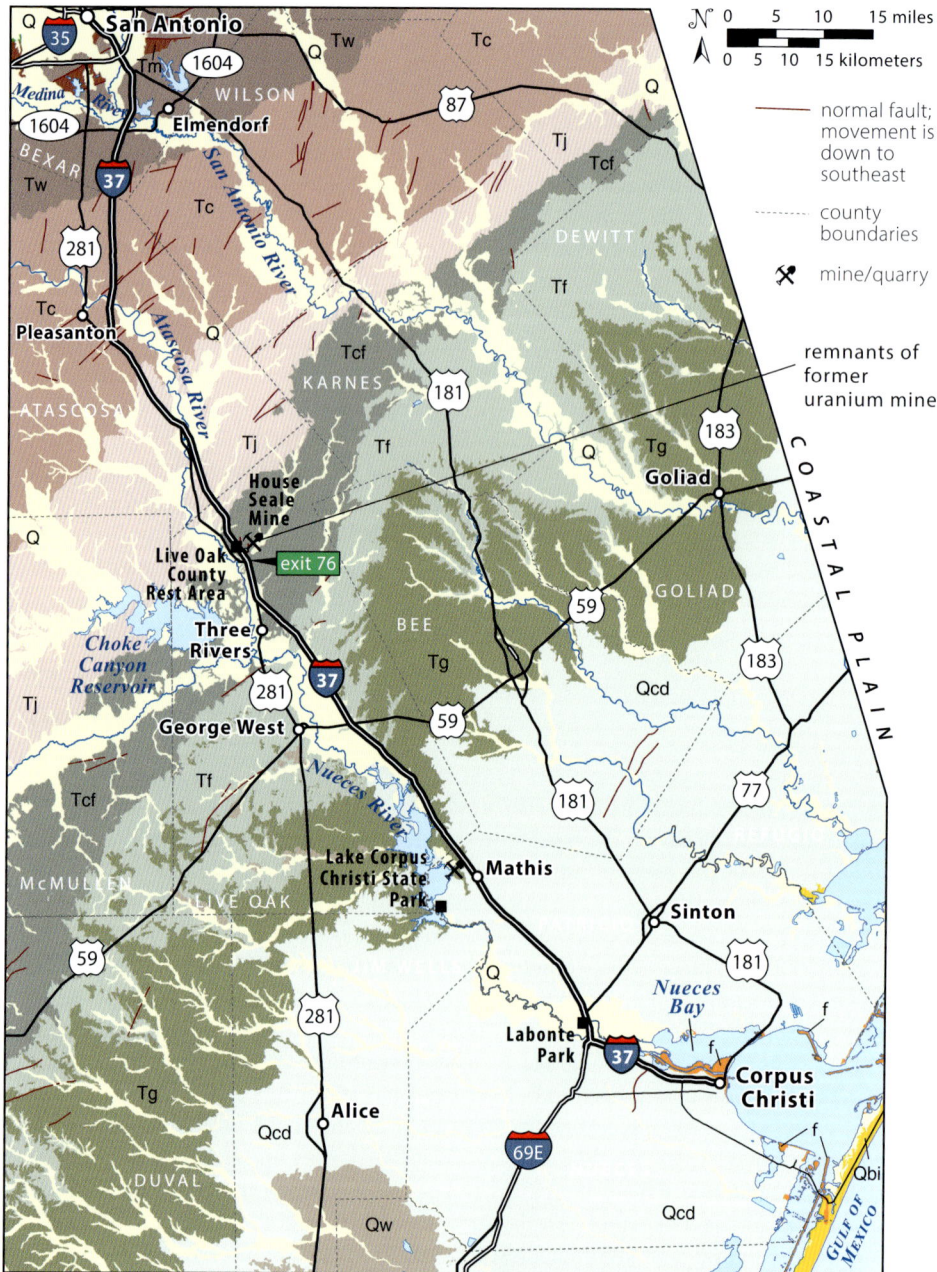

Scale

0 5 10 15 miles
0 5 10 15 kilometers

Legend

— normal fault; movement is down to southeast

···· county boundaries

⚒ mine/quarry

remnants of former uranium mine

Labels on map

San Antonio
35
1604
Medina River
WILSON
Elmendorf
1604
BEXAR
37
Tm
Tw
Tw
Tc
281
Tc
Pleasanton
ATASCOSA
San Antonio River
Atascosa River
Q
Tj
87
Tj
Tcf
DEWITT
Tf
Q
Tc
KARNES
Tcf
181
Tf
Tf
Tg
183
Goliad
House Seale Mine
Live Oak County Rest Area
exit 76
Choke Canyon Reservoir
Tj
Three Rivers
37
281
George West
Tf
Tcf
Nueces River
BEE
Tg
Q
Tg
Qcd
GOLIAD
59
183
McMULLEN
LIVE OAK
Lake Corpus Christi State Park
Mathis
181
77
Sinton
59
181
59
Alice
281
Qcd
Qw
DUVAL
Tg
Q
Labonte Park
37
69E
Nueces Bay
f
f
Corpus Christi
f
Qbi
Qcd
GULF OF MEXICO
COASTAL PLAIN

Legend

CENOZOIC

(QUATERNARY)

- **f** fill material dredged for raising land surface above alluvium and barrier island deposits
- **Q** recent sediments; includes alluvium and terrace deposits (Holocene and Pleistocene)
- **Qbi** barrier island deposits
- **Qw** windblown sand, silt, and sand sheet deposits (Holocene)
- **Qcd** coastal deposits; includes Lissie, Beaumont, and Deweyville Formations (Pleistocene)

(NEOGENE and PALEOGENE)

- **Tg** Goliad Sand (Pliocene)
- **Tf** Fleming Group (Miocene)
- **Tcf** Catahoula and Frio Formations (Oligocene to Miocene)
- **Tj** Jackson Group (Eocene and Oligocene)
- **Tc** Claiborne Group; includes Yegua Formation (Eocene)
- **Tw** Wilcox Group; includes Carrizo Sand (Eocene)
- **Tm** Midway Group (Paleocene–Eocene)

Geology along I-37 between Corpus Christi and San Antonio.

Near exit 56 for George West (US 59), the countryside flattens out somewhat, but the relief is still low and rolling. The Oligocene-to-Miocene-aged sandstone and mudstone beds of this area are not as resistant as the higher-standing Goliad Sand to the south.

Near the town of Three Rivers, just south of exit 69, is a former sand quarry west of the highway, but it is difficult to see from the road. Light-tan to white sandstone, still of Oligocene to Miocene age, was extracted for construction use. Between Three Rivers and Pleasanton, I-37 follows the Atascosa River valley, a tributary of the Nueces River. The headwaters of the Atascosa begin in the hills southwest of San Antonio and south of the Medina River.

Just north of exit 76, a quarry on the east side of the freeway provides a brief look at cuts in whitish tuffaceous sandstones of the Catahoula Formation that mark the northern edge of the Miocene-Oligocene rock band. The abundant tuff fragments in this sandstone tell of Oligocene volcanic eruptions hundreds of miles to the west and northwest.

Between exit 76 and Pleasanton, flatter terrain lies first on the Catahoula Formation, followed by the marly beds of the Eocene Jackson Group, and not much in the way of rock exposures can be seen other than gray mudstones in grassy roadcuts. Large earthen dumps, now vegetated, can occasionally be seen along the highway from former uranium mines in the Catahoula Formation and Jackson Group. The piles from one of these mines, the House Seale Mine, can be seen from the northbound Live Oak County Rest Area, 2 miles north of exit 76. These, and many other mines in the area, are part of the South Texas uranium belt, an approximately 250-mile-long and 15- to 25-mile-wide strip of land extending from east-central Texas across the Rio Grande into Mexico. The belt nearly parallels the Gulf Coast but is about 80 miles

Large vegetated mounds of uranium mine piles, as well as the water-filled pit created during mining at the House Seale Mine, are visible in this satellite image, as well as from the northbound rest area north of exit 76. —Satellite image courtesy Google Earth, December 2024

inland. Volcanic ash and bentonite tuff occur throughout these rock sequences and have long been considered the source for this uranium.

Beginning at and north of the US 281 interchange (exit 103) southeast of Pleasanton, reddish, sandy soils indicate iron-rich sandstones of the Eocene-aged Claiborne Group lie just beneath the surface. Watch for even redder soils and sand in grassy roadcuts just north of the Bexar County line. This bright-red unit is the Carrizo Sand, which forms a distinctive red band across the south half of the state.

Just east of I-37 and north of Loop 1604 are clay pits (not visible from the interstate) dug in the uppermost Wilcox Group claystones, just below the Carrizo Sand. East of the freeway, Elmendorf hosted a large brick industry beginning in the late 1800s until the factory closed in 1989. Not many rocks are to be seen on the southeast side of San Antonio until you stand and study the historic limestone walls of the Alamo.

US 59
HOUSTON — VICTORIA — I-37
190 miles

The drive between Houston and Victoria is entirely across the flat expanse of the coastal plain. Rich soils from overbank flood deposits and a rich vegetation produce much organic humus in the humid climate. Two of Texas's major rivers are crossed in this stretch: the Brazos River just west of Sugar Land and the Colorado River at Wharton. Good views from the bridges show sandbars on the inside bends and the opposing, steep, outside banks. The Colorado and Brazos Rivers slice diagonally across the entire state, originating in the high plateau of the northwestern Texas Panhandle and then reaching the Gulf Coast a mere 20 miles apart. The color of these rivers varies from clear to brown and even red (particularly the Brazos), depending on sediment load, which in turn is associated with upstream rainfall and erosion. The Brazos generally runs redder because it crosses and erodes red Permian rocks in north-central Texas, whereas the Colorado tends to run clearer in its lower reaches, largely because its many dams trap sediment upstream. The highway crosses many other small streams and creeks that roughly parallel each other as they flow southsoutheast toward the Gulf.

Just west of the Brazos River crossing, Crabb River Road (FM 762) heads south 17 miles to Brazos Bend State Park. The park preserves two oxbow lakes created when the meandering Brazos changed its course, leaving parts of its channel orphaned from the main stream. The park is also a haven for migratory birds traveling from the north (or south) to reach their destinations, as well as a prime location to see an occasional alligator or two sunning itself on the shore.

Southwest of Ganado, US 59 crosses the Navidad River. In most places, the river is small, but here it is substantially wider because this is the upper end of Lake Texana, created by a dam downstream.

West of the historical town of Victoria, US 59 passes over the Guadalupe River, whose headwaters are found in the Hill Country northwest of San Antonio. Most of the road between Victoria and Goliad is on flat coastal plain, until a few miles east of Goliad, where the road climbs onto the upland margin of the coastal plain and the

Pliocene-aged Goliad Sand forms a low ridge. The Goliad was named for exposures around the town of Goliad, and several good roadcuts display this gray, cemented, pebbly sandstone. Look for the roadcut near the former railroad overpass on US 183, about four blocks south of US 59. Continue on US 183 to Goliad State Park and Historic Site, where Mission Espiritu Santo and Presidio la Bahía are located.

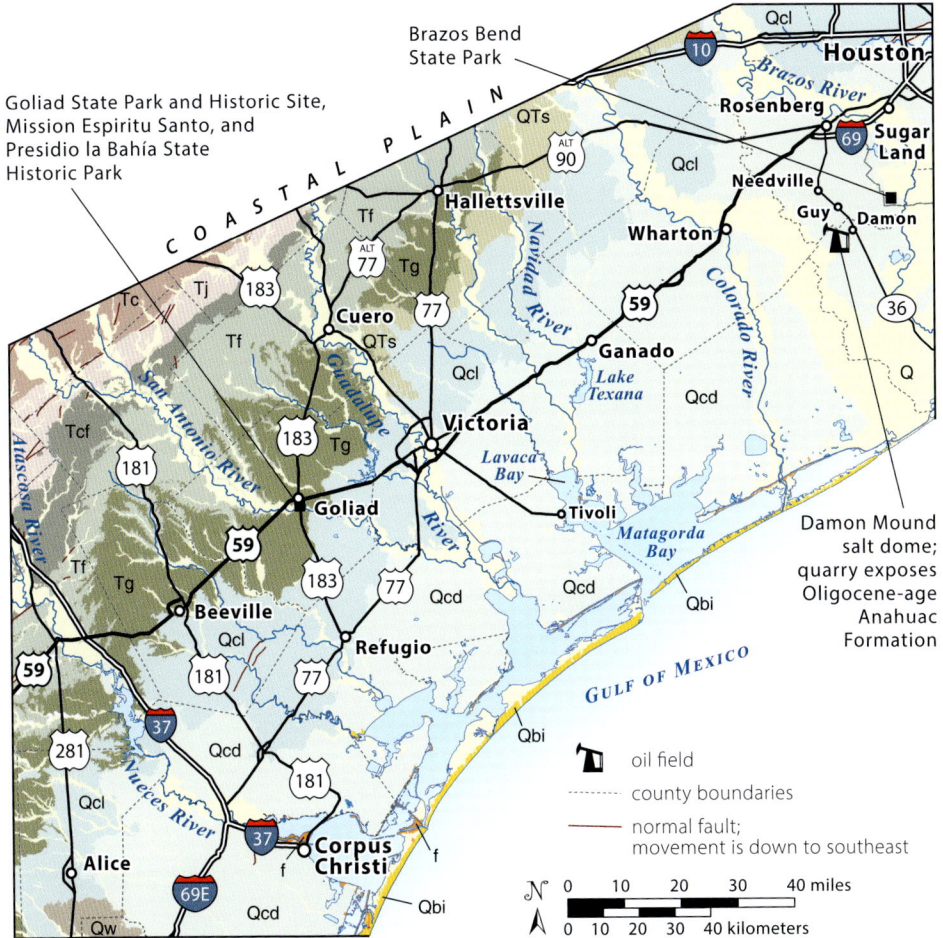

Geology along US 59 between Houston and I-37.

CENOZOIC

QUATERNARY

f	fill material dredged for raising land surface above alluvium and barrier island deposits
Q	recent sediments; includes alluvium and terrace deposits (Holocene and Pleistocene)
Qbi	barrier island deposits
Qw	windblown sand, silt, and sand sheet deposits (Holocene)
Qcd	coastal deposits of the Beaumont and Deweyville Formations (Pleistocene)
Qcl	coastal deposits of the Lissie Formation (Pleistocene)

QTs	Willis Formation (Pleistocene–Pliocene)

NEOGENE and PALEOGENE

Tg	Goliad Sand (Pliocene)
Tf	Fleming Group (Miocene)
Tcf	Catahoula Formation (Oligocene)
Tj	Jackson Group (Eocene and Oligocene)
Tc	Claiborne Group; includes Yegua Formation (Eocene)

The Navidad River and Lake Texana southwest of Ganado. (29.0245, -96.5553)

Roadcuts of Goliad Sand before and after the old railroad overpass south of Goliad give a close look as to what stones were used in constructing buildings of the area. (28.6637, -97.3871)

DAMON MOUND

To reach Damon Mound, head 19 miles south on TX 36 from US 59 in Rosenberg. TX 36 crosses the flat coastal plain, where cotton, sorghum, and formerly much sugarcane are grown on the rich, brown soils. After passing through the small communities of Needville and Guy, a large hill looms on the southwest skyline near the town of Damon. Looking oddly out of place because it stands 83 feet above the flat coastal plain, Damon Mound is one of the best surface expressions of a salt dome on the Gulf Coast. The salt dome pushed its way upward from the Jurassic Louann Salt layer thousands of feet beneath the surface. Pleistocene-aged rocks are tilted on its flanks, which is evidence for the upward migration of salt quite recently in geologic time.

A quarry carves out the northwest corner of the mound, where the caprock of Oligocene-aged coral reefs (Anahuac Formation) is mined for building stone, crushed gravel, and at one time, sulfur. A variety of corals, clams, and one-celled foraminifera are the reef-building organisms found here. Oil wells once surrounded the mound, outlining the underlying salt dome. More than 10 million barrels of oil were produced from the Damon Mound oil field between 1917 and 1952.

Damon Mound was a favorite campsite for Karankawa indigenous people, as evidenced by their burial sites, arrowheads, pottery, and stone implements found on the mound. The town of Damon, on the east side of the mound, was established in 1831 with the construction of a blacksmith shop out of limestone from a natural outcrop. The town cemetery has a number of 1830s headstones and is a registered Texas Historical Commission Landmark.

The gentle upward slope on the otherwise flat coastal plain is a great indication of an underground salt dome, such as at Damon. Note the oil well on the left. (29.3019, -95.7355)

Our Lady of Loreto Chapel, part of the Presidio la Bahía complex south of Goliad and site of the infamous Goliad Massacre, was constructed from nearby quarried Goliad Sand. (28.6483, -97.3824)

Mission Espiritu Santo was established first on the coast at Lavaca Bay in 1722, then relocated to a site north of Victoria, and finally moved to its present site on the San Antonio River in 1749. The stonewalls around the beautifully restored church are Goliad Sand, quarried from the local stone near the mission. Presidio la Bahía was the site of the Goliad Massacre so dear to Texas history, where Colonel James Fannin and 351 of his men were executed by Mexican troops in 1836 during the Texas Revolution, only a few weeks after the ill-fated Alamo fight. The battle cry at San Jacinto, where Sam Houston finally defeated Santa Anna, was: "Remember Goliad! Remember the Alamo!"

From the San Antonio River bridge on US 59 west of Goliad, the topographic relief produced by the resistant Goliad Sand is noticeable. Between Goliad, Beeville, and I-37, the underlying bedrock is all Goliad Sand. The only exception is a low area encountered a few miles east of I-37, where softer Miocene rocks are carved by the Atascosa River.

I-69E/US 77
BROWNSVILLE — CORPUS CHRISTI
152 miles

The Rio Grande, where this road guide begins, is not so grand at the border crossing at Brownsville between Mexico and the United States. It is little more than an entrenched, channelized ditch here, in contrast to its broad expanse and lush, green floodplain near El Paso. Considering the amount of Rio Grande water that is stored upstream behind dams and in irrigation channels, it is a wonder that any Rio Grande water reaches Brownsville. In and around town are a number of long, narrow,

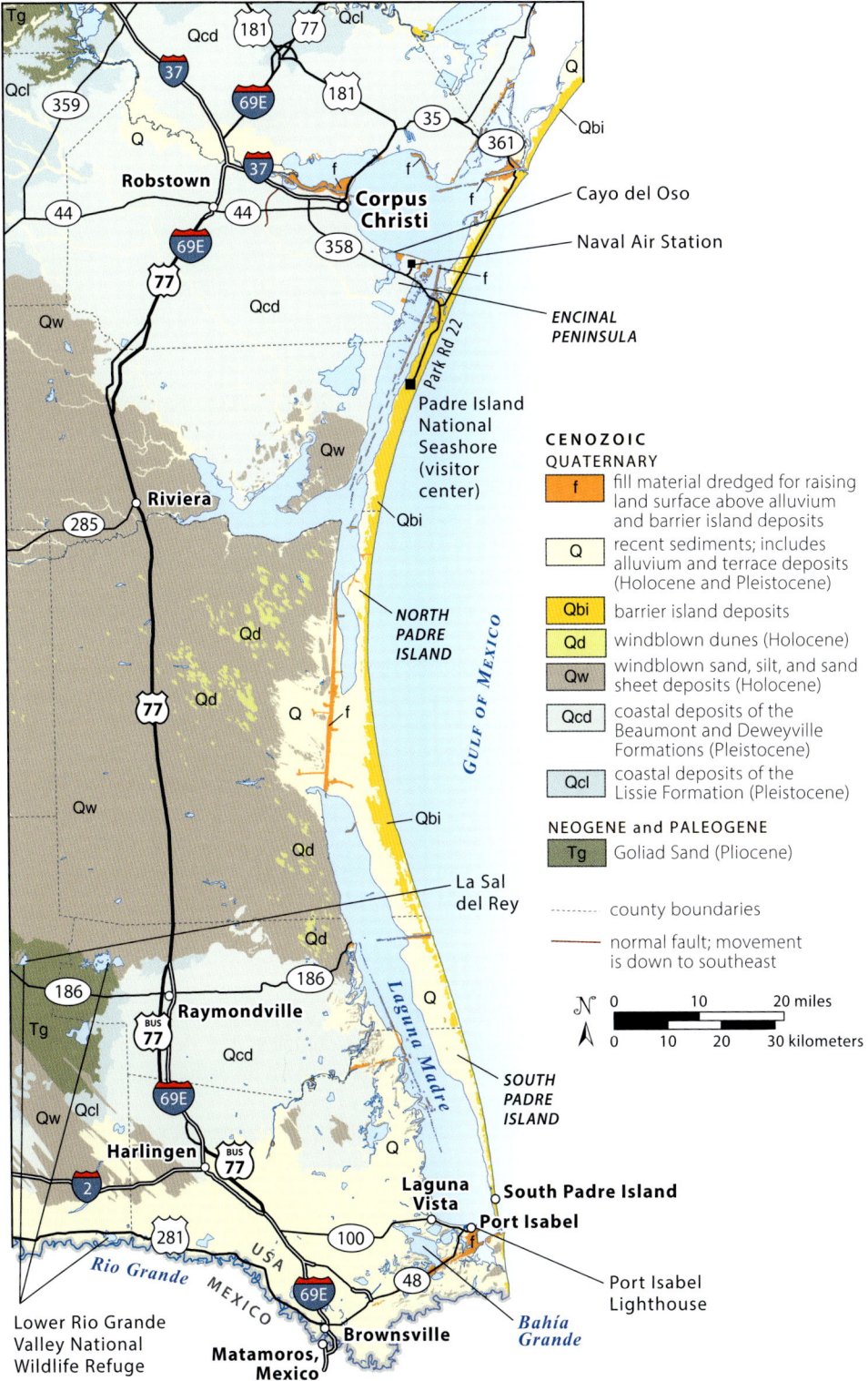

COASTAL PLAIN

Tg

Qcl

181 · 77 Qcl

37 Qcd

69E

359 Q

181

35

Q

361

Qbi

Robstown

37 69E

44

77 44

Corpus Christi f · f'

f

358 · f

f

Cayo del Oso

Naval Air Station

ENCINAL PENINSULA

Qcd

Park Rd 22

Qw

Qw

Padre Island National Seashore (visitor center)

Qbi

Riviera

285

NORTH PADRE ISLAND

Qd

Qd

Qd

Qw

Q · f

Qbi

GULF OF MEXICO

La Sal del Rey

77

Qw

Qd

Qd

Q

186 186

Tg

BUS 77

Raymondville

69E

Qcd

Q

Laguna Madre

SOUTH PADRE ISLAND

Qw Qcl

Harlingen

BUS 77

Q

2

Laguna Vista

South Padre Island

Port Isabel

281

100

48

Port Isabel Lighthouse

Rio Grande

USA

MEXICO

69E

Bahía Grande

Brownsville

Matamoros, Mexico

Lower Rio Grande Valley National Wildlife Refuge

CENOZOIC
QUATERNARY

f	fill material dredged for raising land surface above alluvium and barrier island deposits
Q	recent sediments; includes alluvium and terrace deposits (Holocene and Pleistocene)
Qbi	barrier island deposits
Qd	windblown dunes (Holocene)
Qw	windblown sand, silt, and sand sheet deposits (Holocene)
Qcd	coastal deposits of the Beaumont and Deweyville Formations (Pleistocene)
Qcl	coastal deposits of the Lissie Formation (Pleistocene)

NEOGENE and PALEOGENE

Tg	Goliad Sand (Pliocene)

- - - - county boundaries

———— normal fault; movement is down to southeast

N

0   10   20 miles

0   10   20   30 kilometers

Geology along I-69E/US 77 between Brownsville and Corpus Christi.

meandering lakes, called resacas, which are the oxbow lakes and abandoned channels of the Rio Grande. These are distinctive features of the Rio Grande delta, which occupies the immense area of flat land bordering the river in the vicinity of Harlingen and Brownsville.

PORT ISABEL AND SOUTH PADRE ISLAND

Port Isabel, on the Rio Grande delta east of Brownsville, was an important port in the nineteenth century. To reach Port Isabel, head east on TX 100 (exit 14 from I-69E/US 77) or east on TX 48 from Brownsville. TX 100 crosses the flat floodplain of the Rio Grande. The rich, brown soils support extensive agriculture. At Laguna Vista, the road turns right, and you get the first view of Laguna Madre, the lagoon that lies between the coast and South Padre Island, the barrier island. The lagoon is noticeably calm because the water is only a few feet deep. In the ocean, the wind presses on the water surface, creating large waves, which are circular motions of water that extend many feet downward from the surface. In a shallow lagoon, large waves don't form because the water is too shallow to allow the development of the deep circular motion inherent in big waves. Instead, the wind produces lots of choppy little waves in shallow lagoons. Because big waves don't churn the bottom, sand simply isn't piled along the lagoon's edge in most places, so there is little beach.

The old Port Isabel Lighthouse, located right in town, still stands as a monument to nineteenth-century coastal trade and commerce. The lighthouse was erected by the US government in 1852, extinguished during the Civil War, then discontinued from 1888–1895. The light was permanently extinguished in 1905 but is now maintained as a Texas State Historical Structure by the Texas Parks and Wildlife Department.

The causeway between Port Isabel and South Padre Island gives a high vantage point to see the modern coastal environments of lagoon, tidal flats, sand dunes, and barrier island. Note on the map how narrow the South Padre barrier island is compared to North Padre Island near Corpus Christi. Longshore drift of sand along this southern part of the Texas Gulf Coast moves northward, driven by the predominant southeast to northwest winds. The major sand source is the Rio Grande, augmented by the smaller Texas rivers. Sand drifting longshore from the north and from the south converge near Corpus Christi to build a wide barrier island system there. Hence, the North Padre Island is wider than South Padre Island.

Drive north from the beach town of South Padre Island to see a marvelous stretch of virtually untouched beach, dune, and tidal flat complex. The development on South Padre Island has (so far) been confined to the southern tip of the island, leaving plenty of open space along a 10-mile stretch north of town.

It is easy to spend hours among the dunes looking at their internal cross-bedded structures, seeing how sand moves across the dune's steep faces in low ripples, or observing myriad tracks formed as insects dash across the dry sand. Vegetation plays some role in trapping sand in dunes, though the dunes here are not stabilized by plants nor even extensively covered by the vegetation. Lots of free sand blows around. Note how the high edge of the dunes is along the beach/Gulf side and how the dunes trail off into the lagoon. The source of the sand is therefore the beach, and onshore winds tend to pile up the sand directly behind the beach.

The tidal flats on the lagoon side of the barrier vary in extent, but the wide expanse of tidal flats in this area is somewhat amazing considering the tidal range in the Gulf

Even with vegetation covering the dunes in places, sand still moves and, on occasion, covers State Park Road 100 on South Padre Island. Note the vegetation stabilizing the dunes in the distance. (26.2081, -97.1793)

Ripple marks form as winds blow across the sands. Someday, these ripples may get preserved as sedimentary features in solid rock. (26.2462, -97.1858)

of Mexico is low—only 1 or 2 feet. A lot of low-relief land is inundated, then dried, as the low tidal waters sweep in and out across the flats twice daily.

The best tidal flats are seen a few miles south of Port Isabel along TX 48. Sand dunes are also abundant and easily discernible along this stretch of road. In fact, the first 10-mile stretch south from Port Isabel on TX 48 crosses a delightful area of near-wilderness, where coastal environments in their natural state can be viewed and enjoyed from the highway. The vastness of Bahía Grande, aptly named as the Great Tidal Flat, is impressive. The dunes surrounding the Bahía stand up clearly, their long linear shapes forming ridges above the flat terrain. Note the tall yuccas on sandy slopes of dune ridges, evidence of the dry climate.

TX 48 continues into Brownsville, following the intercoastal waterway, which connects Brownsville with sea traffic in the Gulf. Fishing boats stand above the dry surrounding landscape, 10 miles from Port Isabel, looking oddly out-of-place this far from the ocean.

COASTAL PLAIN NORTH OF BROWNSVILLE

I-69E/US 77 between Brownsville and Corpus Christi crosses classic coastal plain consisting of flat expanses of farmland on rich, brown soils overlying floodplain clay and sand. Between Raymondville and Riviera, however, the roadside scenery is markedly different, with dry sandy soils, scrub trees, lots of cactus, and sparse grass. Wind has blown the sand inland from sandy shores for thousands of years. Flat sand sheets

LA SAL DEL REY

West of Raymondville on TX 186 is the Lower Rio Grande Valley National Wildlife Refuge, a popular destination for bird watchers during the winter months because the area lies along a major bird migratory route in North America. The refuge is also home to a unique geologic feature with an equally unique history called La Sal del Rey (The King's Salt Lake). Early Spanish writings told of Aztec peoples trading obsidian artifacts for salt from the lake. In the mid-1700s, Spaniards claimed the rich salt lake for their king and began mining salt. Later, Mexicans loaded salt into carts and shipped it south, not only to season and preserve food but also to assist in refining rich silver ore.

La Sal del Rey was created as wind scoured out sediment to create a depression called a blowout. Early scientists thought the salt was from a subsurface salt dome, which are numerous in this region of Texas. The source, however, was discovered to be an extremely saline and shallow aquifer. Evaporation further concentrates the salt on the surface, producing white halite crystals around the lake that glisten in the sun. At times, the lake has as much as 10 feet of water in it, while other times the depression is completely dry. Because it is a closed lake, salinity levels (when there is water present) can be more than 400 parts per thousand. For comparison, typical seawater has on average 35 parts per thousand. To reach La Sal del Rey, follow TX 186 west for 17.5 miles from I-69E to Brushline Road and turn right (north). Follow the road for 2 miles to the parking area for East Trailhead. If the gate is open, follow for another one-half mile to another parking area and proceed to the lake.

La Sal del Rey was an important salt source for Aztec peoples, as well as Spaniards and Mexicans.
(26.5321, -98.0543)

Sand blown in from the seashore thousands of years ago created flat sand sheets between Raymond-ville and Riviera, forming dunes in some places. (27.0672, -97.7924)

cover the ground, nearly everywhere, punctuated infrequently by sand dunes that rise above the surrounding surface. The dunes are mostly vegetation-covered and stabilized, though a bright patch of sand can be seen here and there on the flanks of a dune where the wind has somehow wormed its way through the plant cover to form a blowout. The scrub tree and grass region of blowing sand is mostly on the extensive property of the world-famous King Ranch, where Santa Gertrudis cattle were developed as a distinctive North American breed.

Another interesting landform in this desert landscape is the dry lake. Look for large patches of low, grayish ground where plants don't seem to want to grow. These clay patches are lakes when there's enough rainwater to fill them. Distinctive tussocks of grass around the edges of these dry pans apparently get their feet wet frequently enough to hang on for dear life at the water's edge.

PADRE ISLAND NATIONAL SEASHORE

Access to Padre Island National Seashore is from Corpus Christi via TX 358, a divided highway that splits off I-37 and heads southeast through the city. The road crosses two water bodies en route to Padre Island. After passing over Oso Bay, a narrow inlet also called Cayo del Oso, the road then crosses over the Encinal Peninsula on which the Corpus Christi Naval Air Station is located. The peninsula is the remnant of a preserved Pleistocene barrier island built during an interglacial period when sea level was higher than it is today. South of the highway and the Naval Air Station is the Flour Bluff neighborhood, the highest point on Encinal Peninsula. The highway becomes Park Road 22 as it arches over the second water body, the wide expanse of Laguna Madre, on the JFK Memorial Causeway. From the causeway's high vantage point, you can see the lagoon extending to the horizon to the south and get a feel for the shape and size of North Padre Island to the east. Note also the straight line of the intracoastal waterway and its adjacent spoil piles, debris from when the channel was dug. A vast array of goods and materials move up and down the Texas coast in barges plying the protected waters of the lagoons via the intracoastal waterway. The waterway segment near North Padre Island was dredged in 1949 and is maintained at 12 feet deep and 125 feet wide by the US Army Corps of Engineers.

Follow Park Road 22 to enter the national seashore. The two-lane paved highway winds southward down the center of the island, where it ends as a beach access road about 5 miles within the park. Southward beyond this point, car travel is on an unpaved beach road that requires four-wheel-drive vehicles for extended travel down the island.

Padre Island National Seashore preserves a great section of barrier island, a long, sandy island that protects inland areas from the ravages of storms. (27.4353, -97.2929)

North Padre Island is a wonderland of natural geologic processes. Wander down the beach and watch the waves move sand via longshore drift. Look for layers of dark, heavy minerals, such as magnetite, which are moved at a different rate than the lighter quartz sand grains. Note how shells change in size and type along the beach.

Walk inland from the seashore to explore the island. Watch for ghost crab holes as you approach the edge of the dunes. While in the dunes, see if you can spot wind-eroded dune edges where large cross-beds are exposed to show the dune's internal structure. Observe how the dunes are controlled by the hardy vegetation—sea oats, morning glory, sea purslane, beach tea, and panicum, to name the most common. In a few places, the wind has broken sand free from the plants' tight grip to form free-standing sand dunes that take on their own shape and identity. On the land-ward side of the dunes, you can see how the dunes decrease in height and grade into a vegetated barrier flat, where tall grasses and low shrubs cover a surface that is not infrequently inundated by water as storms and high tides churn over and around the barrier island. Ponds and small lakes occupy low spots on the flats.

Another set of dunes is located on the edge of the lagoon where sand is blown from the free-sand edges of the lagoon's inner shore. Windswept flats around the lagoon are the product of tidal variations of water levels in the lagoons. Tides are spawned both by the moon's pull and by wind surges that pile up water in the lagoons, only to subside, leaving behind the flat-exposed surface. Algal mats anchor the tidal flats at many locales, appearing as dark areas among the lighter sand. You might also be lucky enough to find gypsum rosettes in sediments of the tidal flats. These crystals of calcium sulfate form as highly saline water seeps into the sediments, then dries, leaving behind the gypsum as a precipitate. With successive wetting and drying cycles, the crystals grow into rose-shaped clusters.

Close-up of a stabilized dune along North Padre Island. The plants and their roots enhance the dune's resistance to erosion. (27.4353, -97.2929)

View looking inland (westward) across a marshy area toward dunes on the horizon. (27.4437, -97.2999)

US 83/I-2
LAREDO—MCALLEN—HARLINGEN
178 miles

US 83, and eventually I-2, follows the Lower Rio Grande Valley from Laredo to Harlingen, but the river is rarely in view. As the road bends southward in Laredo, note the small exposures of tan sandstones on the south-facing banks of Chacon Creek. These beds belong to the Laredo Formation, a sandstone and claystone unit of the Claiborne Group that was deposited in the shallow sea margin of the early Gulf Coast in Eocene age. The road follows the Laredo beds from Laredo to Zapata near Falcon Reservoir.

The countryside is sandy and dry with only 18 to 20 inches of rainfall per year. The dryness results in a pleasant climate year-round, and consequently, the Rio Grande

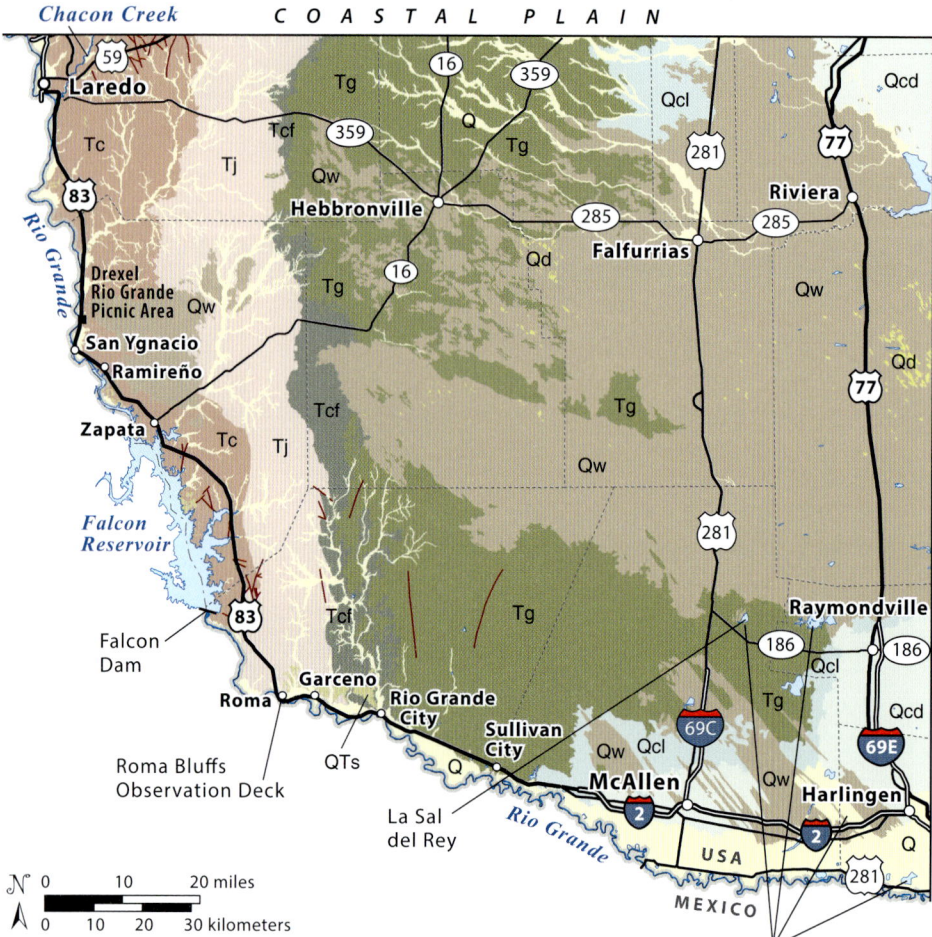

Geology along US 83 and I-2 between Laredo and Harlingen.

CENOZOIC

QUATERNARY

Q	recent sediments; includes alluvium and terrace deposits (Holocene and Pleistocene)
Qd	windblown dunes (Holocene)
Qw	windblown sand, silt, and sand sheet deposits (Holocene)
Qcd	coastal deposits of the Beaumont and Deweyville Formations (Pleistocene)
Qcl	coastal deposits of the Lissie Formation (Pleistocene)

QTs	Uvalde Gravel (Pliocene or Pleistocene)

NEOGENE and PALEOGENE

Tg	Goliad Sand (Pliocene)
Tcf	Catahoula Formation (Oligocene)
Tj	Jackson Group (Eocene and Oligocene)
Tc	Claiborne Group; includes Laredo and Yegua Formations (Eocene)

------- county boundaries

—— normal fault

Lower Rio Grande Valley National Wildlife Refuge

Valley is attracting more and more snowbirds and Texans each year, who seek relief from colder, northern climates.

About 10 miles south of Laredo, the sharp mountain profile of the Sierra Madre Oriental can be seen on the Mexican skyline to the west. Also watch for patches of badlands erosion where watery rills have etched headward into banks of local drainages after thunderstorm downpours.

A rare view of the Rio Grande along this route can be seen at the Drexel Rio Grande Picnic Area north of San Ygnacio. (27.0931, -99.4268)

At the Drexel Rio Grande Picnic Area, about 31 miles south of Laredo, the road ascends a knoll of Laredo Formation, where a decent roadcut displays sandy channels and cross-bedding in the sands. In addition, a great view of the Rio Grande can be observed in the picnic area. Two additional excellent outcrops can also be seen about 5 miles south of San Ygnacio near the town of Ramireño, that shows more of the Laredo Formation cross-bedding with spectacular gypsum beds.

A scenic view of Falcon Reservoir, a dammed portion of the Rio Grande, shows up as the road curves east on a high point north of Zapata. Near Zapata, the highway crosses the upper reaches of Falcon Reservoir. Falcon Dam is operated jointly by the United States and Mexico, under supervision of the International Boundary and Water Commission. Falcon Reservoir attracts boaters and bass fisherman and provides irrigation water and power to the region.

US 83 swings southward at Zapata and follows the next younger Eocene unit, the Yegua Formation of the Claiborne Group, for about 15 miles. A wide variety of crops are grown on the flat, fertile, brown-soil fields situated on the Rio Grande floodplain. Mild weather combined with irrigation water from the Rio Grande make the Lower Rio Grande Valley one of the richest produce regions in the nation.

A great roadcut of Laredo Formation sandstone and claystone lies on the north side of US 83, 0.8 mile north of Ramireño. (27.0141, -99.3942)

Close-up of cross-bedding and gypsum layers within the Laredo Formation, indicating changing conditions during deposition. (27.0141, -99.3942)

Eocene sandstones and claystones continue to form the surrounding bedrock between the Falcon Dam turnoff and the town of Garceno. In Roma, turn right onto Lincoln Avenue to reach the Roma Bluffs Observation Deck. About 37 million years ago in the Eocene, the Gulf Coast shoreline was located around Roma. During this time, a barrier island formed along the coast that protected a shallow lagoon from the ravages of storm waves, allowing giant oysters to happily live. The sandstone around Roma, part of the Jackson Group, formed on that barrier island, while fossiliferous zones represent the lagoon deposits. While exploring Roma, notice that many of the historic buildings in town were built from quarried sandstone, including the John Vale/Noah Cox and Manuel Guerra houses near the observation deck.

Between Rio Grande City and Sullivan City, ridges and mesas are quite prominent on the side of the highway away from the river. A hard, pebbly, gray sandstone holds up the top of the mesas above pinkish claystones and siltstones. Quarries and natural exposures are common along this stretch. The ridge-forming sand is the Pliocene Goliad Sand, which forms a nearly continuous ridge from here to Goliad, its namesake north of Corpus Christi. The road between Rio Grande City and nearly to McAllen is either along the edge of the Goliad ridge or on top of it, before dropping down to the flat, level surface of the coastal plain just north of McAllen. The coastal plain is flatter, grassier, and moister, with browner soils and citrus groves.

Cliffs of Jackson Group sandstone loom over the Rio Grande in Roma. (26.4057, -99.0189)

US 290
Houston — Brenham — Austin
152 miles
See map on page 80.

US 290 traverses pleasant rolling country as it crosses sandy ridges and shale-based valleys cut into the tilted package of Cenozoic sediments that lie adjacent to the coastal plain. US 290 heads northwest out of Houston in a straight line for nearly 20 miles from its junction with West Loop 610, traveling on the flat surface of coastal plain sands and muds. At Cypress, US 290 bends more toward the west, then between the Mason Road exit and the Grand Parkway (TX 99) interchange, the road climbs a slight hill marking the subtle boundary between the flat coastal plain and a more rolling topography (but still on the coastal plain).

Hills around Prairie View and Hempstead are held up by Pleistocene sandstones of the Willis Formation, into which the Brazos River is entrenched. The road crosses the Brazos River valley a few miles west of Hempstead, then climbs through river-edge topography to the high hill on which Brenham is built. Sandy ridges west of Brenham form the Oakville Escarpment, which extends for many miles to the southwest as a recognizable topographic ridge. The Miocene-aged Oakville Sandstone is hard enough to preferentially resist erosion and stand up as a ridge.

Southwest of Brenham, on TX 237, is the historic town of Round Top, where the nearby Winedale Historical Complex, operated by the University of Texas, is located. Outcrops of Oakville Sandstone are well exposed on the ridgetop in and around Round Top. Historic buildings, stone fences, and foundations in both Brenham and Round Top are built of the native sandstone.

Between Brenham and Elgin, US 290 bobs up and down as it traverses progressively older sandstone and shale. Younger Miocene bands are found around Brenham, whereas older Eocene rocks surround Elgin. Roadcut exposures are not common on this vegetated stretch of roadway, but if you watch carefully, you will see a sandstone ledge poking out here and there from the roadside grass.

BASTROP STATE PARK

In Bastrop State Park, near the town of Bastrop, is a geologic oddity. Loblolly pines (*Pinus taeda*), a common sight in the Piney Woods of eastern Texas, also grow here and are known as the Lost Pines. Loblolly pines need plenty of water to survive, but the climate at Bastrop is dry enough that the pines should not be able to grow. During the Pleistocene Epoch, the climate across Texas was much cooler and wetter, allowing pines to cover much of the state. As the climate warmed and began to dry, pines slowly died off. This westernmost stand of loblollies survived here due to the underlying bedrock.

The Carrizo Sand consists of sediments from the Rocky Mountains that were carried by rivers and deposited in channels and deltas along the ancient Gulf Coast. Over time, the rock weathered and eroded into a sandy soil that is great at retaining moisture. Water-loving loblolly pines thus thrive in this soil. Plus, they evolved to be more drought resistant than their more eastern relatives.

In 2011, a major wildfire swept through the Lost Pines. Burning for 55 days, it became the most destructive wildfire in Texas history, burning more than 34,000 acres. The fire killed two people and destroyed almost 1,700 structures. Much of Bastrop State Park also burned in the fire. From destruction, however, came rebirth as new loblolly pines sprouted from seeds, bringing life to the Lost Pines once again.

View of the Lost Pines from Bastrop State Park. In 2011, fire ravaged this area, but growth of new pines has regenerated the forest. (30.1114, -97.2692)

The Brazos River carries a large sediment load at this point in its journey near the US 290 bridge west of Hempstead. (30.1297, -96.1861)

Hills of strikingly bright-red sandstone a few miles west of the junction of US 290 and TX 21 and at Bastrop State Park (southwest of US 290 on TX 21) are exposures of the Carrizo Sand. This red band of iron-rich Eocene sandstone is distinctive for many miles across southeast and southwest Texas. Pine trees, which prefer this moisture-holding substrate, are localized on the sand ridges along TX 21.

Between Elgin and Austin, US 290 rides on Late Cretaceous rocks. These marly limestone units are not exposed along this segment, but look for the glint of white limestone fragments, plowed up to the surface in farm fields and during construction projects.

CENTRAL PLAINS AND HILL COUNTRY

A distinct line curves across Texas, separating the flat, rich farmland of the Gulf Coast from the drier hills, plains, and plateaus to the west. This line generally follows the Balcones fault zone, a series of faults that dropped rocks down to the east relative to the uplifted rocks to the west. The land steps up to the west, and because it looks like a series of balconies, it was called the Balcones Escarpment by Spanish-speaking settlers. Today's steep rise is not an actual fault scarp in which an earthquake broke and lifted the modern land surface; the escarpment formed along the fault zone because softer rocks lie on the east side of the fault and have been eroded more than the harder rocks to the west. The fault zone extends from Del Rio and Uvalde north through San Antonio and Austin to Waco, north of which it becomes less prominent and fades away. The zone follows an ancient suture in the deep crust that formed more than 1 billion years ago and then was reactivated during the forming of the Ouachita Mountains in late Paleozoic time. See the following section on the Llano Uplift for more on this ancient history.

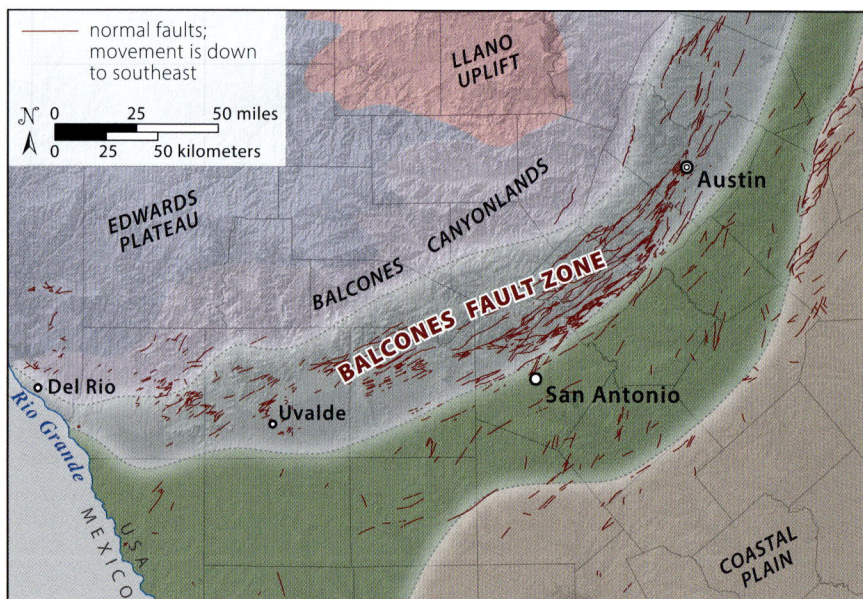

Map of the Balcones fault zone and surrounding areas.
—Modified from Ewing, 2016; faults from Geologic Database of Texas

The Edwards Plateau, the high-standing, flat-surfaced limestone bench of central Texas west and north of the escarpment, stands about 2,000 feet above sea level. These uplifted marine sandstones, limestones, shales, and dolomites were originally deposited in the ocean below sea level in Cretaceous time. Many geologists think this uplift along the Balcones fault zone occurred about 20 to 10 million years ago during Miocene time, as part of a regional upwarping that occurred across the western United States. Other geologists argue that the plateau was first elevated long before the Miocene, perhaps in Paleogene time, and then movement along the Balcones fault zone in Miocene time caused the area on the southeast side of the fault to drop and leave the Edwards Plateau standing high. Whether the uplift occurred in one or two phases, the thick layers of Cretaceous sedimentary rocks have been elevated essentially undeformed; the uplift did not fold or contort the rocks. They are as flat lying today as they were when originally deposited in the ocean 100 million years ago.

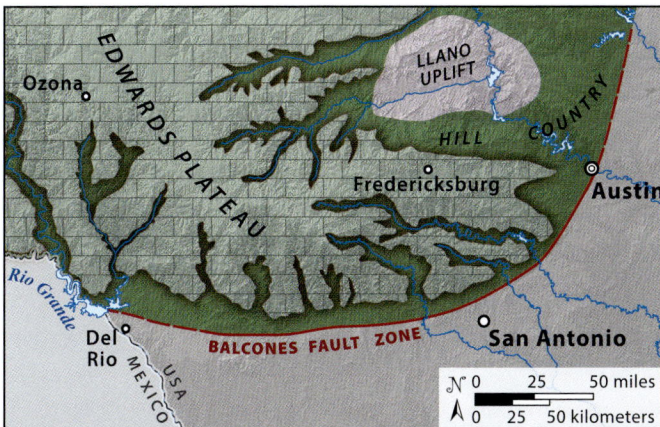

Over the last 10 million years, erosion by streams attacked the margins of the plateau, and the streams continually cut headward as well as downward in an erosional frenzy to flatten this highland. In the span of a mere 10 million years, the job is about half completed. The west half of central Texas remains a high, flat plateau, while the east half, called the Hill Country, is deeply eroded.

North-central Texas was not uplifted as much, and the erosive power of the Brazos River system, which cuts diagonally across the center of north-central Texas, removed an impressive amount of overlying rock to uncover Permian and Pennsylvanian rocks. These older rocks manifest on the geologic map as north-oriented, parallel bands. Looking like a tipped stack of cards, these bands are among the oldest sedimentary rock exposures in the state. The bands are regularly stacked, oldest to youngest, from

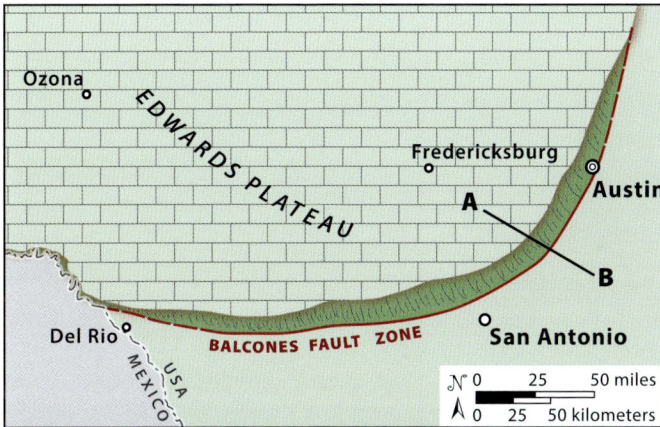

10 million years ago

Present Time

About 10 million years ago, the Edwards Plateau was uplifted along the Balcones fault zone. Since then, streams carved into the Edwards Plateau, exposed the Llano Uplift, and created the Hill Country.

NOTABLE ROCKS IN CENTRAL TEXAS

GEOLOGIC AGE			ROCK NAME	ROCK DESCRIPTION
CENOZOIC	MIOCENE–PLIOCENE		Ogallala Formation	caliche, siltstone, and conglomerate
				———— unconformity ————
MESOZOIC	CRETACEOUS	LATE	Taylor Group — Anacacho Limestone	limestone
			Austin Group — Austin Chalk	limestone and chalk
			Eagle Ford Group — Eagle Ford Shale	oil-producing shale
		EARLY	Washita Group — Buda Formation	white limestone
			Washita Group — Del Rio Clay	shale, clay, and yellow mudstone
			Fredericksburg Group — Edwards Limestone	limestone and dolomite
			Fredericksburg Group — Comanche Peak Limestone	limestone
			Fredericksburg Group — Walnut Formation	marl, yellow-brown claystone, and limestone
			Trinity Group — Paluxy Formation	brown sandstone, mudstone, and limestone
			Trinity Group — Glen Rose Formation	marl, yellow-gray limestone and dolomite, claystone, and sandstone
			Trinity Group — Travis Peak Formation	conglomerate, sand, bluish shale
			Trinity Group — Hensell Sand	limestone and sandstone; contains oyster fossils and geodes
			Trinity Group — Antlers Formation	quartz sand, sandstone, clay, and conglomerate
			Trinity Group — Cow Creek Limestone	limestone
PALEOZOIC	PERMIAN		Whitehorse Formation	red shale, sandstone, dolomite, and gypsum
			Pease River Group — San Angelo Formation	shale, sand, and conglomerate of dolomite and limestone pebbles
			Wichita Albany Group — Lueders Formation	alternating beds of limestone and shale
			Wichita Albany Group — Talpa Formation	
			Wichita Albany Group — Grape Creek Formation	
			Wichita Albany Group — Elm Creek Formation	
			Wichita Albany Group — Admiral Formation	
	PENNSYLVANIAN		Strawn Group — Mineral Wells Formation	limestone, sandstone, and shale
			Strawn Group — Mingus Shale	shale and sandstone
			Bend Group — Smithwick Shale	shale
			Bend Group — Marble Falls Limestone	limestone; known for its fossils
	MISSISSIPPIAN		Barnett Shale	black oil shale

Notable groups and formations in central Texas. For older rock units of the Llano Uplift, see the stratigraphic column on page 167.

east to west, with Permian red beds to the west. Rivers flowing through this land take on the eroded sediments' hue, so much so that the Red River on the northern border of Texas is perfectly named. A high rim of younger Cretaceous rock to the east and south is testimony to Cretaceous sediments having once extended across north-central Texas.

The bands of Pennsylvanian and Permian-aged rocks are the result of progressive filling of a deep crustal basin in southwest Texas, created by the sinking of the crust in that area, while the crust rose in central Texas and adjacent Oklahoma. About 300 million years ago, the Wichita, Arbuckle, and Ouachita Mountains were pushed upward along the southern margin of North America as it collided with Africa and South America to form the supercontinent Pangea. Southwest of this chain of mountain ranges, the crust collapsed to form a deep basin known as the Permian Basin, centered on West Texas. A broad, shelf-like area existed between the Ouachita Mountains to the east, which lay along what is now the Balcones fault zone, and the deep basin to the west. Rivers and streams flowed west across this shelf, carrying and depositing their eroded sedimentary load etched from the mountains. On the shelf, sand was deposited in river channels and mud in floodplains, while red mud, sand, and evaporative gypsum were deposited farther west, where the climate was dry, and shallow water quickly evaporated.

The orderly stack of sediments in north-central Texas is more than 5,000 feet thick, deepening to more than 10,000 feet in the center. The weight tilted the basin about one-half degree to the west. Once the basin was filled, and the stance of the mountains greatly reduced, a rather flat plain, near sea-level, was all that remained. On the late Paleozoic riverbanks and floodplains of north-central Texas, a dramatic biologic story unfolded, as early amphibians and reptiles first walked, then strode mightily into the Mesozoic to spawn a great race of reptilian giants—the dinosaurs.

By Jurassic and Cretaceous time, the continents drifted apart once again, new seas opened up, and sea level rose across the flat plains. A thick, widespread layer of limestone was deposited from the marine waters, while sandstone and mudstone were laid down on the coastal margins. Along this seaway, mighty beasts shook the ground as they marched across firm, sandy flats. Their tracks are well-preserved in the Cretaceous beds of north-central Texas, and Dinosaur Valley State Park preserves some of the best. At its highest level, in the middle of the Late Cretaceous, the shallow sea extended from the Gulf of Mexico to the Arctic Ocean, earning the name Western Interior Seaway.

TEXAS CAVERNS

In the Texas Hill Country and Edwards Plateau, limestone is everywhere. Thousands of square miles of Cretaceous limestone are exposed in streambanks, hillsides, roadcuts, and cliffs. The hard, limy rock resists surface erosion better than softer, surrounding shales, so the generally flat-lying limestone beds stand out as ridges and cliffs and form ledges on top of hills and mesas.

As hard and indestructible as limestone appears at the surface, underground it is riddled with holes and cracks, endless caverns, and intricate subterranean passageways. Limestone, you see, is soluble. Light raindrops pattering on a flat limestone outcrop are not gentle to the rock. Rainwater contains dissolved carbon dioxide from its passage through the air, which makes even natural rain slightly acidic. This slightly acidic rainfall (called carbonic acid) is normal. Pollutants add even more carbon dioxide, sulfur dioxide, or nitrous oxide to the water in the air to create acid rain, a much more acidic and worrisome problem.

For millions of years, this weak acid has percolated downward through tiny fractures and along openings between bedding planes in the subterranean limestone of the Edwards Plateau. The groundwater slowly dissolves the limestone along these microscopic passageways, and bit by bit, the passageways are enlarged. The water percolates downward until it reaches the water table, then moves laterally toward a

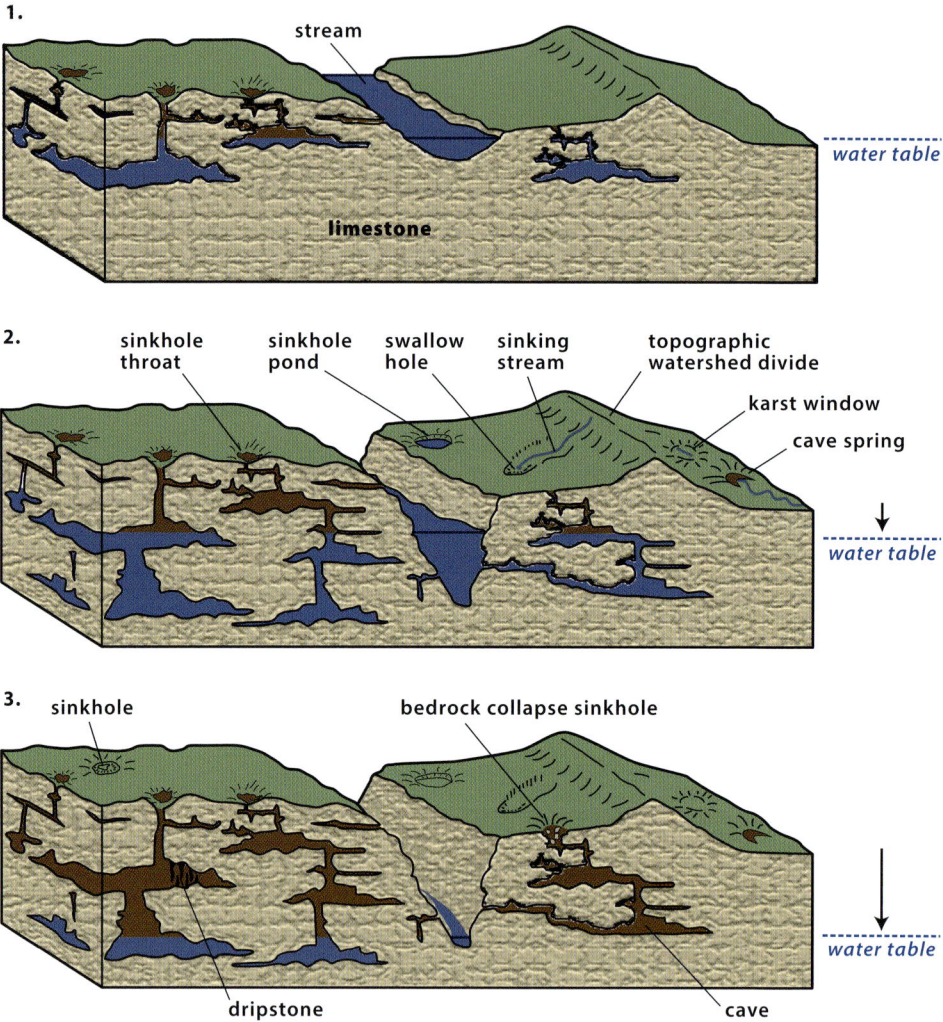

How a typical Texas cave forms. First, percolating groundwater enlarges fractures in limestone as it seeks its way downward toward the water table. Second, passageways enlarge, and water flows laterally through the subterranean cavern system at the level of the water table toward streams. Springs emerge from banks at or above the water table level, and streams continue downcutting. Finally, the streams cut even deeper, the water table lowers, and new, deeper passageways are created. Higher caverns are left dry, except for minor dripping water that builds dripstone formations.

nearby surface river, always dissolving the limestone. Passageways are enlarged until a subterranean channel, or cave, is created at the water table level.

Meanwhile, all has not been quiet at the surface either, for rivers and streams have been doing their job of cutting downward. As the rivers and water table reach a new, lower level, an equivalent level is sought in the subterranean system, and a new, lower, main passageway is created, leaving the upper, former main passage dry. Rainwater continues to fall on the plateau above, and groundwater continues to percolate downward through passageways. Some water still leaks downward through tiny fractures to drip from the ceiling of the abandoned, upper-level caves. This water is the same as always; it continues to be slightly acidic and dissolves limestone from the walls of fractures and carries away the dissolved calcium in a watery solution. But, the story now becomes intriguing because a little chemistry trick occurs just as the water droplet emerges from the tiny crack in the cave ceiling to hang motionless for a moment before dropping to the ground floor far below. In the crack above the ceiling, the drop is carrying a full load of calcium as well as a balanced load of carbonate (carbon dioxide) from the dissolving action on the limestone (which is calcium carbonate). As the drop hits the crack opening in the cave ceiling, it is carrying more carbon dioxide than is in the cave air, so a bit of carbon dioxide is released to the cave air to even things up, chemically speaking. As soon as that happens, the remaining water in the hanging drop becomes instantly supersaturated with calcium, and just before

Stalactites, like these in Caverns of Sonora, form where calcium-charged water drips from a low spot on the cave ceiling. Stalagmites build up from the spot of water-impact on the cave floor.

the drop lets go, the extra calcium (along with some carbon dioxide) is left behind on the ceiling as a thin, precipitated ring or film of pure calcium carbonate (the mineral calcite), and a stalactite is born.

Falling through black cave space, the drop loses more carbon dioxide to the air, and by the time it splashes on the floor, it is again supersaturated with calcium. Another bit of calcite is left on the cave floor at the point of the droplet's impact, and a stalagmite starts to form. Repeating the process drop by drop for countless millennia creates the wonderful, artful, and awe-inspiring dripstone formations we love to see in caves. Stalactites (with a "c") hang from the *ceiling*, stalagmites (with a "g") *grow* from the ground, and where they meet, solid columns form. Delicate soda straws (hollow, tubular stalactites) form when one drop at a time exits the ceiling from a tiny hole, leaving a perfect calcite ring behind; successive rings simply grow into a tube. Droplets emerging from side walls deposit calcite in irregular patterns to form odd-shaped twigs, fans, butterflies, and fishtails—all called helictites. Some speleologists (cave scientists) have even suggested air currents may blow droplets awry, causing the many angles at which helictites grow. Water emerging in a continuous line along a crack lays down a line of calcite, and continued growth results in the marvelous drip curtains seen in many caves. Though pure calcite is white, some cave dripstone is colored yellow, tan, brown, and even red. These warm colors are derived from varying amounts of iron carried in groundwater and precipitated along with the calcium.

Large passageways are frequently enlarged to caves as unstable roofs collapse, creating sinkholes above and piles of collapse debris below at the cave floor. Sinkholes are common over much of the Edwards Plateau and are significant openings where rainwater enters the underground piping system to form the aquifers that provide drinking water for much of the Hill Country and the San Antonio area. The Edwards Aquifer is carefully watched and managed by the citizens of this region.

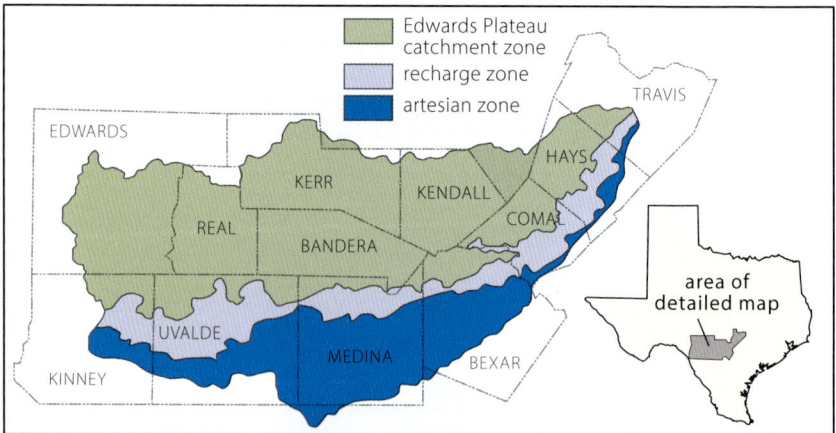

The Edwards Aquifer system includes three zones. The catchment zone is the area where water comes from (for example, the drainage basins of rivers that flow into recharge zone). The recharge zone is where water refills the aquifer, for example, by seeping into the ground or flowing into a sinkhole. The artesian zone is an area where water flows freely from the aquifer and onto the land surface, emerging as springs.

I-10
SAN ANTONIO — SONORA — FT. STOCKTON
312 miles

I-10 passes many roadcuts and natural outcrops as it traverses the Hill Country near San Antonio and tops out on the Edwards Plateau farther west. This mainly limestone terrain, miles and miles of flat-lying but uplifted Cretaceous rocks, handsomely illustrates the broad extent of the Western Interior Seaway, which covered the region. Fossils of marine snails, clams, oysters, and sea urchin spines are especially abundant, and a stop at virtually any roadcut will turn up fossils.

As you drive north out of San Antonio, I-10 crosses the Balcones fault zone, a parallel series of faults. Just north of the I-410 crossover, look north to see the high hills of the uplifted "balcony" ahead. A large quarry can be seen to the northeast of the Loop 1604 interchange where the Edwards Limestone, a major unit in the Fredericksburg Group, is mined to make cement and road gravel. The quarry exposes the thick-bedded character and flat-lying nature of the Edwards Limestone.

Immediately north of Loop 1604 between exits 556 and 552, I-10 climbs uphill, and thereby crosses the main Balcones fault. Several highway cuts along I-10, and the abandoned limestone quarry that has been converted to Six Flags amusement park to the west, display the yellowish, thin-bedded limestone and marl character of the Early Cretaceous Glen Rose Formation of the Trinity Group. I-10 follows the Glen Rose for about 50 miles to just west of Kerrville.

The area between San Antonio and Kerrville, known locally as the Hill Country, represents the deeply dissected edge of the Edwards Plateau. Streams have carved away at the uplifted sedimentary rock for the last several million years. Watch for

The Glen Rose Formation limestone is prominent in the walls of a former quarry, now converted into an amusement park near the Loop 1604 interchange. (29.6008, -98.6045)

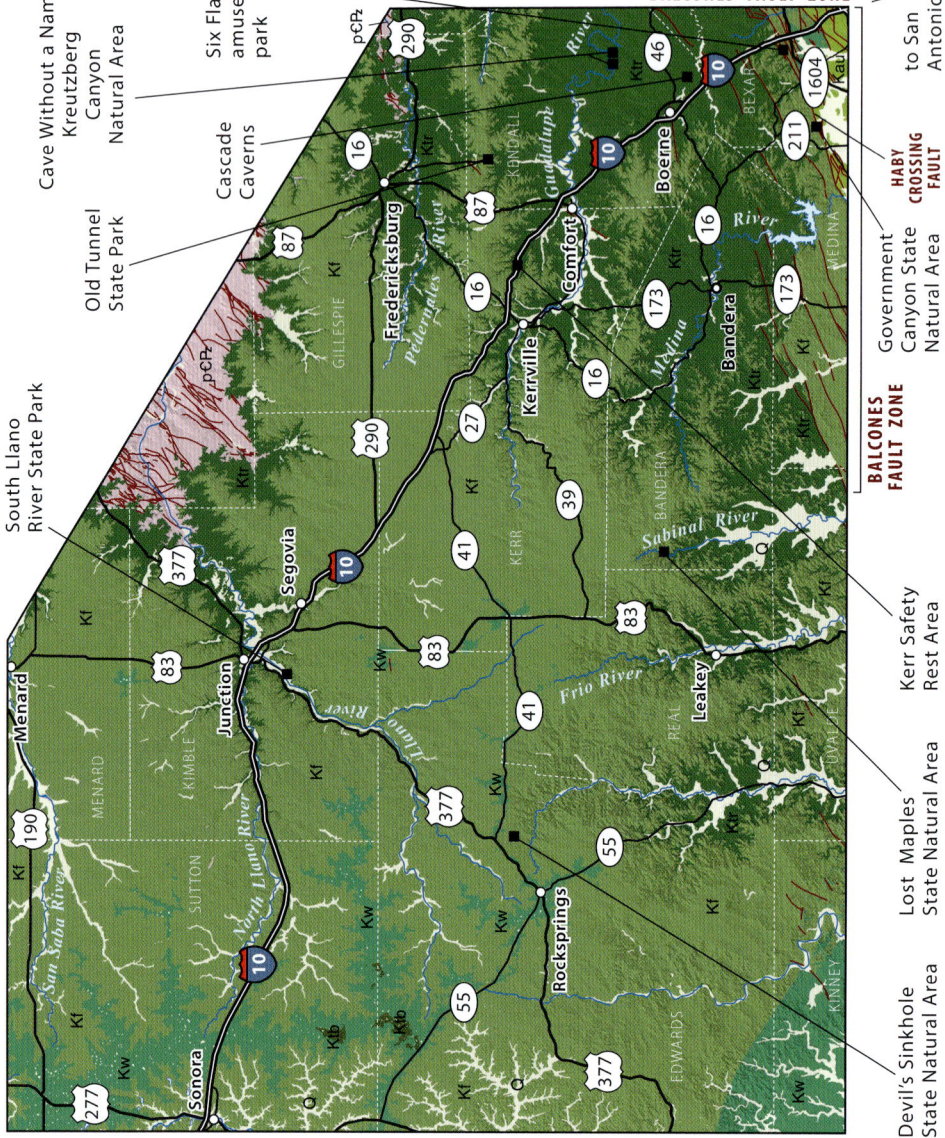

BALCONES FAULT ZONE

Cave Without a Name; Kreutzberg Canyon Natural Area

Six Flags amusement park

Cascade Caverns

Old Tunnel State Park

South Llano River State Park

to San Antonio

HABY CROSSING FAULT

Government Canyon State Natural Area

BALCONES FAULT ZONE

Kerr Safety Rest Area

Lost Maples State Natural Area

Devil's Sinkhole State Natural Area

pCPz
290
River
46
10
1604
211
BEXAR
Kau
10
Boerne
16
Kf
Ktr
KENDALL
87
Guadalupe
River
Kf
Fredericksburg
Pedernales River
16
Comfort
173
Ktr
16
River
Medina
Bandera
173
GILLESPIE
pCPz
87
290
16
27
Kerrville
KERR
39
Kf
Sabinal River
Q
Kf
Ktr
377
10
Segovia
41
83
Kw
Frio River
83
REAL
Menard
83
Junction
Kf
41
Kw
Leakey
UVALDE
190
MENARD
KIMBLE
Llano River
377
Kw
Kf
San Saba River
SUTTON
Kw
Kw
Rocksprings
55
Kf
10
Sonora
Ktb
Ktb
EDWARDS
Kf
KINNEY
277
Kw
Q
55
377
Kf
Kw
North Llano River

N

CENOZOIC
Q recent sediments; includes alluvium, playa deposits (small dots), and older terraces (Holocene and Pleistocene)

MESOZOIC (CRETACEOUS)
Kau Austin Chalk; includes some Navarro and Taylor Groups
Ktb Boquillas Formation of the Terlingua Group
Kw Washita Group
Kf Fredericksburg Group; includes Edwards Limestone, Segovia Member of the Edwards Limestone, and the Devils River Limestone
Ktr Trinity Group; includes Glen Rose Formation and Hensell Sand
pCPz Paleozoic and Proterozoic rocks of the Llano Uplift

— normal fault
county boundaries shown as white dotted lines

0   10   20 miles
0 10 20 30 kilometers

Geology along I-10 between San Antonio and Sonora. See the map on page 127 for the geology between Sonora and Ft. Stockton.

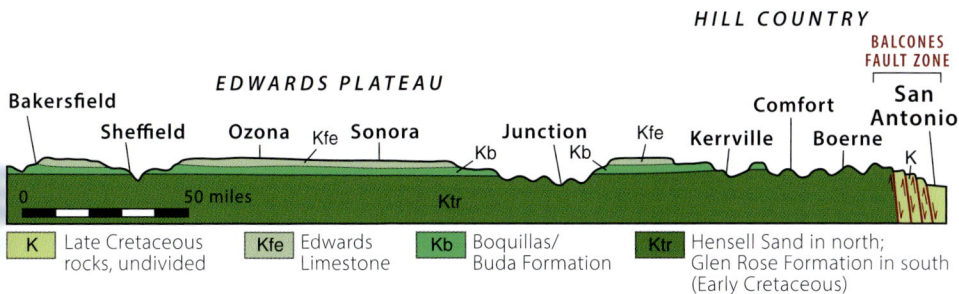

Cross section along I-10 between San Antonio and Bakersfield.

stair-stepped hills typical of Glen Rose terrain. The alternating hard limestone and soft marl beds erode unequally, so stairsteps are produced.

At Boerne (pronounced "Bernie"), take exit 543 and head 5 miles southeast on Cascade Caverns Road to see Cascade Caverns, with its stalactites, stalagmites, and a 90-foot-high, underground waterfall.

Also near Boerne is Cave Without a Name, a national natural landmark near the Guadalupe River, designated in 2009 for its biologic and geologic features such as stalactites, stalagmites, and soda straws. Take FM 474 northeast of town for about 6 miles and turn right on Kreutzberg Road. The nearby Kreutzberg Canyon Natural Area is a gorge cut into Glen Rose limestones on the Guadalupe River.

Near Comfort (exits 524 and 522) is Old Tunnel State Park, where an old railroad tunnel through the Glen Rose limestones provides a seasonal habitat for around three million bats. To visit, take FM 473 northeast 5 miles, then continue straight on Old 9 Road.

Near Kerrville, the upper parts of hills are composed of light-gray, thick-bedded Edwards Limestone that rests on top of the Early Cretaceous Glen Rose Formation. The yellowish limestone beds of the Glen Rose show up in the lower parts of hills in the low drainage area of the Guadalupe River around Kerrville.

GOVERNMENT CANYON STATE NATURAL AREA

Government Canyon State Natural Area, located northwest of San Antonio, is a beautiful tract of land that preserves features associated with the Balcones fault zone, as well as numerous dinosaur trackways. The park is divided into two main areas; the Frontcountry, which is an area of relatively flat topography, and the Backcountry, which is hillier and craggier. These areas are separated by the Haby Crossing fault, a normal fault within the Balcones fault zone. The relief at this location is around 600 feet, which is dramatically observed throughout the park.

Numerous dinosaur tracks and trackways can be found in the Glen Rose Formation limestones in the Backcountry. More than three hundred footprints of *Acrocanthosaurus*, *Sauroposeidon*, and an unknown dinosaur have been found. While the Glen Rose Formation gets most of the attention, the most abundant rocks within the Backcountry are related to the younger Edwards Limestone. Its many caves and numerous other karst features allow surface water to enter and recharge the Edwards Aquifer, a major water supplier to such cities as San Antonio and Austin. To protect this important source of water, no vehicles are allowed in the Backcountry parts of the park.

Southwest of Kerrville and west of Boerne, TX 16 and other paved backcountry roads wind through canyons along streams, offering the traveler a different pace from the maddening freeway rush. Deep, enfolding canyons, pattering streams, high mesas, and curving cliff walls of limestone upon limestone are the brush strokes in the geologic painting of this intriguing terrain.

All the rocks in this area are Early Cretaceous in age, deposited over millions of years in warm, shallow seas that once covered Texas. The Glen Rose Formation, a collection of limestone, shale, marl, and siltstone beds, was deposited along the shifting margins of the sea where dinosaurs roamed in great numbers, leaving their footprints in the sands. The Western Interior Seaway then spread over Texas in earnest, depositing thick layers of solid marine limestone, called the Edwards Limestone, over the Glen Rose beds. This sequence of rock units—Glen Rose below, Edwards above— is found throughout this area. An excellent roadcut showing this relationship can be seen on both east- and westbound lanes of I-10 just east of the Kerr Safety Rest Area east of Kerrville.

The gray Edwards Limestone (top half of roadcut) overlies yellowish limestone of the Glen Rose Formation (bottom half) in a roadcut along I-10 east of the Kerr Safety Rest Area east of Kerrville. Red Swiss Army Knife for scale. (30.0532, -99.0224)

The whole region was uplifted several thousand feet about 10 million years ago, which caused streams to erode intensely and cut back the margins of the Edwards Plateau. Evidence of the deep stream etching can be seen on the geologic map, where the intricate, feather-like patterns of the Edwards Limestone outline hilltops adjacent to drainages, while the underlying Glen Rose is exposed in the etched-out river bottoms.

Most of the roads around Kerrville, such as TX 16, TX 39, and TX 173, follow low areas along stream courses. Watch for yellowish, thin-bedded, marl and limestone beds of the Glen Rose Formation. Eroded material from the mesas commonly forms talus piles on the lower slopes, through which some roads are cut. The chaotic mix of sand, gravel, and boulders is a sure indicator of these talus deposits. In many places in the streams and rivers, low, stair-step waterfalls are found. This phenomenon is a direct result of the bedding character of the Glen Rose Formation wherein hard limestone beds alternate with softer marl or silt beds. Where roads climb over ridge crests, the thick, gray beds of solid Edwards Limestone show up in high-level roadcuts.

Southwest of Kerrville, TX 39 and TX 16 top out on the flat, high terrain of the Edwards Plateau. This is a good place to get a feel for the nature of the plateau prior to erosion and where one can also get an appreciation for just how much rock has been removed from the area by stream erosion.

West of Kerrville on I-10, close to the Gillespie County line near milepost 500, are a series of wonderful roadcuts where broken-up limestone, called breccia, can be seen along with local collapse folds. Layers of the Edwards Limestone have collapsed into underlying caverns, also in the Edwards.

LOST MAPLES STATE NATURAL AREA

In a remote part of the Texas Hill Country lies Lost Maples State Natural Area, a remnant of an ice age forest. During the Pleistocene, the climate of North America was vastly different than today. Glacial ice covered most of the northern parts of the continent, and although the glaciers did not reach Texas, they still had a major influence on the state's climate and ecosystems. Northern trees and plants, such as maples, thrived here in the cooler climate. Once the ice retreated and the climate warmed, most of these northern species disappeared from Texas except in places like Lost Maples State Natural Area.

The Sabinal River flows through the park, carving its way through Cretaceous-aged Edwards Limestone and Glen Rose Formation at the southern edge of the Edwards Plateau. Rocky cliffs protect the canyons from the hot sun, and springs emerge in the lower reaches, providing the perfect conditions for a remnant stand of bigtooth maples to grow. Other northern trees, such as red oak and sycamore, also inhabit the park, making for a brilliant autumn display of colors that rival New England or the Great Lakes region.

Large blocks of Edwards Limestone tumble from the adjacent cliffs into the Sabinal River canyon bottom in Lost Maples State Natural Area. (29.8185, -99.5707)

About 12 miles west of Kerrville, the road finally tops out on the flat countryside of the Edwards Plateau, which is the upper surface of the Edwards Limestone. To the east, intensive erosion in the Hill Country removed the Edwards Limestone and carved hills and valleys down into the underlying Glen Rose rocks. Erosion continues today as flash flooding events scour the bedrock channels and transport sediment. The interstate heads down through the Edwards into the Glen Rose again along the erosional drainage of the Llano River near the town of Junction. Natural, river-cut outcrops are all around Junction and at South Llano River State Park, 3 miles south of Junction along US 377.

A collapse breccia feature within the Edwards Limestone at the Kerr/Gillespie County line about 10 miles west of Kerrville. Note the lack of uniform layers and blocky nature because the jumble of rocks fills an old cavern. (30.1304, -99.2256)

This roadcut along Loop 481 south of Junction includes bright-red Hensell Sand, along with an overlying, yellowish shale unit, also of the Hensell. (30.4829, -99.7596)

Bright-red beds of the Cretaceous-aged Hensell Sand, another unit in the Trinity Group, are also particularly noticeable along I-10 and around Junction, especially in the deeper valley cuts. The sand was deposited close to shore at the same time the Glen Rose limestone was deposited farther out to sea.

I-10 follows the North Llano River drainage for about 20 miles west of Junction, then climbs to Edwards Plateau level again on the way to Sonora. Watch for roadcuts on hilltops about 8 miles east of Sonora that expose dark-gray and brown beds of nodular marl of the Buda Formation, which overlies the Edwards Limestone and is a unit in the Washita Group. It caps hilltops over much of the Edwards Plateau country.

DEVIL'S SINKHOLE STATE NATURAL AREA

Devil's Sinkhole State Natural Area, located near Rocksprings, is one of the Edwards Plateau's many karstic features. The earliest known visitor to the bottom of the sinkhole was H. S. Barber, who carved his name and the date (1889) into a rock at the bottom. The sinkhole is around 65 feet wide at the top and extends 350 feet in depth. In the center is a large pile of collapse debris approximately 200 feet high called a breakdown mountain. Groundwater gradually dissolved a portion of the Edwards Limestone as the water flowed through cracks in the rock, creating a large, water-filled cavern. Over time, rivers eroded downward, deepening their drainages and lowering the water table, so the cavern drained. Once dry, the cavern roof could not support itself and collapsed, forming the sinkhole we see today.

Since its discovery, people have tried to profit from the sinkhole. In the 1920s, attempts to extract bat guano for fertilizer use were abandoned due to being uneconomic. An ambitious and unsuccessful project during the Second World War involved American soldiers capturing bats to be used to deliver small napalm bombs to Japanese cities. Today, the sinkhole is the seasonal home to as many as three million Mexican free-tailed bats. During the summer months, the bats leave at nightfall in a spectacular display to feast on insects. Because of its geologic features and unique life forms, Devil's Sinkhole was declared a national natural landmark in 1971 and as such, the area can be accessed only via a guided tour.

Devil's Sinkhole State Natural Area near Rocksprings preserves one of many sinkholes on the Edwards Plateau. (30.0576, -100.1092)

Nodular marl beds of the Buda Formation (top one-third) overlie more blocky Edwards Lime-stone (bottom two-thirds) in this hilltop roadcut east of Sonora. (30.5683, -100.5522)

CAVERNS OF SONORA

One of the most beautiful caves in the country, the private Caverns of Sonora offers a unique look at a cave system formed in a much different way than usual. Typically, caves form from groundwater containing dissolved carbon dioxide from the atmosphere, which creates carbonic acid, a weak acid that dissolves carbonate rocks. At Caverns of Sonora, fluids containing dissolved hydrogen sulfide rose from deep underground and combined with oxygenated groundwater to form sulfuric acid, a powerful acid. Geologists think a fault provided the pathway for the deep fluids to move upward late in the Miocene Epoch. This process, called hypogenic cave formation, is also how the much larger Carlsbad Caverns in New Mexico formed. Beginning around 3 to 1 million years ago, the groundwater drained from the cavern, allowing calcite to precipitate on the walls and grow into speleothems. The deepest part of the cavern reaches about 120 feet below the surface with as much as 6 miles of passages spread out over four levels in the Cretaceous-aged Edwards Limestone.

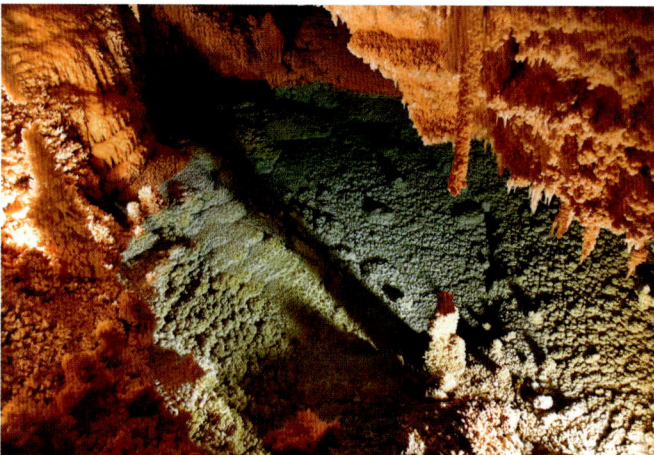

This pool of water, deep in the Caverns of Sonora, is surrounded by dripstones, flowstones, and popcorn. (30.5552, -100.8120)

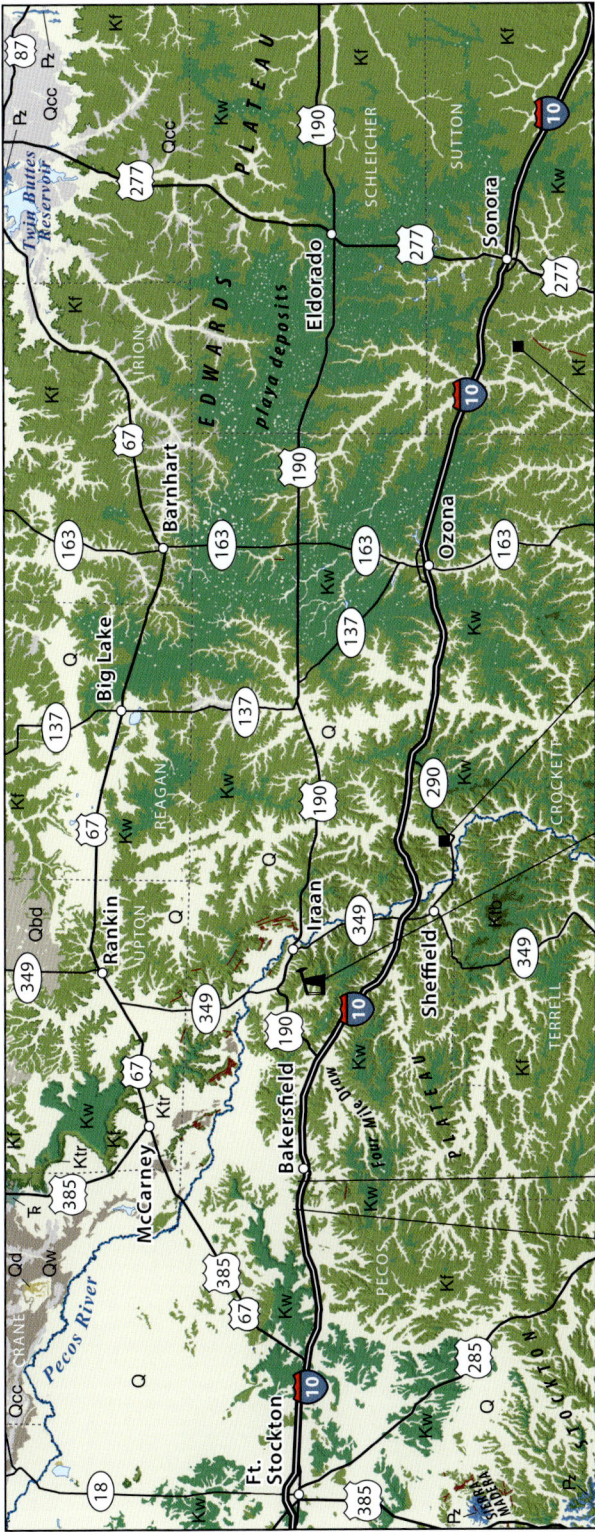

Geology along I-10 between Sonora and Ft. Stockton.

CENOZOIC

QUATERNARY

Q   recent sediments; includes alluvium, alluvial fan deposits, playa deposits (small dots), and older terraces (Holocene and Pleistocene)

Qcc   caliche

Qw   windblown sand, silt, and sand sheet deposits (Holocene)

Qd   windblown dunes (Holocene)

Qbd   Blackwater Draw Formation and windblown cover sand (Pleistocene)

MESOZOIC

CRETACEOUS

Ktb   Boquillas Formation of the Terlingua Group

Kw   Washita Group; includes Buda Formation

Kf   Fredericksburg Group; includes Edwards Limestone and Segovia Member of the Edwards Limestone

Ktr   Trinity Group

TRIASSIC

Tr   Dockum Group

PALEOZOIC

Pz   sedimentary rocks, undivided; includes Ocho, Clear Fork, and Pease River Groups (Permian)

oil field

normal fault

county boundary

20 miles

30 kilometers

West of Sonora, the steep-sloped, rocky, poorly vegetated mesas look different from the rounded, well-vegetated hills near San Antonio, even though flat-lying limestone dominates both regions. The difference in rainfall accounts for the change in vegetation and weathering. Average rainfall near San Antonio is 28 inches per year, but it is quite desert-like at 18 inches per year west of Sonora, and even less rain falls farther west. In moist regions, chemical weathering causes exposed rocks to disintegrate to pebbles and soil, which creep downslope, rounding off slopes and making soft-looking hills. In drier regions with little continuous moisture, chemical weathering is minimal, so soil is thin. Rocks simply break from cliff edges, and the main agent of erosion is high-intensity run-off from water supplied by thunderstorms. Sharp-edged mesas of exposed rock are thus a common landform in deserts. The junipers, low shrubs, and mesas are starting to look like West Texas cowboy country.

Several good roadcuts in Edwards Limestone near Ozona interrupt the otherwise flat terrain between Sonora and Ozona. See the photo on page 13 in this book's introduction for a good look at the Edwards roadcut at exit 368 east of Ozona. Low mesas and isolated buttes become more and more common west of Ozona.

An interesting side trip on TX 290 (exit 343) takes the traveler over typical Texas mesa country and past old Ft. Lancaster in the Pecos River valley. Established in the 1850s on the San Antonio–El Paso Road, the fort played a part in an experiment to use camels for transportation. West of exit 328, I-10 crosses the Pecos River, which begins high in the Sangre de Cristo Mountains of New Mexico and flows generally south for 926 miles before reaching the Rio Grande near Del Rio. The Pecos also served as the boundary for Native American tribes of the area. The phrase "West of the Pecos" was used to describe the rugged desolation of the Wild West during the late-nineteenth century.

TX 349 (exit 325) leads north to Iraan past the Yates oil field, one of the giant fields of North America. Discovered in 1926, the field has produced more than 1 billion barrels of oil from cavernous Permian limestones only 1,000 feet below the surface.

Tunas Peak, a prominent hill of Edwards Limestone capped by the Buda Formation, stands above the barren, flat landscape of central Texas. (30.8909, -102.3038)

West of Sheffield, I-10 follows the 1920s national highway called the Old Spanish Trail along the bottomlands of Four Mile Draw, a tributary of the Pecos River. Hillside exposures of Edwards Limestone, capped by the Buda Formation, surround the roadway. Note how the bottomlands are broader and the mesas more widespread as I-10 heads west into Bakersfield (exit 294). Look northwest from this exit to see Tunas Peak.

Near the junction of US 67 with I-10, about 20 miles west of Bakersfield, the mesas near the highway have limestone caprocks of Early Cretaceous age. What you are seeing here, as the mesas get farther apart and are finally gone by Ft. Stockton, is the western, eroded edge of the Edwards Plateau.

I-20
SWEETWATER — ABILENE — FT. WORTH
191 miles

This stretch of I-20 between Sweetwater and Ft. Worth is a perfect transect of the geology of north-central Texas. The highway runs perpendicular to the colorful red bands of Permian rocks in the western half between Sweetwater and Cisco, then cuts across the southern part of the older Pennsylvanian band of rocks between Cisco and the Brazos River crossing, and finally traverses younger, limy Cretaceous rocks between the Brazos River and Ft. Worth. Though rocks are not exposed at the surface over much of this area, the vegetation is frequently a clue to the underlying rock units. Mesquite commonly grows over claystones, while post oaks favor sandstone terrain. Junipers and live oaks prefer limestone country, so if you see green trees and shrubs in the winter, you are on calcareous soils and rocks.

East of Sweetwater at exit 249, a gypsum plant is located on the south side of the freeway. The gypsum is mined from the Whitehorse Formation, deposited during the Permian Period where extensive areas of shallow, evaporating water left behind bed after bed of drying gypsum along with bright-red, oxidized sand and mud layers. Gypsum (calcium sulfate) is used in making Plaster of Paris and is the white sandwich filling in the wallboard, or sheetrock, used in building construction.

Between Sweetwater and Trent, red-bedded sediments of slightly older Permian rocks, the Pease River Group, are exposed north of the highway. South of the road along the skyline is a high, flat-topped ridge called the Callahan Divide. This ridge is

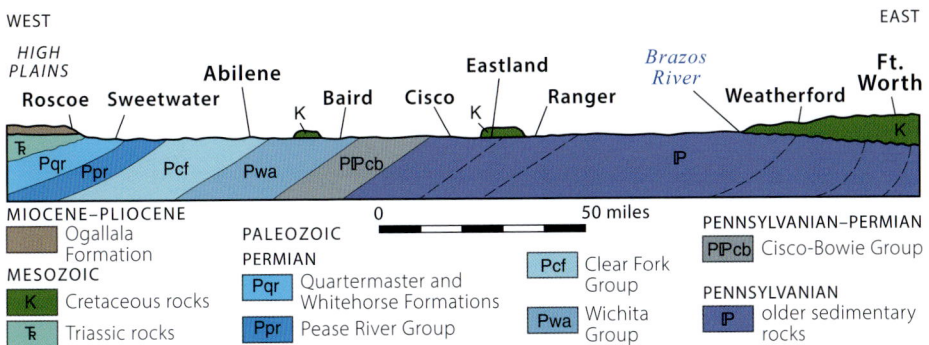

Cross section along I-20 between Ft. Worth, Abilene, and Sweetwater. The tilt of the rocks is greatly exaggerated.

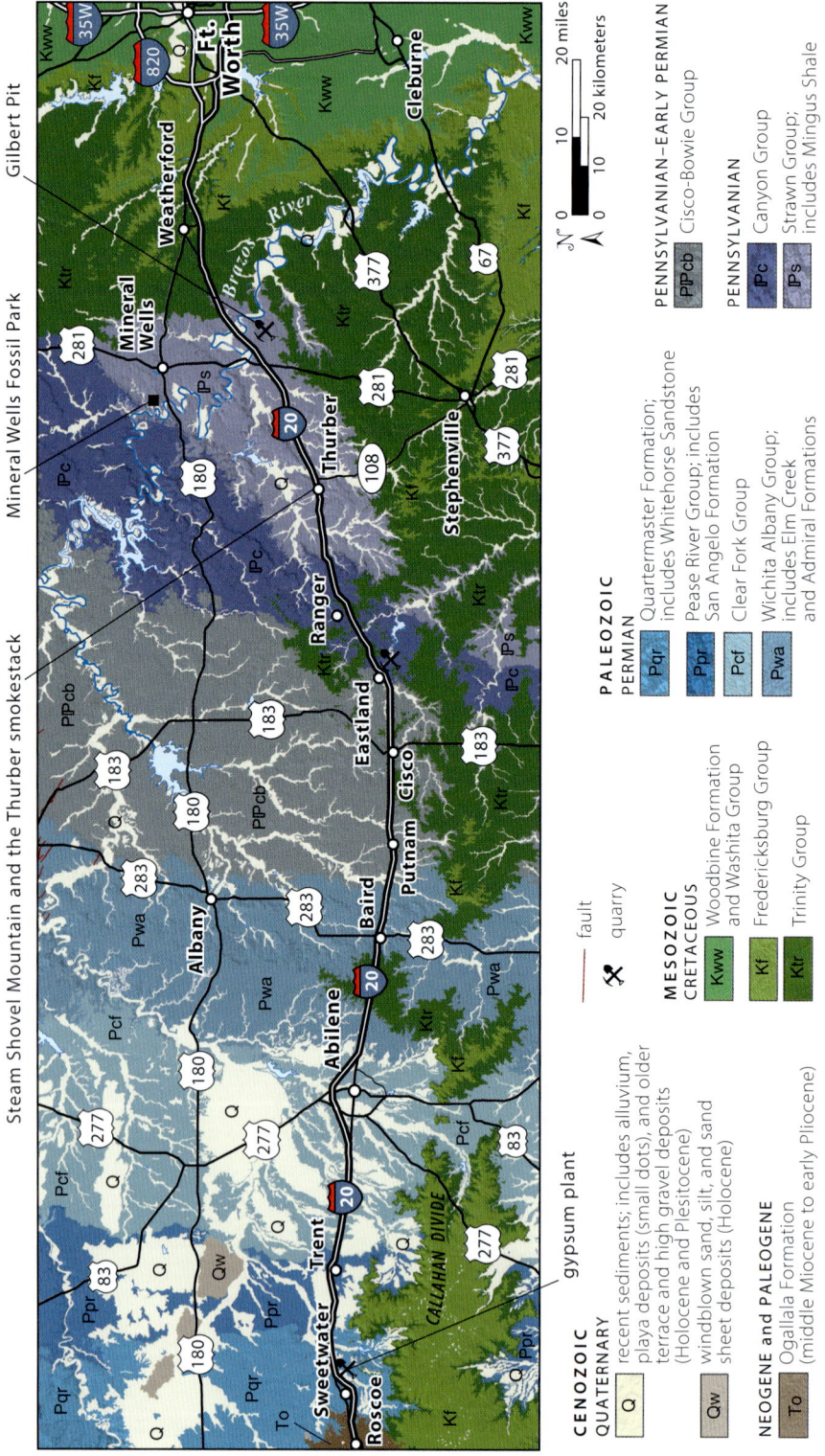

Geology along I-20 between Sweetwater and Ft. Worth.

Steam Shovel Mountain and the Thurber smokestack

Gilbert Pit

Mineral Wells Fossil Park

PENNSYLVANIAN–EARLY PERMIAN
IPIPcb Cisco-Bowie Group

PENNSYLVANIAN
IPc Canyon Group
IPs Strawn Group;
     includes Mingus Shale

PALEOZOIC
PERMIAN
Pqr Quartermaster Formation;
     includes Whitehorse Sandstone
Ppr Pease River Group; includes
     San Angelo Formation
Pcf Clear Fork Group
Pwa Wichita Albany Group;
     includes Elm Creek
     and Admiral Formations

MESOZOIC
CRETACEOUS
Kww Woodbine Formation
     and Washita Group
Kf   Fredericksburg Group
Ktr Trinity Group

— fault
✗ quarry

gypsum plant

CENOZOIC
QUATERNARY
Q   recent sediments; includes alluvium,
    playa deposits (small dots), and older
    terrace and high gravel deposits
    (Holocene and Pleistocene)
Qw windblown sand, silt, and sand
    sheet deposits (Holocene)

NEOGENE and PALEOGENE
To Ogallala Formation
    (middle Miocene to early Pliocene)

Ft. Worth
Weatherford
Cleburne
Mineral Wells
Thurber
Stephenville
Ranger
Eastland
Cisco
Putnam
Baird
Albany
Abilene
Trent
Sweetwater
Roscoe

Brazos River
CALLAHAN DIVIDE

N
0   10   20 miles
0   10   20 kilometers

A close-up of gypsum (in a roadcut along the frontage road east of Sweetwater near exit 249) reveals bright-red sand and mud layers that have oxidized prior to being covered by more gypsum. (32.4853, -100.3502)

an erosional remnant of a once-extensive cover of Cretaceous rocks that overlay the red Permian rocks. The Callahan Divide is a northern outlier of Cretaceous limestone beds that lie at the surface of the Edwards Plateau and Hill Country in central Texas to the south.

Just east of exit 256 is a highway rest stop where you can get good views of the surrounding countryside from the elevated topography. At exit 261, west of Trent, are red, inclined sandstone beds of the Permian-aged San Angelo Formation, part of the Pease River Group.

ERA	PERIOD		AGE (mya)*	ROCK NAME	ROCK DESCRIPTION	GRAPHIC COLUMN
PALEOZOIC	PERMIAN	late		Quartermaster Formation	red shale, sandstone, dolomite, and gypsum	
				Whitehorse Formation		
		middle		Pease River Group	red dolomite, sandstone, and gypsum	
				Clear Fork Group	red shale, siltstone, and some sandstone	
				Wichita Albany Group	reddish-brown shale and sandstone; cuestas	
		early		Cisco-Bowie Group	gray-tan limestone, sandstone, and shale	
	PENNSYLVANIAN		299			
		late		undivided	sandstone, limestone, and shale	

*mya = millions of years ago

[shale] shale [sandstone] sandstone [limestone] limestone [dolomite] dolomite [gypsum] gypsum

Stratigraphic sequence of Permian rocks in north-central Texas.

West of Trent at exit 261, sandstone beds of the Permian-aged San Angelo Formation crop out on the east side of the overpass. (32.4945, -100.1414)

East of Trent, low ridges called cuestas extend out for many miles on either side of the expressway. These cuestas are formed where slightly harder, west-dipping sandstone beds come to the surface. Erosion works a bit faster on the softer, surrounding mudstones, so the sandstones are left standing higher as cuestas. This sandy part of the Permian is in a slightly older, thicker band of rocks, the Clear Fork Group, which extends to the east side of Abilene.

The escarpment of Cretaceous rocks—the Callahan Divide—can still be seen on the skyline south of Abilene. The road between Trent and Abilene is quite flat, as is the surrounding terrain. The smooth, flat ride continues about 15 miles east of Abilene. The interstate crosses onto another belt of Permian rocks, the Wichita Albany Group, near exit 292A, but it is difficult to distinguish because of the flat terrain and few roadcuts or outcrops.

East of Baird between mile markers 309 and 310, the roadway climbs through the Elm Creek and Admiral Formations, both of the Permian-aged Wichita Albany Group. Note the thin beds of limestone, sandy limestone, and mudstone that are tan and gray, distinctively not reddish like other surrounding Permian rocks. The

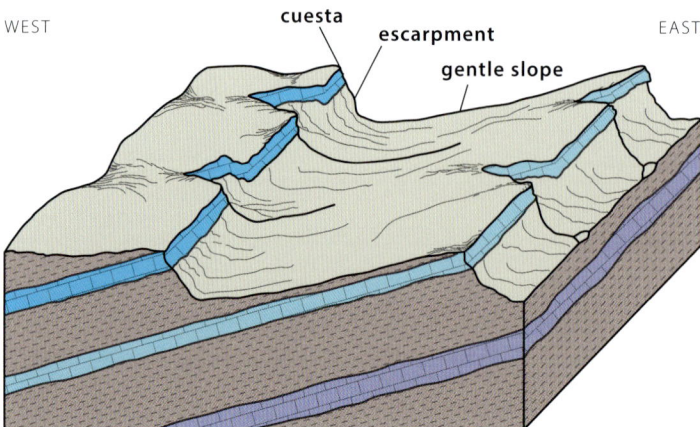

Cuestas, ridges where edges of inclined rock layers are differentially eroded at the surface, are common along I-20.

Roadcuts of Permian-aged limestone, mudstone, and sands of the Wichita Albany Group about 1 mile east of Baird. (32.3947, -99.3612)

limestones have abundant clams and snails—fossils that indicate deposition in clear, shallow marine water.

The roadcuts east of Baird through the Wichita Albany Group form a line of sandy cuestas extending away from the highway. These rocks, which include sandstones, mudstones, and shales, were originally sediments eroded from the Ouachita Mountains that lay to the east in Permian time. Streams from these mountains meandered their way westward across a vast floodplain, depositing these sediments along a developing sea. A second set of cuestas in Putnam are part of the Permian–aged Cisco Group, the oldest deposits found in north-central Texas. The brown sandstones and multicolored shales of this group were deposited on deltas that were built along the same seashores. The town of Cisco is the demarcation within the Cisco Group between Permian rocks to the west and Pennsylvanian rocks to the east. Blocks of tan-orange limestones and sandstones are exposed on ridgetops and old roadcuts a few miles east of Cisco.

Between Eastland and Ranger, light-tan limestone outcrops and quarries are in Cretaceous terrain—another spur of younger rocks that extends northward from the main body of Cretaceous rocks to the south. The Cretaceous rocks were deposited directly on an eroded surface of the older Pennsylvanian section. Around Ranger, the ledge-bordered hills and ridges are composed of Cretaceous limestone and sandstone, mainly in the Trinity Group.

In the 40 miles between Ranger and the Brazos River, I-20 crosses a series of prominent cuestas formed by resistant beds of limestone and sandstone, mainly of Pennsylvanian age. Softer claystone and coal beds are found in the low, eroded areas between the cuestas. Near Thurber (exit 367), look for the smokestack as a marker to identify nearby weathered coal beds. The depositional environment during the Pennsylvanian was quite interesting because of its cyclic nature and variety of sediments that led to economic accumulations of coal, limestone, clay, and hydrocarbons.

Pennsylvanian-aged orange-to-yellow sandstone beds overlie reddish-brown shale beds in eroded cuesta country near the TX 108 crossover (exit 370). Sandstone and shale cuestas are common between TX 108 and the Brazos River crossing. A large quarry, the Gilbert Pit, can be seen south of the highway west of the Brazos River.

The Brazos River forms one of the major drainage systems across Texas. It begins in the high plateau country of the Panhandle and flows across the caprock edge, where its tributaries have cut magnificent canyons. The river then flows across the plains of north-central Texas, before it crosses limestone terrain of the Cross Timbers on its way southeast. The river's last run takes it through the coastal plain before it enters the Gulf of Mexico at Freeport, Texas. Stephen F. Austin thought the Brazos

THURBER

Today, Thurber is a sleepy little town along I-20. However, during a fifty-year period, Thurber was the center of a major coal mining industry. During the Pennsylvanian Period, Thurber was located where rivers flowing from the nearby Ouachita Mountains created a delta at the edge of an ocean. Plants growing on that delta and in nearby coastal marshes were later buried under sediments from the rivers and turned into coal deposits. The coal at Thurber was unlike the typical lower-grade lignite coal found in other parts of Texas. Here, higher-grade bituminous coal formed in a bed 2 to 3 feet thick in the Mingus Shale of the Strawn Group.

Extraction of this coal began in 1886, and two years later, the Texas & Pacific Coal Company was formed to manage the mines. The company built Thurber and furnished it with homes, stores, saloons, an opera house, and an electric plant. The coal that was extracted was sold to railways, but smaller pieces called "pea coal" could not be sold. In 1897, a brick plant was built that used the pea coal to fire its kilns. Bricks from Thurber were used from the Galveston Seawall to paving streets in Austin.

Steam Shovel Mountain, east of Thurber, was an important source of clay required to make bricks. (32.5070, -98.4069)

By 1917, the company was looking to expand its operations into the lucrative petroleum industry when it discovered oil west of Thurber in the Ranger Field. Petroleum from Thurber was soon being used in oil-burning steam locomotives and for making asphalt and concrete for roads. By the later 1920s, coal mining ceased at Thurber with the brick company closing in 1933. Piles of waste rock from the mining days can be seen around Thurber, along with piles of brick. Steam Shovel Mountain, to the east of town, exposes the clay that was used to make bricks. However, the most noticeable feature of the area is the tall, red, brick smokestack from the electric plant.

would be the chief waterway of commerce for Texas, and indeed, steamboats and flatboats plied up and down the brown Brazos until railroads took away the trade later in the nineteenth century. The Brazos River now provides power, water, and recreation via the many dams and lakes along its length.

The easternmost extent of the Paleozoic rock outcrops is just east of the Brazos River. East of there, Cretaceous rocks cover them. It is difficult to see this important change from the interstate because there is no real topographic edge and no prominent outcrops. A careful look at soil color might distinguish the transition: Cretaceous limestones and sandstones are light-colored and weather nearly white, whereas the Pennsylvanian limestones are gray and weather to buff-tan colors. Vegetation cover, limestone rubble in fields, rolling topography, and a few erosional hills with ledges of nearly white limestone constitute the Cretaceous geologic scene between the Brazos River and Ft. Worth.

MINERAL WELLS FOSSIL PARK

For the most part, fossil collecting is not allowed in state parks or other state-owned property in Texas. Mineral Wells Fossil Park, located west of Mineral Wells, provides an opportunity to not only observe Pennsylvanian-aged fossils in situ but also collect these fossils legally. Around 300 million years ago, this area was located along a shallow sea where the Mineral Wells Formation of the Strawn Group was being deposited. Rivers from the Ouachita Mountains flowed into the sea, bringing sand and building deltas. The shallow, nearshore waters were teeming with abundant life that are now preserved as fossils in the shale unit.

When walking around the park, it is impossible to not notice the abundance of fossils, with the most common being crinoids, or "sea lilies." Crinoids look like large plants with roots and stems (hence, the name sea lily), but they are animals related to modern-day starfish. When crinoids die, their stem segments are preserved and look like small poker chips or shirt buttons. Other fossils in the park include brachiopods, clams, bryozoans, snails, nautiloids, and petalodus shark teeth. For more information, refer to *Texas Rocks! A Guide to Geologic Sites in the Lone Star State.*

Mineral Wells Fossil Park, located west of the town of Mineral Wells, provides an opportunity to legally collect fossils on public land in Texas. (32.8262, -98.1913)

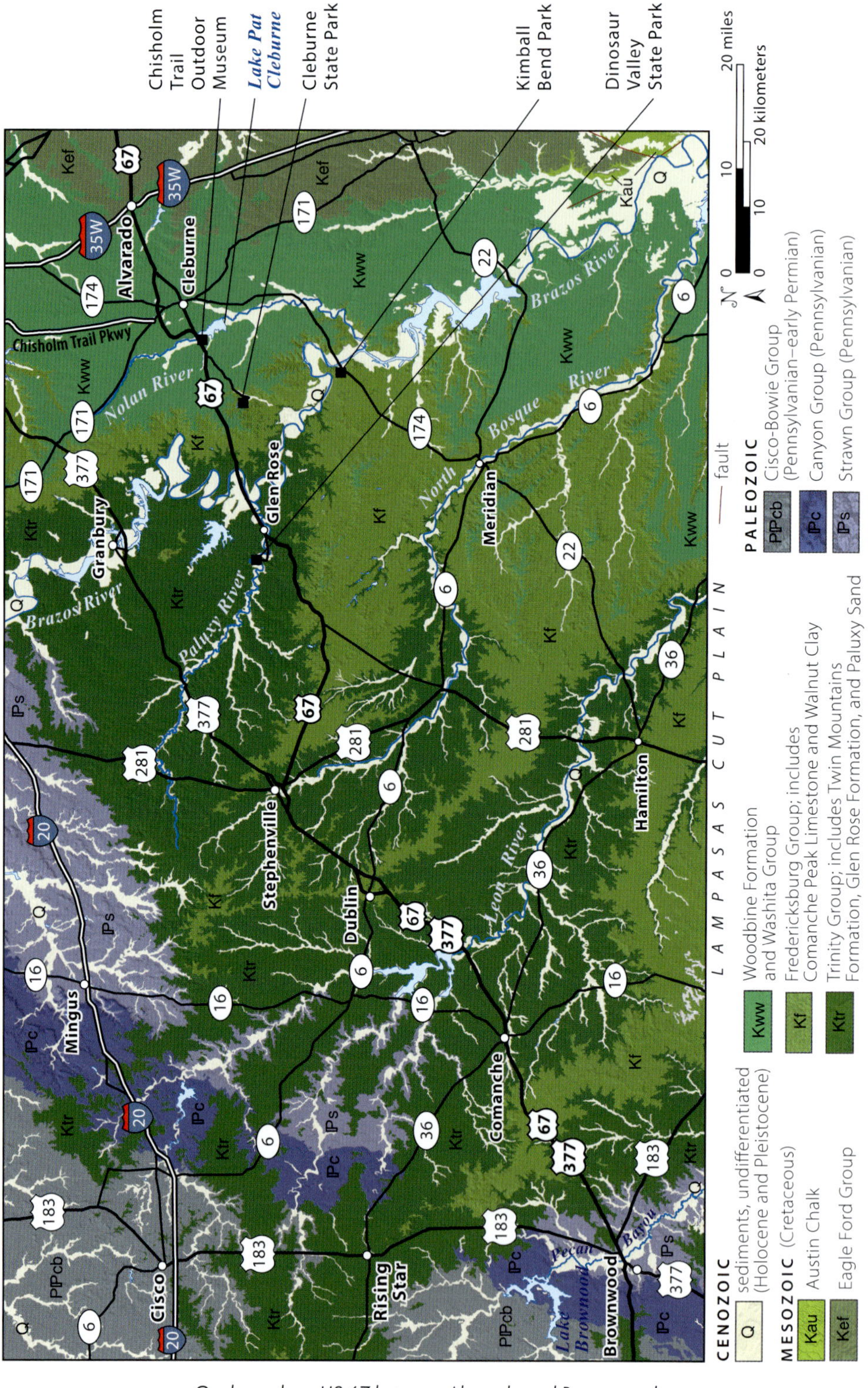

Geology along US 67 between Alvarado and Brownwood.

CENOZOIC

Q — sediments, undifferentiated (Holocene and Pleistocene)

MESOZOIC

Kau — Austin Chalk

Kef — Eagle Ford Group

Kww — Woodbine Formation and Washita Group

Kf — Fredericksburg Group; includes Comanche Peak Limestone and Walnut Clay

Ktr — Trinity Group; includes Twin Mountains Formation, Glen Rose Formation, and Paluxy Sand

PALEOZOIC

ₚIₚcb — Cisco-Bowie Group (Pennsylvanian–early Permian)

IPc — Canyon Group (Pennsylvanian)

IPs — Strawn Group (Pennsylvanian)

— fault

Chisholm Trail Outdoor Museum
Lake Pat Cleburne
Cleburne State Park
Kimball Bend Park
Dinosaur Valley State Park

LAMPASAS CUT PLAIN

Brazos River
Bosque River
North Bosque River
Nolan River
Paluxy River
Leon River
Pecan Bayou
Lake Brownwood

Alvarado
Cleburne
Glen Rose
Granbury
Meridian
Stephenville
Dublin
Hamilton
Comanche
Mingus
Cisco
Rising Star
Brownwood

Chisholm Trail Pkwy

US 67
ALVARADO — BROWNWOOD
125 miles

US 67 crosses through the northern section of the Texas Hill Country, passing many of the rocks that also crop out in other parts of the Hill Country. West of Alvarado, the roadway traverses the barren, mostly treeless landscape of the Grand Prairie, a physiographic region delineated by nearly flat plains, rolling hills, and limestone uplands. The underlying limestone rock layers are interlayered with clays that create shallow, slightly acidic soils. While much of the Grand Prairie has been transformed into cropland, the area still supports one of the largest native grasslands in Texas.

Just west of Cleburne is the Chisholm Trail Outdoor Museum, which greets you with an 80-foot-long wall and the message "Welcome to Cleburne on The Chisholm Trail." The museum preserves the trail stop at the former town of Wardville on the Nolan River and Lake Pat Cleburne. The Chisholm Trail was used after the Civil War to drive cattle overland from the large ranches in southern Texas (like the famous

CLEBURNE STATE PARK

About 6 miles southwest from US 67 on Park Road 21 is Cleburne State Park, donated to the state in 1934. Shortly thereafter, the Civilian Conservation Corps (CCC) built a small earthen dam to hold back waters from the 116-acre Cedar Lake. Work continued through 1940 with the construction of a concessions building, a boathouse on the lake, and a bathhouse for day visitors and campers. The CCC workers' craftsmanship is on full display at the still-standing Camp Creek Bridge, a testament to the masonry and carpentry skills these men possessed. The real jewel of the workers' craftsman-ship, however, is a three-tier spillway below the earthen dam. The spillway was hand-dug into the Cretaceous-aged Comanche Peak Limestone, a nodular carbonate unit of the Fredericksburg Group that was deposited in a shallow sea and contains abundant bivalve and coral fossils. To control erosion, workers devised an ingenious idea to lay stone blocks along the shoulders of the spillway set in a stair-step pattern. A short, quarter-mile scenic trail from the parking area takes visitors to the spillway, as well as to the best exposures of Comanche Peak Limestone.

The unique three-level spillway at Cleburne State Park was dug into Comanche Peak Limestone by hand by Civilian Conservation Corps workers in the 1930s. (32.2593, -97.5530)

King Ranch), across the Red River and Indian Territory, eventually ending up in Kansas to meet with the rail line stops. The trail followed an established pathway created by the Native American scout Black Beaver in 1861 and a later wagon road established by Jesse Chisholm in 1864.

South of Cleburne, TX 174 crosses the Brazos River at the former townsite of Kimball, established around 1854 where the Chisholm Trail crossed the Brazos. The town operated an all-important ferry to cross the often deep and swift-flowing river. During times of exceptionally high water, the ferry was used to move cattle across as well. Between the end of the large cattle drives and the railroad bypassing the town in 1881, Kimball rapidly declined. Kimball Bend Park preserves a portion of the townsite today.

In the 17 miles between the turnoff to Cleburne State Park and the town of Glen Rose, for which the Glen Rose Formation is named, US 67 travels over rolling hills composed of Cretaceous-aged clays and carbonates, but outcrops of bedrock are few and far between. The only indication of what lies underneath are some small, rubbly patches along the roadway and the occasional gravel road. Just over 4 miles west of the Park Road 21 intersection, an unusual sight appears: dinosaurs roaming a field! Local businessmen, along with a metal-art designer, created several dinosaurs and placed them on a ranch about 15 miles northeast of Glen Rose. On a hillcrest about 1 mile west of County Road 1119 are outcrops of Fredericksburg Group carbonates, another of the Cretaceous-aged rocks in this area. West of the hill, the highway drops into the broad valley of the mighty Brazos River, which cuts into the Paluxy Sand of the Trinity Group, also Cretaceous in age. Limestone of the Glen Rose Formation, also of the Trinity Group, crops out in the riverbed and valley walls.

Fredericksburg Group carbonates, exposed about 1 mile west of County Road 1119, are favorite growing substrates for junipers and the occasional yucca. (32.2829, -97.6079)

DINOSAUR VALLEY STATE PARK

At Glen Rose, take TX 205 a few miles north along the Paluxy River to reach Dinosaur Valley State Park. Around 113 million years ago, during the Cretaceous Period, this area was a tidal flat at the edge of a shallow sea. Large, carnivorous dinosaurs, *Acrocanthosaurus*, prowled the area looking for food. The wet flats where the animals strode preserved not one but thousands of tracks, scattered at localities across Texas, but the best trackways are preserved at Dinosaur Valley State Park.

Stop at the visitor center to see the excellent displays and get oriented. The displays illustrate the types of dinosaurs that roamed Texas in Cretaceous time, their biology and family trees, and how tracks were preserved in the environments of the time. A particularly well-done display shows track casts with the foot bones of the animals that made the tracks.

At the end of the north road is an overlook where you can peer down on the Paluxy River bed to see several distinct trackways, although sometimes riverbed silt covers the trackways. Both washtub-like tracks of sauropods, most likely *Sauroposeidon*, and three-toed tracks of theropods (mostly the two-legged carnivore *Acrocanthosaurus*) are preserved here. The trackways march across the riverbed away from the overlook.

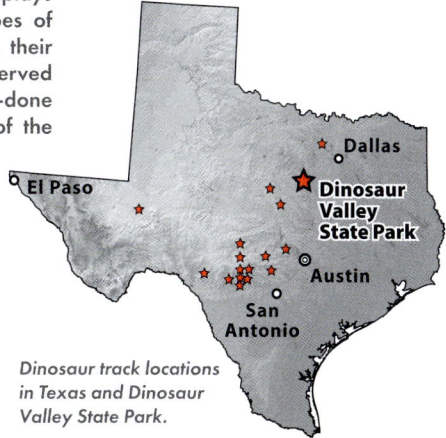

Dinosaur track locations in Texas and Dinosaur Valley State Park.

Many individual tracks are found on the limestone layers at river level along the west loop of the Paluxy River. Wander around for a while to see if you can spot both types of tracks preserved here. Directly across the river from the stairs and northwest parking lot is a set of sharply defined three-toed tracks, recently exposed by the bank-carving action of the river. Note the gray mudstone layer in the bank that once covered the sandy track bed before erosion stripped the mud to expose the track. This little rock sequence graphically shows how the tracks were preserved: the mud was laid down immediately and gently over the track to form a tough, protective layer that lasted for millions of years, until the Paluxy River removed it only a few years ago.

Alternating sandstone, limestone, and shale layers of the Glen Rose Formation are exposed in the cliff face along the tight river bend. These alterations are typical for the Cretaceous sedimentary rocks in this area and tell us a lot about the environment at the time the dinosaurs walked around here. The Cretaceous sea lay to the south and southeast, roughly where the Gulf of Mexico is now. The Glen Rose area was part of the flat Cretaceous seacoast, which was not unlike today's Texas coastal plain. During the Cretaceous, however, the sea was shallow for a long distance between the seacoast and deeper ocean water. Shorelines moved

The three-toed tracks of the two-legged dinosaur Acrocanthosaurus. (32.2480, -97.8190)

A large track, complete with imprints of stubby toes, is from the four-legged dinosaur Sauroposeidon, which could grow to be 110 feet long and weigh as much as 44 tons. (32.2480, -97.8190)

back and forth across this shallow pan as sea level bobbed up and down, driven by the amount of seawater that was being displaced at any one time by gigantic lava eruptions in all the world's mid-ocean ridges. Recall that seafloor spreading, the process in which magma rises to the surface and creates new ocean floor, was in high gear during the Cretaceous.

During periods of high lava extrusion into the oceans, sea level rose and inundated the continental margins. Shorelines retreated for many miles across the flat pans, marine organisms moved in, and their shells accumulated as limestone. The limestone was deposited directly over the old shoreline sand beds and river silts and muds that once occupied the flat pan. During quiescent episodes of seafloor spreading, the ocean surface settled back to lower levels, and seawater evacuated the flat pans of the continental margins. Shoreline sand and river silt and clay were now spread over the limestone.

Dramas of daily life are told by the trackways. One famous tale is unfurled at the visitor center, wherein a chase involving a predator and its plant-eating prey is written in a long trackway uncovered in the 1920s. It was removed for display at the American Museum of Natural History (where it is today) and the Texas Science and Natural History Museum in the 1940s.

As you enter the town of Glen Rose, it is obvious that the locals love their dinosaurs and with good reason. The Glen Rose Formation is famous for dinosaur trackways, and many tracks, both real and imitation, are scattered around town and Texas in general. While the statues, paintings, and dino art are great entertainment to look at, the real treasures lie within Dinosaur Valley State Park west of town.

West of Glen Rose, US 67 follows the northern edge of the Texas Hill Country. Several roadcuts are visible along the roadway, first the Glen Rose Formation, then alternating roadcuts of Fredericksburg Group carbonates and the Walnut Clay, also of the Fredericksburg. All of these rocks are Cretaceous aged and represent changing sea level conditions.

Clay, limestone, and shales of the Cretaceous-aged Walnut Clay outcrop in fantastic fashion at the intersection with County Road 200. (32.1514, -97.9498)

About 17.5 miles west of Glen Rose, the topography changes dramatically from the typically hilly landscape of the Texas Hill Country to one of flat, wide-open spaces with few small hills and no roadcuts. Based upon drilling data, this area from Stephenville to Dublin and Comanche is underlain by more Glen Rose Formation and Walnut Clay. Even though roadcuts are nonexistent, one can still get an idea of local geology because rock from nearby quarries was used to construct buildings that, in many cases, still stand today.

The former county jail in Comanche was built in 1888 of locally sourced sandstone of the Twin Mountains Formation. (31.8973, -98.6063)

Between Comanche and Brownwood, US 67 continues across a relatively flat landscape, with the occasional small hill or broad valley. About the only interesting views of the underlying bedrock are four prominent roadcuts beginning about 2.5 miles west of Comanche. The first is of limestone of the Glen Rose Formation. About 1 mile farther, a more substantial roadcut of Paluxy Sand with interbedded limestone occurs on both sides of the road. One-half mile farther to the west, another substantial roadcut, this time of Walnut Clay, lines both sides of the road, with another roadcut of the Walnut about one-half mile farther. One possible explanation for having three different sedimentary rock units in quick succession is that in the Cretaceous, a large river system was emptying into the sea in this part of Texas, so many sedimentary environments existed in close proximity.

Drill core data shows that the underlying bedrock is mostly the same as we've seen throughout this road guide: Glen Rose Formation, Paluxy Sand, and Walnut Clay. Closer to Brownwood, additional sandstones of the Travis Peak Formation, another Cretaceous unit in the Trinity Group, and claystones of the Middle Pennsylvanian–aged Strawn Group, are known from drill cores, but these rock units do not outcrop along the highway.

The yellowish Paluxy Sand is interbedded with limestone of the Glen Rose Formation in this roadcut about 3.5 miles west of Comanche. (31.8802, -98.6652)

US 84
ABILENE — COLEMAN — BROWNWOOD
82 miles

Between Abilene and Coleman, US 84 crosses three bands of Permian rocks and passes by high hills of younger Cretaceous rocks. South of Abilene, the prominent hills on either side of the road are part of the Callahan Divide. Rivers on the south side of the drainage divide spill into the Colorado River, while water from the north side of the Callahan Divide flows northward to join the Brazos River. The divide is a remnant, erosional plateau of Cretaceous rocks that once covered much more terrain in north-central Texas, until erosion removed the rocks from most of the area.

South of Abilene and about 5 miles west of US 84 is Buffalo Gap, a small historic town, located where buffalo once ran through a narrow break in the Callahan Divide. The pioneer town is now operated as a historic village where visitors are treated to a bit of the Old West.

Fifteen miles south of Abilene, the road runs through a wider gap in the Callahan Divide, where high hills frame both sides of the highway. Yellow-tan Edwards Limestone of Early Cretaceous age forms the resistant cap on top of the ridges. Pink and gray shales and sandstones of the Antlers Formation on the lower slopes, also Early Cretaceous in age, rest directly on red Permian rocks at road level. The red soils are indicative of the Permian bedrock here. Another set of roadcuts south of Lawn exposes the Lueders Formation, a Permian-aged series of tan-gray to red, interbedded shale and limestones in the Wichita Albany Group.

US 84 passes through several good roadcuts in Permian rock nearer Coleman. Near the western intersection of FM 702 (about 18 miles north of Coleman), fossils of brachiopods, clams, crinoids, and fossil hash are found in thin-bedded, gray

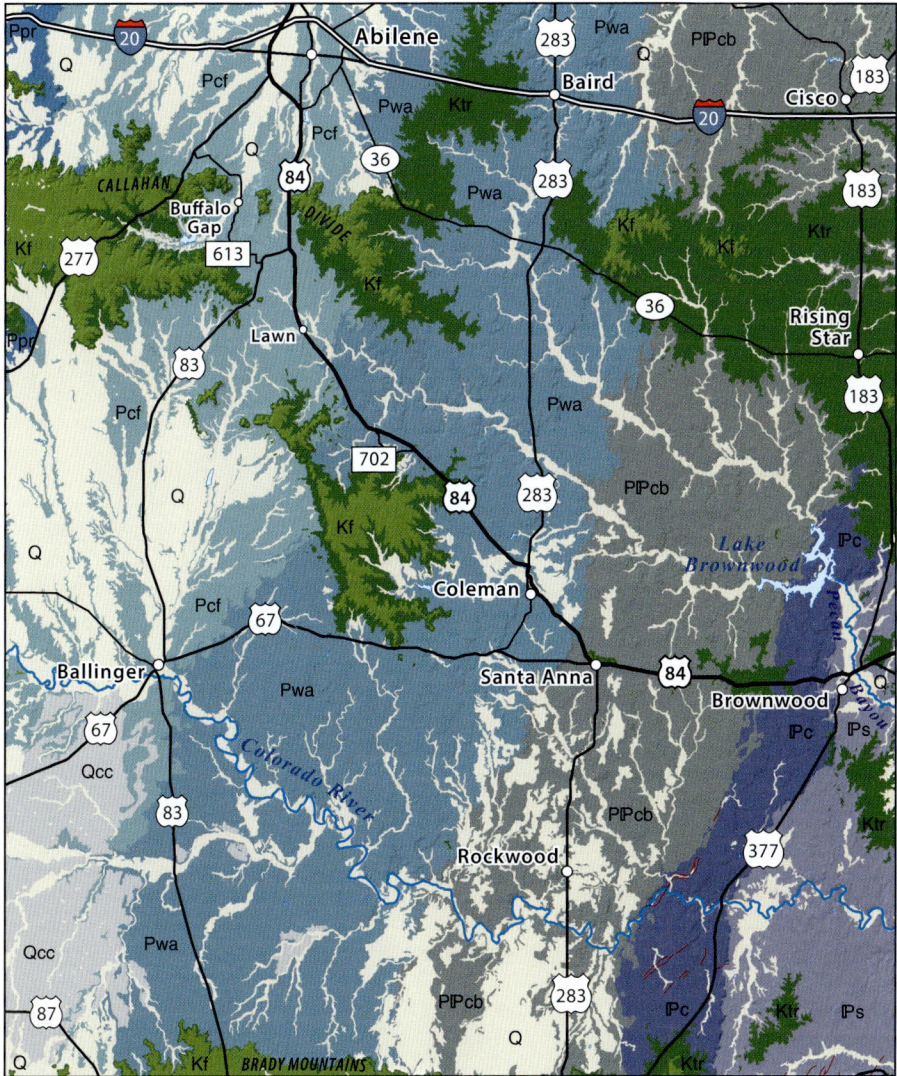

Ppr
Q
I-20
Abilene
Pcf
Pwa
Q
283
Pwa
PⅢPcb
Q
183
Baird
Cisco
I-20
Pwa
Ktr
Pcf
Q
84
36
283
183
CALLAHAN
Buffalo
Gap
DIVIDE
Pwa
36
Kf
277
613
Kf
Ktr
Kf
Ppr
Kf
Lawn
Rising
Star
183
83
Pcf
Pwa
702
PⅢPcb
Q
84
283
Ⅲc
Lake
Brownwood
Q
Pecan
Q
Kf
Coleman
Pcf
67
Santa Anna
84
Brownwood
Q
Ballinger
PⅢPcb
Ⅲc Ⅲs
67
Pwa
Ktr
Qcc
Colorado River
Rockwood
377
83
PⅢPcb
Qcc
Pwa
283
Ⅲc Ⅲs
87
PⅢPcb
Ⅲc
Ktr
Q
Q
Kf
BRADY MOUNTAINS
Ktr

L A M P A S A S   C U T   P L A I N

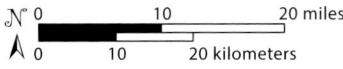

to
Brady

CENOZOIC

Q surface sediments, undifferentiated; includes alluvium, playa, older terrace and high gravel deposits (Holocene and Pleistocene)

Qcc caliche

MESOZOIC
CRETACEOUS

Kf Fredericksburg Group; includes Edwards Limestone

Ktr Trinity Group; includes Antlers Formation

N° 0   10   20 miles
0   10   20 kilometers

—— fault

PALEOZOIC
PERMIAN

Ppr Pease River Group

Pcf Clear Fork Group

Pwa Wichita Albany Group; includes Lueders, Grape Creek, and Talpa Formations

PENNSYLVANIAN–EARLY PERMIAN

PⅢPcb Cisco-Bowie Group

PENNSYLVANIAN

Ⅲc Canyon Group

Ⅲs Strawn Group

Geology along US 84 between Abilene and Brownwood.

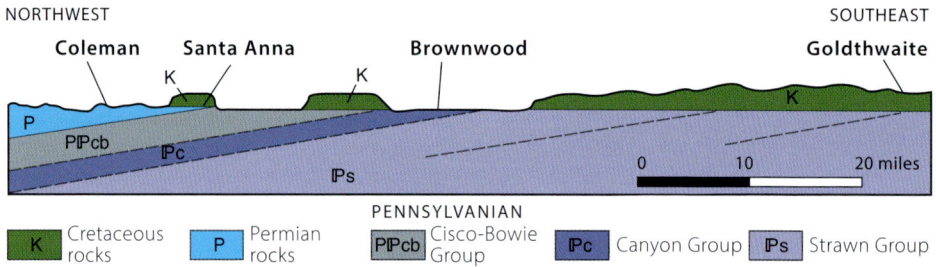

NORTHWEST   SOUTHEAST

Coleman   Santa Anna   Brownwood   Goldthwaite

K   K   K

P

PₚPcb   Pc

Pₚs   0   10   20 miles

PENNSYLVANIAN

| K | Cretaceous rocks | P | Permian rocks | PₚPcb | Cisco-Bowie Group | Pc | Canyon Group | Pₚs | Strawn Group |

Cross section along US 84 between Coleman and Goldthwaite. The tilt of Permian and Pennsylvanian rocks is greatly exaggerated.

Tan-gray to red shales and limestones of the Permian-aged Lueders Formation are eroding in a prominent roadcut 4 miles south of Lawn at the intersection of County Road 166. (32.0880, -99.6983)

Limestones of the Grape Creek and Talpa Formations are interbedded with shale and some sandstone at the west intersection of FM 702 near Novice. (32.0144, -99.6373)

limestones of the Grape Creek and Talpa Formations, both of the Wichita Albany Group. The limestones are interbedded with soft shales and some sandstones. The fossils are indicative of the shallow, clear seawater in which these limestones were deposited. Although the Permian section tilts gently to the west, the half-degree tilt is not discernible at the scale seen in a single roadcut. Thus, the layers in these cuts look quite flat lying.

Between Coleman and Brownwood, US 84 travels on flat country of Permian and Pennsylvanian rocks. Mesas standing above the general level of the countryside are formed of limestone layers of younger Cretaceous age that have been etched and eroded for more than 10 million years by tributaries of the Colorado River. As the Edwards Plateau, south of this area, was elevated along the Balcones fault zone about 10 million years ago, the ancient Colorado River began downcutting through the Cretaceous limestones. The carving continued unabated until the Cretaceous layers were breached and eroded back, exposing the older, tilted Permian and Pennsylvanian rocks. Isolated mesas of Cretaceous limestone were left standing as lone sentinels above the bottomlands. The high, rocky ridge near Santa Anna was part of a continuous Cretaceous layer connected to the Brady Mountains to the south near Brady. All the Cretaceous rock between Santa Anna and the Brady Mountains has been removed by erosion and carried away by the Colorado River system over the last few million years.

The flat area around Brownwood is in the eroded bottomland of the Pecan Bayou, a tributary of the Colorado River. Look carefully in stream cuts, roadcuts, and outcrops right around Brownwood to see tan, yellow, and reddish sandstone and limestone beds of Pennsylvanian age. The hills to the east of Brownwood are another example of the Cretaceous mesas.

US 90
SAN ANTONIO—DEL RIO—SANDERSON
269 miles

West of San Antonio, US 90 crosses the Medina Valley, where rich soils yield abundant crops. The highway runs a few miles south of, and parallels, the main Balcones fault, which separates this lower area from the uplifted Texas Hill Country to the north.

US 90 climbs over a north-south ridge of Uvalde Gravel just west of Castroville. The ridge's mostly grassed-over roadside exposures of gravels and sandstones were deposited by streams in the Pleistocene Epoch. Similar roadcuts are scattered along US 90 for about 10 miles. The heavy rainfall of the cool ice age climate engorged streams, increasing their power to erode and transport sediments. Look for white caliche, a calcareous deposit that formed by precipitation of lime within some of these cherty gravels. The chert, composed of resistant silica, is derived from chert nodules weathered from the Edwards Limestone and transported here from the higher Edwards Plateau country to the north.

Hill Country State Natural Area, located about 26 miles north of Hondo on FM 462 near Bandera, offers an opportunity to see what the Hill Country may have looked like before human settlement. Opened in 1984 after the state acquired nearly 5,400 acres from the Merrick Bar-O Ranch in 1976, the park is a beautiful array of craggy hills, flowing springs, and a small oasis brimming with birds, plants, and reptiles.

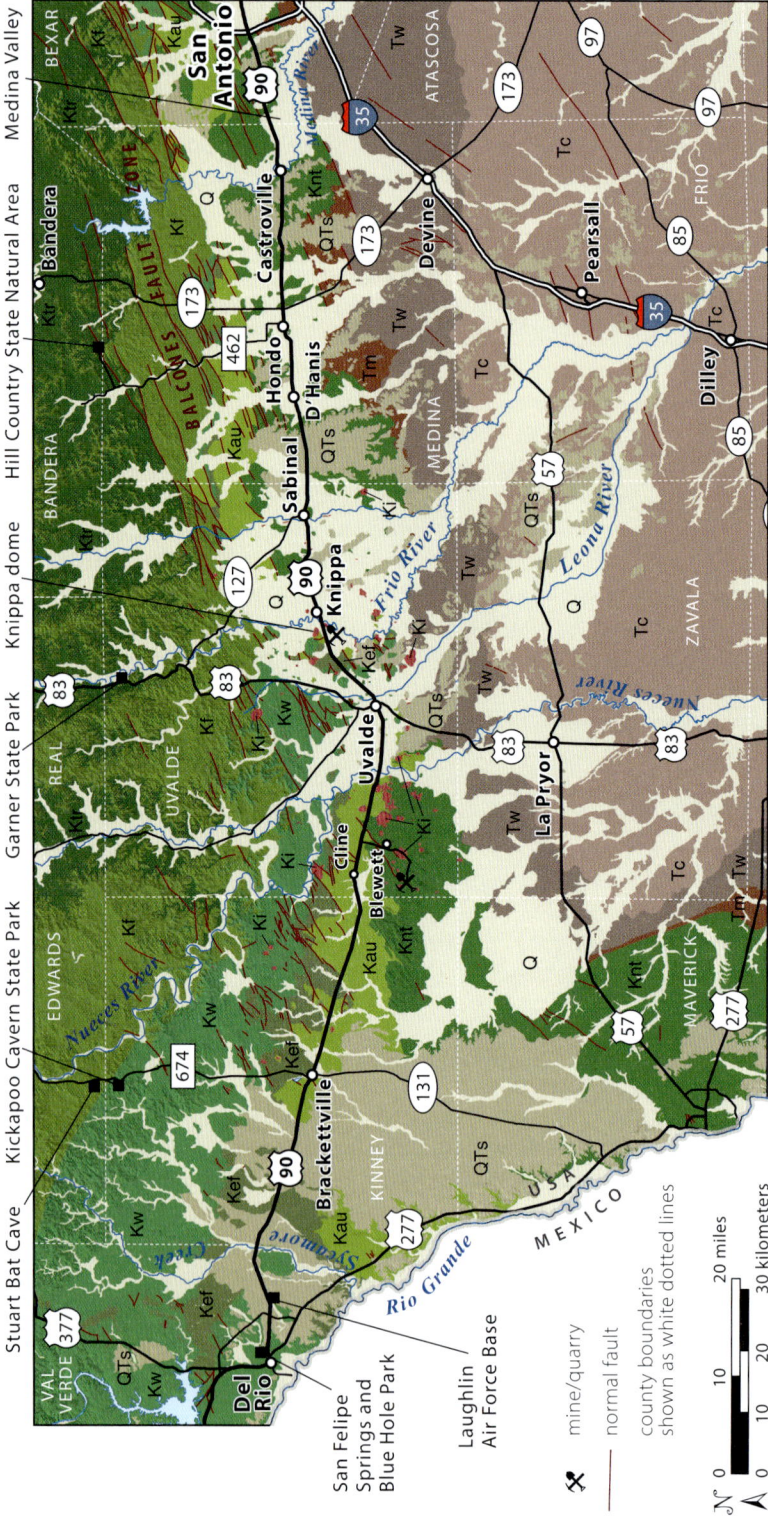

Geology along US 90 between San Antonio and Del Rio.

CENOZOIC

Q — recent sediments; includes alluvium and older terraces (Holocene and Pleistocene)

QTs — Uvalde Gravel (Pliocene or Pleistocene)

NEOGENE and PALEOGENE

Tc — Claiborne Group (Eocene)

Tw — Wilcox Group; includes Carrizo Sand (Eocene)

Tm — Midway Group (Paleocene–Eocene)

MESOZOIC (CRETACEOUS)

Knt — Navarro and Taylor Groups; includes Anacacho Limestone of the Taylor Group

Kau — Austin Chalk

Kef — Eagle Ford Group

Kw — Washita Group; includes Salmon Peak Limestone

Kf — Fredericksburg Group; includes Devils River and Edwards Limestones

Ktr — Trinity Group; includes Glen Rose Formation

VOLCANIC ROCKS

Ki — basalt and pyroclastic rocks in sills, laccoliths, volcanic necks, and dikes (Cretaceous)

✕ — mine/quarry

— normal fault

county boundaries shown as white dotted lines

0 10 20 miles
0 10 20 30 kilometers

At D'Hanis, fossiliferous clay pits, a brick factory, and red brick houses and buildings attest to the importance of clay in our lives. West of D'Hanis, the road climbs out of the valley bottom to a gravel-capped terrace ridge.

A picnic area on the north side of the highway 6.5 miles west of Sabinal lies opposite an outcrop of the Late Cretaceous Anacacho Limestone on the south side of the road. Pleistocene stream gravels, rounded by their transport by running water, overlies the limestone. Elsewhere, cross-bedding and some fossils have been found within this limestone unit, a formation in the Taylor Group.

Pleistocene stream gravels, rounded during transport, overlie a roadcut of Late Cretaceous Anacacho Limestone about 6.5 miles west of Sabinal. (29.3177, -99.5660)

At Knippa, two prominent volcanic domes stand above the otherwise flat terrain north of the highway. A good view can be had from the bridge over the Frio River west of town. Another of these volcanic domes is mined for aggregate west of town in the Knippa quarry. US 90 passes near the edge of the quarry, so you get a good opportunity to see these intriguing rocks from your car. The Knippa domes are part of a swarm of igneous bodies of Late Cretaceous age (average age about 80 million years) that occur in a belt that trends eastward from Del Rio nearly to Waco. The belt roughly parallels the edge of the Cretaceous Edwards Plateau and the Balcones fault zone because the magma rose along faults. The magma fed submarine volcanic centers where eruptions created explosion craters on the Cretaceous seafloor. Rocks in the Knippa quarry are dark-colored nephelinite, a sodium-rich rock similar to basalt. At Knippa, fragments of mantle rock are commonly found, so the magma rose from deep within the Earth. There must have been quite a Cretaceous show when these volcanoes erupted in the sea, spewing steam, hot rocks, and red lava skyward.

Between Knippa and Uvalde, the road continues to traverse relatively flat terrain where occasional Late Cretaceous limestone and claystone exposures peek out from beneath the extensive cover of river deposits. At Uvalde, a prong of Late Cretaceous rocks extends southward toward town from the Edwards Plateau to the north,

Large blocks of nephelinite, a rock similar to basalt, sit in front of the Knippa quarry on the west side of Knippa. (29.2914, -99.6513)

allowing groundwater contained within the Edwards Limestone to emerge as springs near Uvalde. The freshwater is used to irrigate the adjacent agricultural fields.

About 30 miles north of Uvalde on US 83 is Garner State Park, which sits along the crystal-clear waters of the Frio River at the southern edge of Hill Country. Located on the southwest edge of the Edwards Plateau, the park features Cretaceous-aged rocks that are familiar to most people studying geology in the Hill Country: the Glen Rose Formation that formed along the shoreline of an ancient sea and, above the Glen Rose, the Edwards Limestone, another carbonate that was deposited when the ancient sea level rose and covered all of Texas. Visitors to the park are greeted by the Balcones Canyonlands, a region of canyons, high mesas, and naturally sculpted cliffs.

Between Uvalde and Del Rio, US 90 traverses mostly flat-lying, Late Cretaceous limestone uninterrupted by canyons or stream cuts. White limestone rubble in fields adjacent to the highway is all you see in many places. Look for noticeable intrusive volcanic hills about halfway between Uvalde and Brackettville.

At the crossing of the Nueces River about 6 miles west of Uvalde, the meandering river bends expose excellent views of sandy point bars. River sediment is extracted by a sand and gravel operation on the north side of the road. Note another set of quarries where volcanic rocks are mined about 12 miles west of Uvalde on the hills south of the road.

About 15 miles west of the Nueces River, in an area south of Cline and Blewett, are old asphalt mines from the turn of the century. Though unseen from US 90, these were the largest natural asphalt mines in Texas and perhaps in the United States. The asphalt is from natural oil seeps that impregnated a coarse-grained, shelly limestone. The asphaltic limestone was mined, crushed, and laid down—as is—on roads. FM 1022 to Blewett was, at one time, composed of this type of pavement. Just west of FM 1022, piles of dark rock can be seen on a hill to the south; these were also part of the asphalt mine operations.

A noticeable change in vegetation occurs about 20 miles west of Uvalde. Temperate trees, shrubs, and grasses give way to desert topography and plants. Desert terrain is more and more prevalent west from here.

About 22 miles north of Brackettville on Ranch Road 674 is Kickapoo Cavern State Park. The land for the park was acquired in 1986 and was fully opened in 2010. Kickapoo Cavern formed much like every cavern in this area when slow-moving, acidic groundwater dissolved Cretaceous-aged Devils River Limestone. The park provides guided tours of the cavern, but reservations are required. Nearby Stuart Bat Cave is also worth a visit to observe cave swallows and up to a million Mexican free-tailed bats that call the cave home during the summer months. In the evenings, the bats leave in a furious flurry to hunt for insects in the nearby area.

Just east of Del Rio, note the white caliche beds in the roadside exposures near the entrance to Laughlin Air Force Base. Caliche forms in dry, arid conditions and is composed mostly of calcium and magnesium carbonates that bind together sediments, forming an almost cement-like layer. In Del Rio, San Felipe Springs bubble up in the San Felipe Springs Golf Course from the Salmon Peak Limestone, an Early Cretaceous rock that hosts some of the vast Edwards Aquifer. The springs feed San Felipe Creek, which empties into the Rio Grande. US 90 crosses the creek where it passes through Blue Hole Park in Del Rio.

The stretch of West Texas desert country between Del Rio and Sanderson forms the western margin of the Edwards Plateau. The geologic story, while expectedly one of Cretaceous limestone terrain, is decidedly interesting, especially topographically. Thick sequences of limy rocks are deeply dissected by the Devils and Pecos Rivers, which carve their way southward to their inevitable junctures with the Rio Grande. Magnificent cliffs stand tall above these rivers, and scenic highway crossings are found at Amistad Reservoir (Devils River) and at the spectacular high bridge over the Pecos River.

The US 90 roadbed around Del Rio rides on the surface of the Salmon Peak Limestone, a thick, gray Early Cretaceous unit that is beautifully exposed to the west in the Pecos River canyon as the equivalent Devils River Limestone. This widespread

Beds of caliche, the hard, white soil layer composed mostly of calcium and magnesium carbonates in the upper quarter of the roadcut, outcrop near Laughlin Air Force Base east of Del Rio. (29.3697, -100.8062)

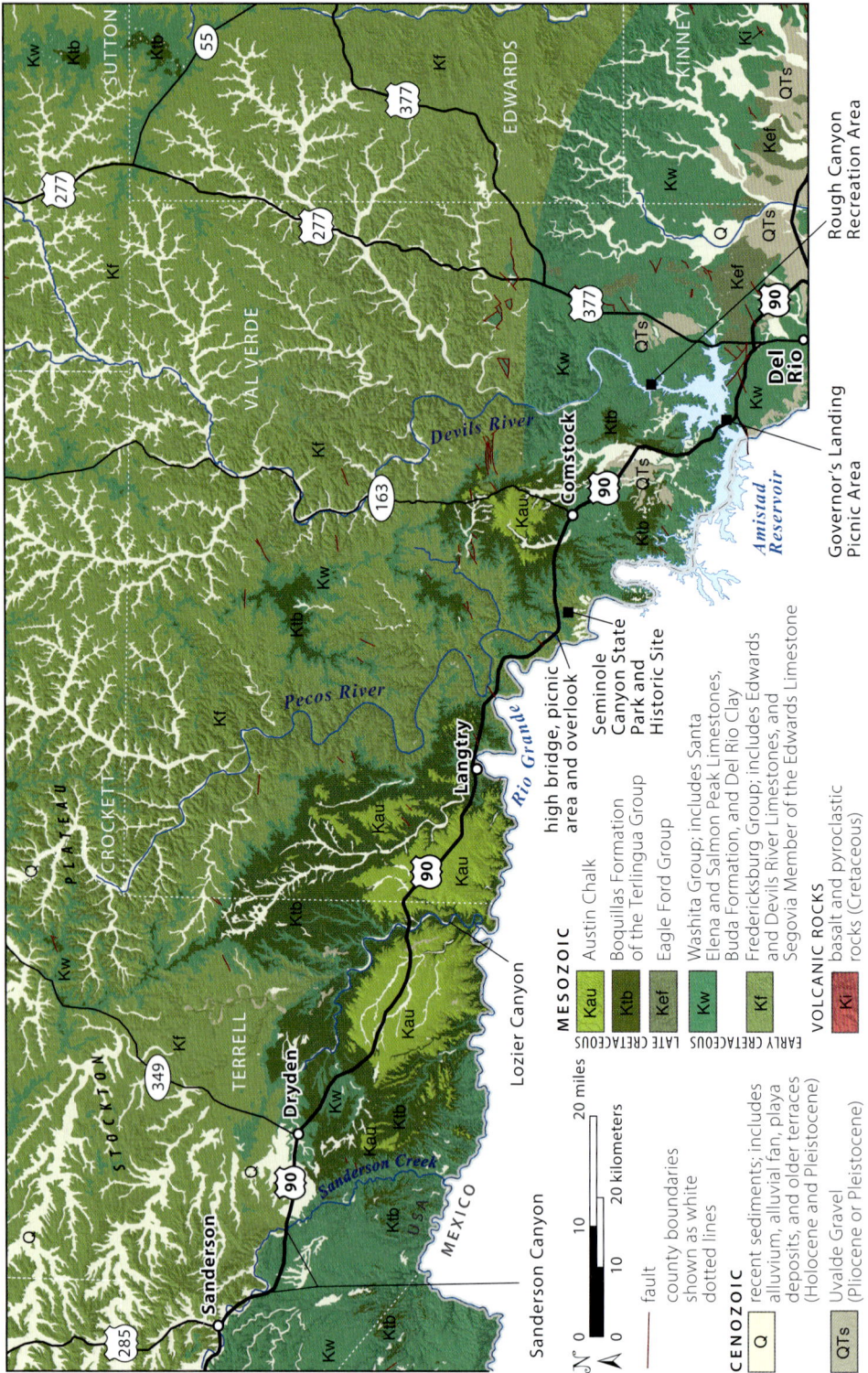

Geology along US 90 between Del Rio and Sanderson.

Map labels:

Counties and regions: SUTTON, EDWARDS, KINNEY, VAL VERDE, CROCKETT, TERRELL, STOCKTON PLATEAU, MEXICO

Towns: Del Rio, Comstock, Langtry, Dryden, Sanderson

Water features: Devils River, Pecos River, Rio Grande, Amistad Reservoir, Sanderson Creek

Features: Rough Canyon Recreation Area, Governor's Landing Picnic Area, Seminole Canyon State Park and Historic Site, high bridge, picnic area and overlook, Lozier Canyon, Sanderson Canyon

Highways: 55, 277, 377, 163, 90, 349, 285

Scale: N, 0 10 20 miles, 0 10 20 kilometers

Legend:

CENOZOIC
- Q — recent sediments; includes alluvium, alluvial fan, playa deposits, and older terraces (Holocene and Pleistocene)
- QTs — Uvalde Gravel (Pliocene or Pleistocene)

MESOZOIC

LATE CRETACEOUS
- Kau — Austin Chalk
- Ktb — Boquillas Formation of the Terlingua Group
- Kef — Eagle Ford Group

EARLY CRETACEOUS
- Kw — Washita Group; includes Santa Elena and Salmon Peak Limestones, Buda Formation, and Del Rio Clay
- Kf — Fredericksburg Group; includes Edwards and Devils River Limestones, and Segovia Member of the Edwards Limestone

VOLCANIC ROCKS
- Ki — basalt and pyroclastic rocks (Cretaceous)

— fault

county boundaries shown as white dotted lines

limestone is also the same one that forms the upper part of the giant cliffs in Santa Elena Canyon on the west side of Big Bend National Park, named the Santa Elena Limestone there. The Santa Elena is equivalent in age to the Salmon Peak and Devils River Limestones but was deposited farther west in a different part of the sea.

Two miles north of Del Rio, US 377 splits off to the right, heading north to the Rough Canyon Recreation Area of the Amistad Reservoir. Between this junction and

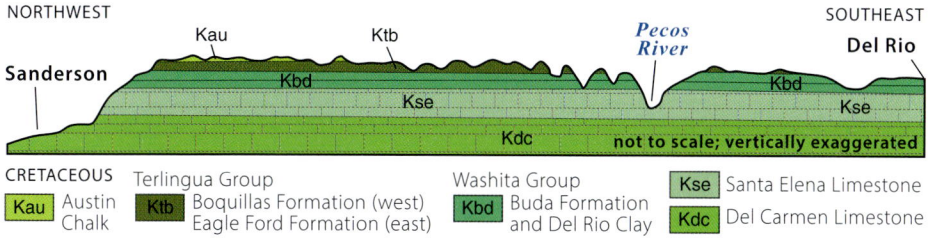

Cross section of the Cretaceous rocks along US 90 between Del Rio and Sanderson.

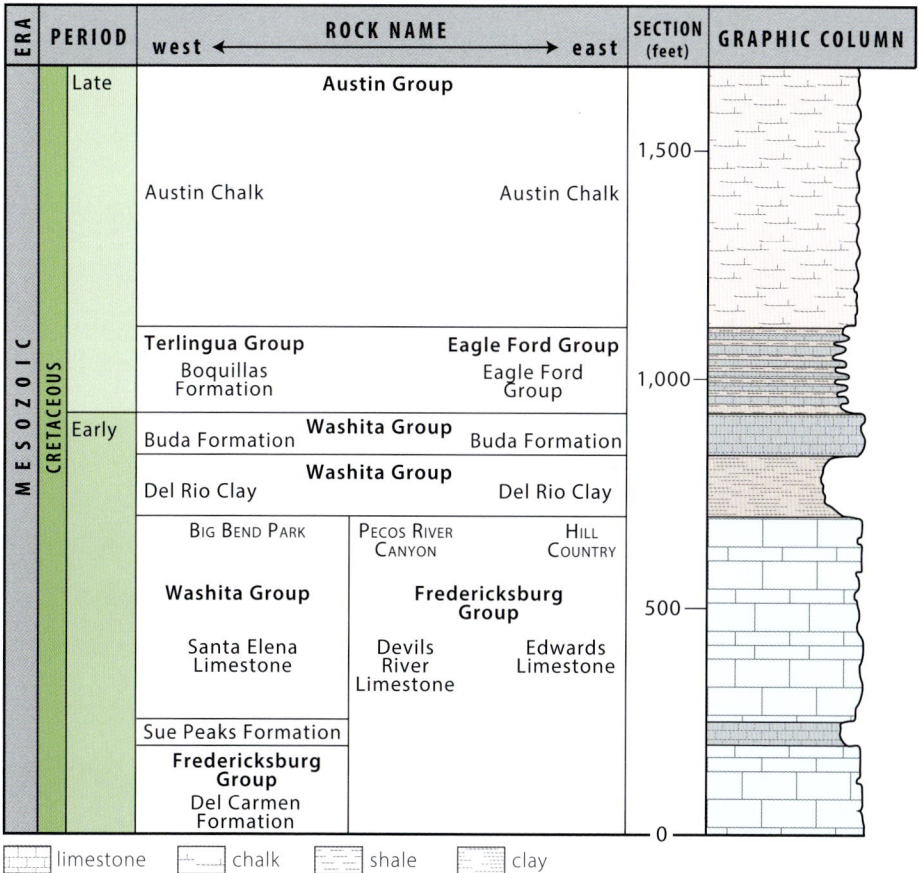

Stratigraphic sequence of rocks between Del Rio and Sanderson area along US 90.

Yellowish, thinly bedded limestone of the Eagle Ford Group overlies white Buda limestone in a roadcut on US 377 about 3.2 miles north of the US90/US377 junction. (29.4907, -100.9109)

the lake, US 377 crosses a ridge where several roadcuts slice through the Eagle Ford Group, Buda Formation, and Del Rio Clay of Late Cretaceous age. The Eagle Ford (called Boquillas farther west of here) is mainly thin-bedded (flaggy) limestone, alternating with shale and siltstone. In a quarry a half mile north of the US 90/US 377 junction, the Eagle Ford is mined for crushed stone and aggregate.

West of Del Rio, US 90 crosses the bridge over Amistad Reservoir. On the southeast side of the bridge, take Spur 349 west 2 miles to see the dam, which impounds the Devils River and the Rio Grande. From the top of the dam, look south to see the broad canyon of the Rio Grande downstream from the dam.

Governor's Landing Picnic Area, along US 90 south of Governor's Landing Bridge, provides a great view of the bridge and the Early Cretaceous Salmon Peak Limestone across the lake. North of the US 90 bridge, and overlying the Salmon Peak, are outcrops of weathered, thinly bedded, gray limestone of the Early Cretaceous Boquillas Formation in scenic cliffs and bluffs.

The Early Cretaceous Salmon Peak Limestone forms the shore of Amistad Reservoir, as viewed from the picnic area below Governor's Landing Bridge. (29.4805, -101.0292)

Beginning about a half mile west of Comstock, several good roadcuts show hard, light-gray, thin-bedded Buda limestone, full of clams and burrows, overlying brown, clayey beds of the Del Rio Clay, which contains abundant fossil hash. Look for ramshorn (*Exogyra*) oysters in the Del Rio Clay here. As the highway goes up and down these hills and through the roadcuts, you pass through this Cretaceous Buda Formation/Del Rio Clay stratigraphic section several times.

Close-up of fossils, including ramshorn (Exogyra) oysters, in the Del Rio Clay. Swiss Army Knife for scale. (29.6995, -101.2009)

About 1.9 miles west of Comstock are beautiful roadcuts of gray limestone of the Buda Formation overlying yellowish-tan Del Rio Clay. (29.6995, -101.2009)

About 2 miles west of the Buda/Del Rio roadcut, large roadcuts appear where the tan Boquillas limestone overlies the gray limestone of the Buda Formation along a sharp contact. (29.7097, -101.2360)

West of the turnoff to Seminole Canyon, watch for a blue sign that says "Picnic Area-Historical Marker" and marks the entrance to a gorgeous scenic overlook on the south side of US 90. From the overlook, the deep, cliff-walled incision of the Pecos River canyon spreads north and southwest, while the Pecos–Rio Grande confluence can be seen on the hazy horizon to the southwest. The Early Cretaceous Devils River Limestone composes most of the cliff walls down to river level. The Late Cretaceous beds on top of the Devils River Limestone form the uppermost cliff edge, the same units we've seen along the highway. One can only marvel here at the power of erosion

SEMINOLE CANYON STATE PARK

Seminole Canyon State Park is cut into the Devils River Limestone. (29.6996, -101.3130)

Nine miles west of Comstock is the turnoff to Seminole Canyon State Park and Historic Site, where a tributary to the Rio Grande has carved deeply into the Devils River Limestone, a Cretaceous unit in the Fredericksburg Group. On the canyon walls and sheltered alcoves, beautifully preserved pictographs provide a snapshot of life at the end of the Pleistocene. Hunters in search of bison, mammoths, and other big game arrived around 12,000 years ago. The earliest paintings were done in what is known as the Pecos River Style between 9,000 and 2,750 years ago. This style is known for its human-like figures, geometric patterns, and wildlife, including its most famous pictograph, the Panther. Around 1,280 years ago, the Red Linear Style with its deep-red stick figures appeared, although the timing is controversial. Some studies shows that Red Linear art has been painted over by Pecos Style art. Next, at around 1,050 years ago, a new style of art, called Red Monochrome, appeared and is known for images of animals and hunters using bows and arrows. The latest style, called Historic Images, was painted sometime after 1600 and depicts churches, cowboys lassoing cattle, and people wearing more modern European clothing.

In addition to the pictographs, Seminole Canyon's Bonfire Shelter is also the southernmost and oldest known buffalo jump in the United States. Beginning around 12,000 years ago and as recent as 800 BCE, people herded bison over the cliff to their deaths. Bonfire Shelter got its name because gases that built up from the decomposing bison spontaneously combusted, heating and calcining the bones. Also at the park are numerous other sites and artifacts including roasting ovens used approximately 1,500 years ago and stone ovens used by Italian railway workers as recently as the 1880s.

and the persistent downcutting by the Pecos River over the last few million years as it chipped away at these hard limestones to create this beautiful canyon. You can reach the canyon bottom by continuing down Park Service Road 3. Others before us have also enjoyed this canyon, as evidenced by abundant living sites, rock art, and lithic fragments of the ancient Pecos River Culture. Boquillas, Buda, and Del Rio rock units are beautifully exposed in cliffy roadcuts on the eastern approach to the US 90 bridge over the Pecos River.

On the east side of Langtry, the colorful town where Judge Roy Bean once held legal sway in the late nineteenth century, watch for crinkly, gray to red, nodular, thin-bedded limestone in roadcuts. These beds are the Late Cretaceous Boquillas Formation, quite spectacular in their odd, nodular (rounded) appearance.

Cliffs of the Devils River Limestone line the Pecos River Canyon, as viewed from the picnic area along US 90. (29.7061, -101.3536)

US 90 climbs westward from Langtry and, in a few miles, tops out on a plateau where excellent roadcuts reveal shockingly white, blocky mudstone and chalk beds of the Late Cretaceous Austin Chalk. It is surprising how recognizable the Austin is; it looks just the same here as it does around Austin and as far away as Dallas. Clams and their thick calcite shells are common in the Austin Chalk.

For the 30 miles between Langtry and Dryden to about 10 miles east of Dryden, the highway rolls along on the Austin Chalk. The plateau-like countryside is punctuated by Lozier Canyon, a dry wash crossed by US 90 about 18 miles west of Langtry. The wash has been cut down into the underlying Boquillas and Buda units, which form the lower walls in a natural river cut south of the road. The Austin-supported plateau gives way westward to valley and mesa country, cut into Boquillas and Buda limestones by satellite creeks and draws of the Rio Grande.

Eight miles west of Dryden, US 90 begins its descent into Sanderson Canyon, a delightful drainage cut deeply into Early Cretaceous limestone by Sanderson Creek, a tributary of the Rio Grande. Rocks north of the road are mainly cherty, fossiliferous Edwards Limestone (Early Cretaceous), whereas south of the highway equivalent

Large roadcuts of Austin Chalk along US 90 about 3.7 miles west of Langtry. (29.8302, -101.6266)

The Santa Elena Limestone rises above a picnic area about 3.4 miles east of Sanderson in Sanderson Canyon. (30.1072, -102.3656)

limestones are called Santa Elena. Such terminology is used in West Texas for this significant assemblage of cliff-forming limestones that we won't see again until we visit the Sierra del Carmen Mountains on the east side of Big Bend National Park and Santa Elena Canyon on the west side of the park in the chapter on West Texas. There, the sequence is seen in its entirety in sheer canyon walls. Sanderson Canyon deepens and darkens as it narrows toward Sanderson.

US 183
BROWNWOOD — LAMPASAS — AUSTIN
141 miles

Southeast of Brownwood, US 183 leaves the Colorado River lowlands and begins a gentle climb into the Lampasas Cut Plain, a hilly, rolling region that is more of a vegetative change than anything else. Beds of Cretaceous-aged Edwards Limestone, the same rock that forms the surface of the Edwards Plateau, form ridges and hilltop edges but do not crop out along the roadway.

In a limestone quarry north of Goldthwaite are marly, yellowish-tan beds of the Glen Rose Formation that dates to approximately 113 million years ago. A unit of the Trinity Group, this hard limestone resists erosion, so it forms ledges. While the quarry cannot be viewed from the road, look for small outcrops of the Glen Rose at the intersection of US 183 and US 84 on the north side of town.

Note old buildings constructed of limestone and sandstone blocks in the small towns, particularly in Lampasas. Pioneer folks liked to build sturdily, but they were smart and didn't drag their heavy building stones from long distances if good stone was locally available. If you want to know what rocks are found nearby, look at building stones used in old houses.

Lampasas lies in the Sulphur Creek drainage, where erosion has cut completely through the Early Cretaceous Edwards Limestone and Glen Rose Formation, exposing Pennsylvanian Marble Falls Limestone and Ordovician Honeycut Formation in the creek bottom west of town. However, these older rocks do not crop out along US 183.

Along US 183 in Lometa, this building is constructed of local orange sandstone and white limestone blocks. (31.2163, -98.3936)

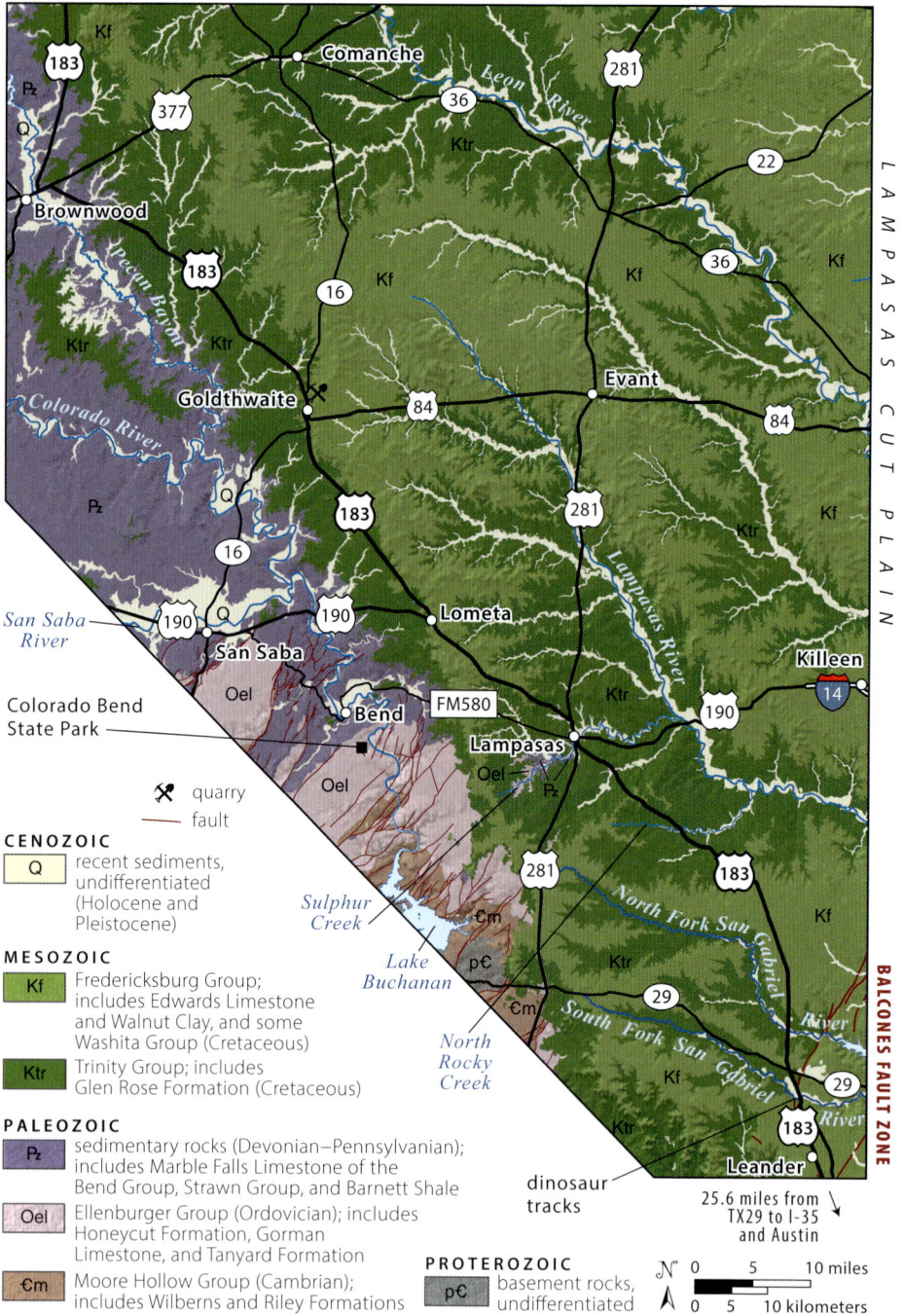

Geology along US 183 between Brownwood and Austin.

Colorado Bend State Park (discussed in depth in the section on page 192) is accessible from Lampasas and features steep cliffs, outcrops, and waterfalls of the Llano Uplift. Follow FM 580 (W. North Ave.) west approximately 23.5 miles to the community of Bend, then head south.

South of Lampasas, US 183 climbs through more yellowish limestone beds of the Glen Rose Formation. Look for exposed rocks in roadcuts and natural outcrops on both sides of the road south of North Rocky Creek. About 20 miles southeast of Lampasas, the road climbs noticeably to the top of a plateau where the Walnut Clay and Edwards Limestone form the flat surface.

The older Glen Rose Formation appears again, however, in a 20-foot-high roadcut south of the bridge over the North Fork of the San Gabriel River. The road again climbs onto the Walnut Clay and Edwards Limestone surface around Seward Junction (intersection of TX 29), then heads downward again to cross the South Fork of the San Gabriel River. A small parking area near the south side of the southbound San Gabriel River bridge provides access to a well-preserved dinosaur trackway in the Glen Rose Formation. Walk a half mile west (upstream) to a series of tracks left by the dinosaur *Acrocanthosaurus*, the top predator of this area 100 million years ago, while it walked along a tidal flat. Much of the time, the trail and trackway are above water and dry. However, after a heavy rain, the trail is slippery, and the trackway may be underwater. Please use caution and check weather forecasts before visiting.

The Glen Rose Formation limestone appears in roadcuts on both sides of the roadway about a quarter mile south of North Rocky Creek on US 183. (30.9424, -98.0262)

A trackway left by the dinosaur Acrocanthosaurus in the dry riverbed of the South Fork of the San Gabriel River near Leander. (30.6169, -97.8684)

Between Leander and Austin, the highway crosses the flat Edwards surface. The road passes over the Balcones fault zone on the northwest suburban outskirts of Austin and crosses the Mt. Bonnell fault, the western main fault of the Balcones fault zone, about where the North Capital of Texas Highway (Spur 360) joins US 183. For more on the Mt. Bonnell fault, see the section about Austin in the North Gulf Coastal Plain.

US 287
CHILDRESS—WICHITA FALLS—FT. WORTH
220 miles

US 287 runs a transect across the bands of gently west-dipping Permian rocks east of Childress, but few outcrops are visible from the highway. Rolling farmland extends in all directions, and the bright-red soils indicate the presence of the underlying red rocks. Two major bands of Late Permian rocks, the Pease River and Clear Fork Groups, are crossed between Childress and Electra. You can see these rocks at Copper Breaks State Park, 12 miles south of Quanah on TX 6.

PALEOZOIC

PERMIAN

Pqr — Quartermaster Formation

Ppr — Pease River Group; includes San Angelo and Blaine Formations

Pcf — Clear Fork Group

Pp — Patrolia Formation

Pwa — Wichita Albany Group

PENNSYLVANIAN–PERMIAN

PlPcb — Cisco-Bowie Group; includes Markley and Archer City Formations of the Bowie Group

PENNSYLVANIAN

lPc — Canyon Group

CENOZOIC

QUATERNARY

Q — recent sediments; includes alluvium, playa deposits, and older terrace deposits (Holocene and Pleistocene)

Qw — windblown sand, silt, and sand sheet deposits (Holocene)

MESOZOIC

CRETACEOUS

Kfw — Fredericksburg and Washita Group, undivided

Ktr — Trinity Group; includes Antlers Formation

⬧ — oil field

Trinity Group sandstone outcrop

40 miles from Decatur to Ft. Worth

GRAND PRAIRIE

Red River

Wichita River

Pease River

Red River

Lake Arrowhead

Lake Kickapoo

Lake Kemp

OKLAHOMA

Medicine Mounds

Copper Breaks State Park

Childress
Quanah
Vernon
Electra
Burkburnett
Wichita Falls
Mabelle
Benjamin
Throckmorton
Bellevue
Bowie
Alvord
Decatur
Jacksboro

287
6
70
82
83
183
277
380
281
59
82
81
82
44

N

0   10   20 miles
0   10   20 kilometers

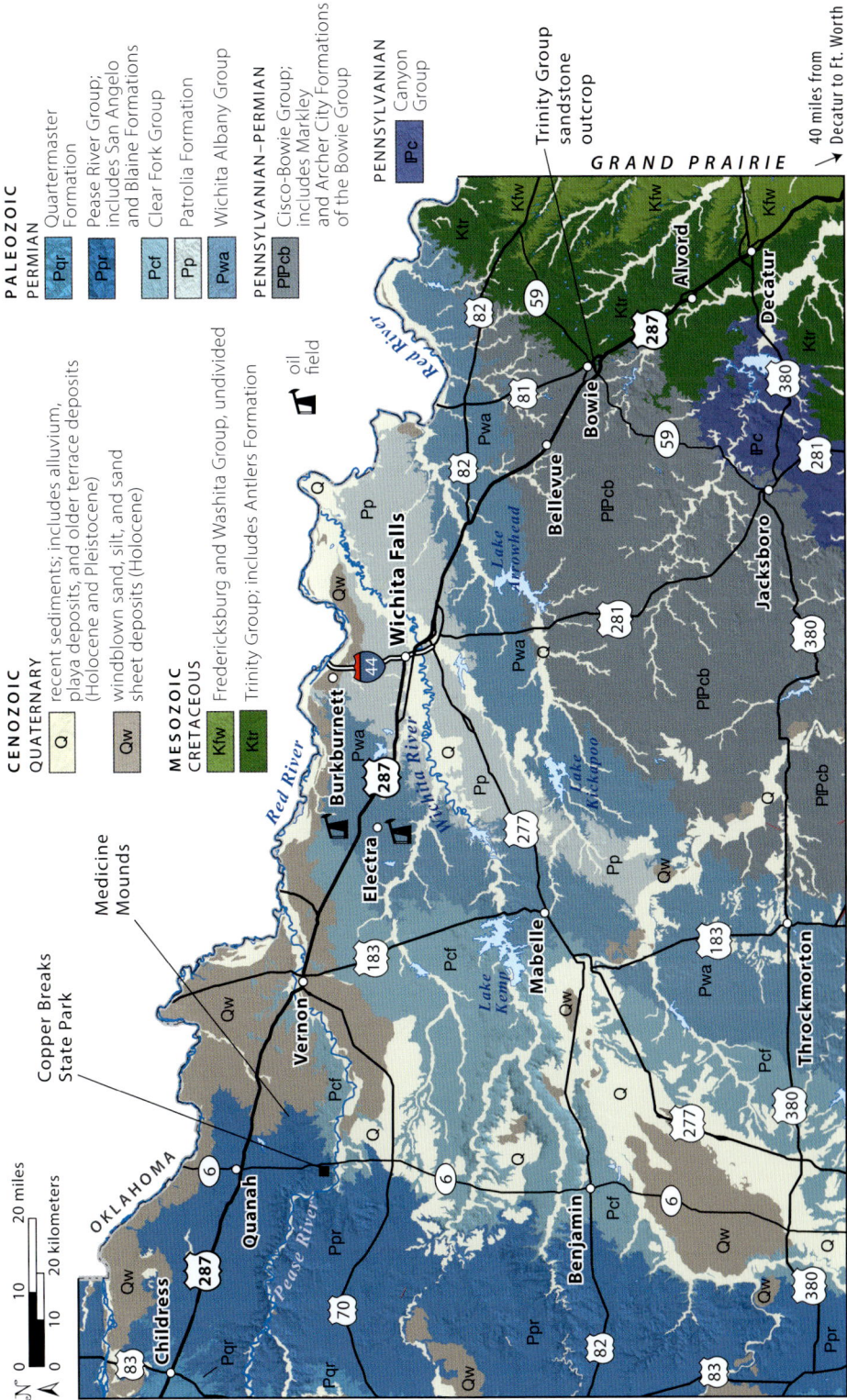

Geology along US 287 between Childress and Decatur.

COPPER BREAKS STATE PARK

Copper Breaks State Park, 12 miles south of Quanah along TX 6, features rugged hills of colorful red and greenish-gray sedimentary beds. The rocks are part of the Permian-aged Clear Fork Group carbonates and the San Angelo sandstone of the Pease River Group. Contained within these rocks is copper ore, giving them their greenish color. The copper was first noted by George McClellan in 1852 while on an expedition searching for the headwaters of the Red River. Many hypotheses and mechanisms exist as to how the copper got concentrated here. Copper-bearing groundwater may have flowed through the sulfide-rich bedrock, or the copper-bearing water moved to the surface through fractures (or evaporation pulled it to the surface) where it encountered hydrogen sulfide from decaying bacteria. Regardless of how the copper sulfide minerals covellite and chalcocite originally formed, they were altered to the mineral malachite later. The malachite gives the copper ore its tell-tale green color.

In 1877, George McClellan returned to purchase large tracts of land around the Copper Breaks. He formed the Grand Belt Copper Company and by 1884 was attempting to mine the copper ore, but he died in 1885. Ore assays proved that the copper content was poor, so the company ceased all activity on the property in 1888.

The greenish-gray rocks of the red San Angelo sandstone at Copper Breaks State Park contain trace amounts of copper ore. (34.1080, -99.7467)

The green color of this copper ore is from the copper carbonate mineral malachite. —Photo courtesy Kelly Nash

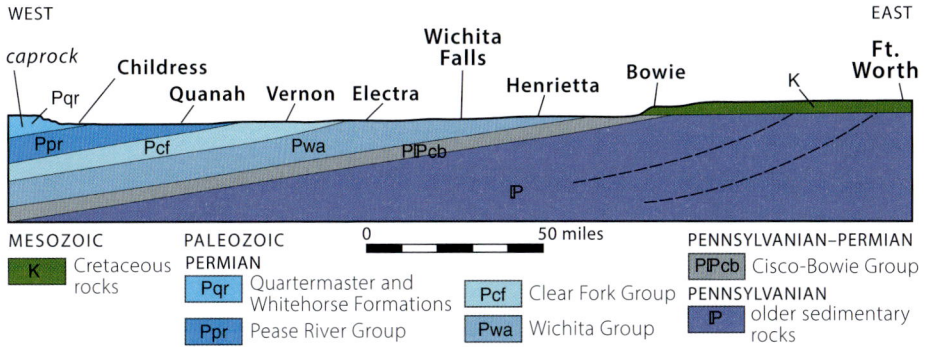

WEST .... EAST

caprock Childress .... Wichita Falls .... Ft. Worth

Pqr ... Quanah Vernon Electra ... Henrietta Bowie K

Ppr ... Pcf ... Pwa ... PIPcb

IP

MESOZOIC	PALEOZOIC		PENNSYLVANIAN–PERMIAN	
K Cretacheous rocks	PERMIAN		PIPcb Cisco-Bowie Group	
	Pqr	Quartermaster and Whitehorse Formations	PENNSYLVANIAN	
	Pcf	Clear Fork Group	IP	older sedimentary rocks
	Ppr	Pease River Group		
	Pwa	Wichita Group		

0 — 50 miles

Cross section along US 287 between Childress, Wichita Falls, and Ft. Worth. The tilt on the rock layers is greatly exaggerated.

From the rest area a few miles east of Quanah, four distinct mounds are visible, framed against the skyline to the south. Known as the Medicine Mounds, these high-standing erosional remnants were sacred ground to the indigenous people who first occupied this area. The mounds indicate the amount of erosion that has taken place in this area, assuming the regional land surface was once level with the top of the mounds. They are part of the Blaine Formation of the Pease River Group.

The Medicine Mounds, south of the highway east of Quanah, are erosional remnants from when the original land surface was higher. (34.2649, -99.6105)

Just west of Vernon, US 287 crosses the Pease River. This tributary of the Red River carries red sediment eroded from the surrounding Permian red beds and contributes significantly to the maintenance of the Red River's image.

Numerous oil fields dot the countryside between Vernon and Wichita Falls. The underlying Pennsylvanian rocks are the source and reservoir for much of the petroleum produced in this area. The Pennsylvanian cycles of coal and limestone deposition produced the rich organic layers to source the oil, and oil rose upward to accumulate in layers of sandstone.

At Wichita Falls, US 287 crosses the Wichita River, another major tributary to the east-flowing Red River. Not unexpectedly, the Permian band of rocks in this area is the Wichita Albany Group, though few rock exposures are visible from the highway. Most rock outcrops are found in steep river embankments, and geologists studying these rocks drive many back roads and hike along streams for miles to get to the best outcrops.

Four miles east of Bellevue, Pennsylvanian- to Permian-aged Bowie Group sediments show up on cuesta hills and in a quarry cut. The rocks exposed are Markley Formation sandstones and Archer City Formation shales, both originally deposited in stream channels and floodplains that bordered the high-standing Ouachita Mountains during this time

Bowie straddles the line between Permian rocks to the west and Cretaceous rocks to the east, though no significant escarpment is seen from the highway to mark the transition. On the east side of Bowie east of the TX 59 crossover, watch for a Cretaceous sandstone outcrop on a ridgetop in the trees above a large concrete retaining

ELECTRA/BURKBURNETT

April 1, 1911, became one of the most historic days in Texas petroleum history when a geyser of oil gushed from the Clayco No. 1 Well near Electra. The well produced 650 barrels of oil per day and sparked a rush of wildcatters and oil speculators to the sleepy farming town. Several hundred wells were drilled in and around Electra, and by 1913, the area had produced around eight million barrels of oil. In the nearby town of Burkburnett, a similar geyser of oil on July 19, 1918, set off another mad rush to strike it rich. The area around the two towns became known locally as the World's Wonder oil field as more productive wells were drilled. By 1920, the area, now officially known as the Wichita County regular field, accounted for almost half of all the oil being produced in Texas. Towering oil drilling derricks were built less than twenty feet apart, in some cases within a foot of people's homes. Forests of pumpjacks covered the area. The pumpjack, also known as the nodding donkey or rocking horse, was a device invented in the 1920s to pull oil from the subsurface.

The oil being recovered from this area is from a layer of sandstone within the Cisco Group, a formation deposited in deltas during the Pennsylvanian Period. Over time, oil production decreased, and the boom times became bust during the Great Depression. The 1940 film *Boomtown* starring Clark Gable and Spencer Tracy was inspired by the events surrounding the Burkburnett oil rush. Electra is known as the Pumpjack Capital of Texas, and today, upward of five thousand pumpjacks in a 10-mile radius bring oil to the surface.

Pumpjacks on the outskirts of Electra still produce oil. (34.0418, -98.9482)

wall. The sandstone, a unit of the Trinity Group, is more resistant to erosion than adjacent shales, so it holds up the ridge and stands tall above the surrounding landscape. Rainwater soaks into this exposed sandstone and flows down dip into the subsurface, where the sand forms an important aquifer in this part of the state. The Cretaceous sandstones were deposited in streams and along shorelines at the edge of a sea to the southeast.

The road between Bowie and Ft. Worth rides on an elevated plateau surfaced by flat-lying, resistant Cretaceous limestone and sandstone beds. Watch for oil pumps and a red, white, and pink sandstone and shale outcrop of the Antlers Formation at the highway crossover in Alvord. The Antlers, a unit of the Trinity Group, was deposited in deltas and alluvial fans and is the coalescence of two other sandstone units in this part of Texas, the Hensel and Paluxy.

Cretaceous Trinity Group sandstone exposure just east of the TX 59 interchange. (33.5502, -97.8580)

Just south of the north exit to Alvord, a multicolored outcrop of Antlers Formation sandstone and shale on the right (west) side of the highway appears behind a pumpjack. (33.3678, -97.7061)

A former Texaco station built with locally sourced stone at the corner of Washburn and Walnut Streets in Decatur. (33.2349, -97.5934)

In Decatur, a former Texaco service station built completely of locally sourced petrified wood, sandstone, and limestone can be viewed at the corner of Washburn and Walnut Streets. Careful examination of the blocks reveals fossils as well. Southeast of Decatur, the countryside flattens out on the Grand Prairie, and fertile farmland becomes the main scenery from here to Ft. Worth.

LLANO UPLIFT

A unique and spectacular group of rocks is found in the center of the state, north of Fredericksburg, surrounding the town of Llano. Here, hard granite knobs reflect pink in the harsh, highland sunlight, while tortured, black schists absorb the sun. This assemblage of ancient, glinting, crystalline rocks is more expected in the jagged crags of Colorado than in the soft, rolling, calcareous canyon country of central Texas. Buried, squeezed, melted, faulted, then uplifted, the granite and schist formed the heart of once-lofty mountain ranges that time has eroded flat, only to be later born again, raised to near surface, and finally exposed in central Texas by erosion.

The oldest Llano rocks formed deep in the geologic past, about 1.35 billion years ago, when the area that is now central Texas lay in an ocean basin off the coast of North America. Sediments poured into this sea from adjacent mountains, building a coastal plain and continental shelf, not unlike the Texas coast today. First, a mix of rhyolitic volcanic rocks and associated tuffaceous sediments were shed from the young mountain ranges to accumulate a wedge of poorly sorted sediments in the sea. For this book, we'll call it the Valley Spring wedge. As the mountain range wore down over the next few million years, a second wedge of dark, fine-grained, well-sorted muddy sediments (herein called the Packsaddle wedge) was laid over the top of the Valley Spring wedge.

EON	ERA	PERIOD	ROCK NAMES IN THE LLANO UPLIFT	GRAPHIC COLUMN

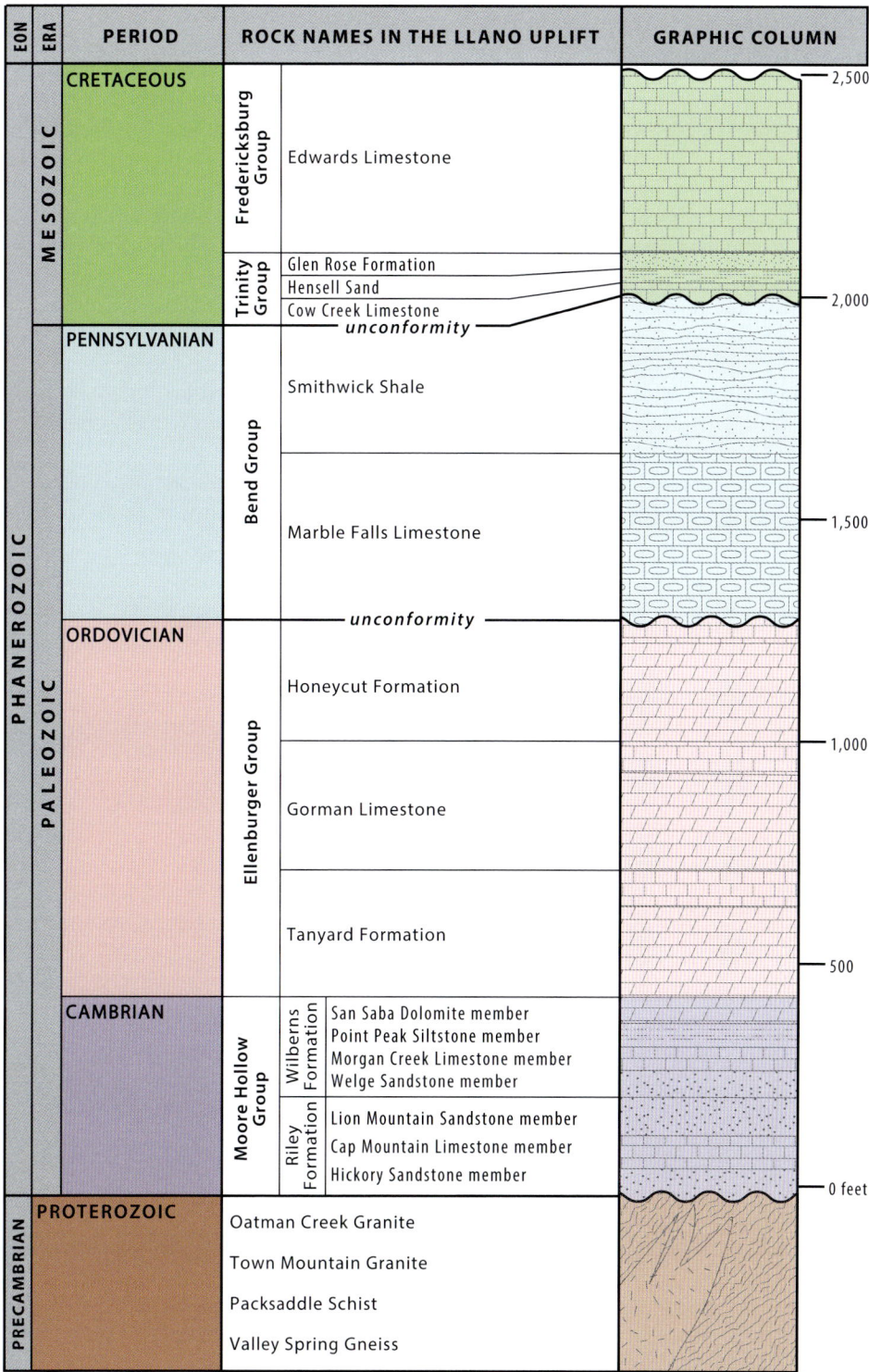

Stratigraphic sequence of rocks in the Llano Uplift.

Table contents:

PHANEROZOIC

MESOZOIC
- CRETACEOUS — Fredericksburg Group: Edwards Limestone — 2,500
- Trinity Group: Glen Rose Formation, Hensell Sand, Cow Creek Limestone — 2,000
  - *unconformity*

PALEOZOIC
- PENNSYLVANIAN — Bend Group: Smithwick Shale; Marble Falls Limestone — 1,500
  - *unconformity*
- ORDOVICIAN — Ellenburger Group: Honeycut Formation; Gorman Limestone — 1,000; Tanyard Formation — 500
- CAMBRIAN — Moore Hollow Group:
  - Wilberns Formation: San Saba Dolomite member, Point Peak Siltstone member, Morgan Creek Limestone member, Welge Sandstone member
  - Riley Formation: Lion Mountain Sandstone member, Cap Mountain Limestone member, Hickory Sandstone member — 0 feet

PRECAMBRIAN

PROTEROZOIC
- Oatman Creek Granite
- Town Mountain Granite
- Packsaddle Schist
- Valley Spring Gneiss

Legend: siltstone · sandstone · limestone · dolomite · granite · schist

One billion years ago, the edge of North America collided with another landmass. This collision, known as the Llano event in Texas, was part of the larger Grenville mountain building event that served to consolidate the supercontinent Rodinia. Trapped in the subduction zone between the colliding continents, the thick Packsaddle and Valley Spring sediment wedges were squeezed, folded, and heated with such intensity that all the sedimentary minerals changed—metamorphosed—into new crystalline forms. The Valley Spring Gneiss and Packsaddle Schist thus were born.

While the rocks metamorphosed at depth, a mountain range rose in the melded zone of the new continent. Rocks buried even deeper beneath the range melted, and the ensuing magmatic mixture, lighter in weight than surrounding rocks, rose through the overlying rock column as large spherical bodies of red-hot liquid. Eventually, subduction ended, the magma solidified into granite, and the metamorphic rocks cooled to surface temperatures.

The Earth doesn't like topographic irregularities, so for the next few hundred million years, erosion wore the mountains down to a preferably flatter surface. By the time

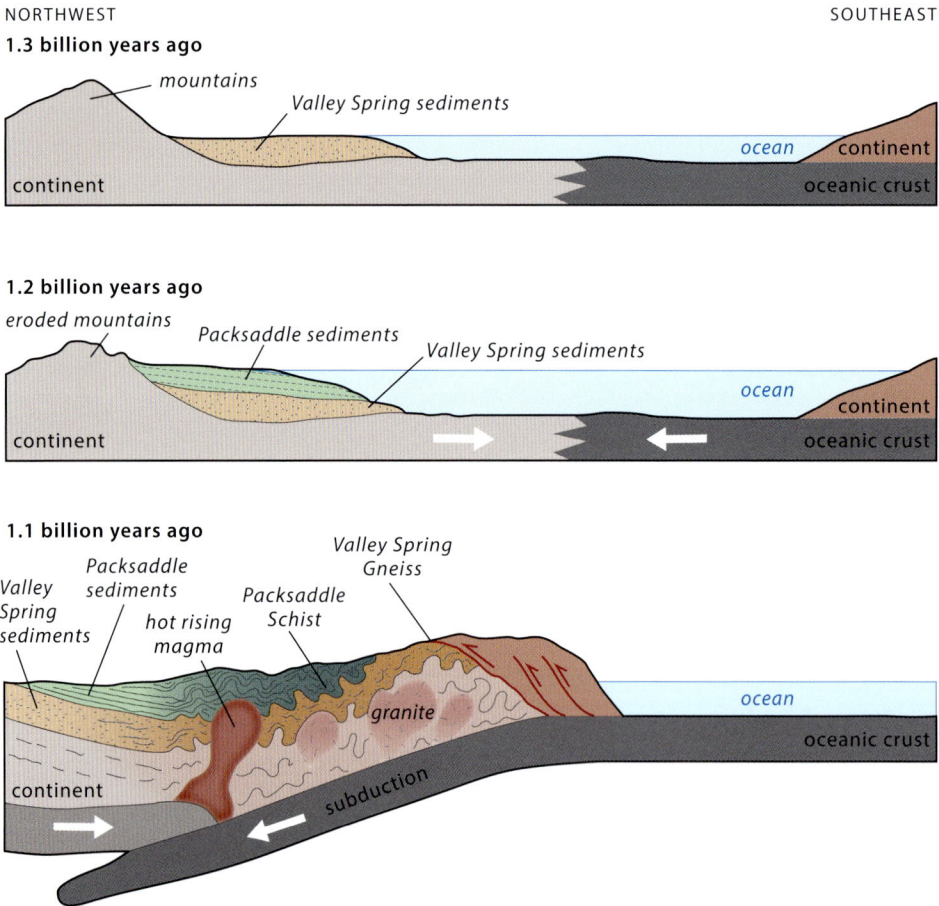

How the rocks of the Llano Uplift came to be formed in the Proterozoic Eon.

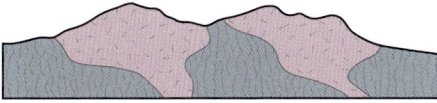

A. About 1 billion years ago, mountains rose, rocks were metamorphosed, and granitic batholiths were emplaced.

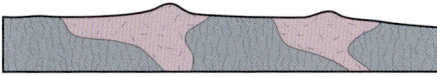

B. In 400 million years, erosion reduced the range to a nearly flat plain, though 800-foot hills were left here and there.

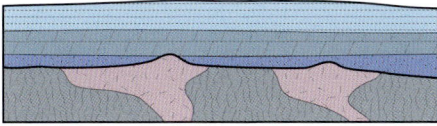

C. Sediments were laid over the plain during the Paleozoic as seas invaded the land.

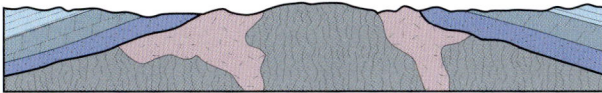

D. In a burst of Mesozoic mountain building, the Llano was tilted, faulted, and eroded.

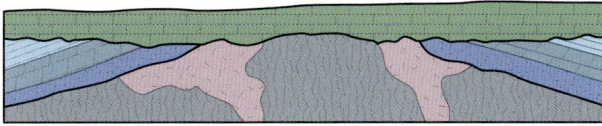

E. Cretaceous sedimentary rocks covered the Llano as seas, once again, invaded the land.

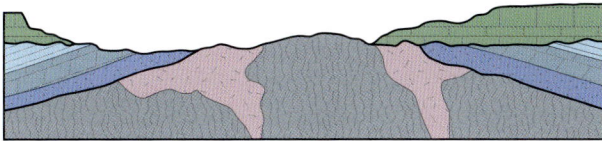

F. In the Cenozoic, uplift caused renewed erosion; Cretaceous rocks were removed to reveal the Paleozoic and Precambrian rocks beneath.

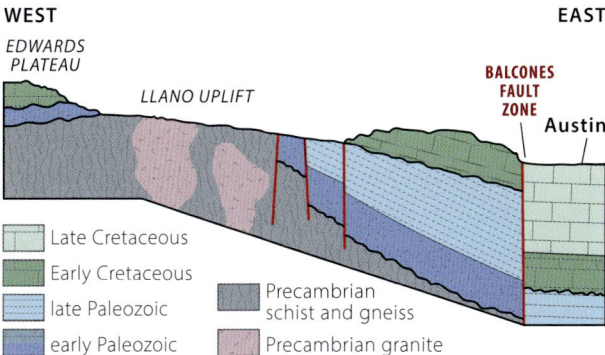

WEST

EAST

EDWARDS PLATEAU

LLANO UPLIFT

BALCONES FAULT ZONE

Austin

G. This cross section is specific to the Llano country west of Austin.

The Llano Uplift is structurally high but topographically low.

Late Cretaceous

Early Cretaceous

late Paleozoic

early Paleozoic

Precambrian schist and gneiss

Precambrian granite

Much has happened to the rocks of the Llano Uplift since their formation during the Proterozoic Eon.

Ediacaran life was first abundantly preserved in the fossil record (around 600 million years ago), the old range was gone, worn to a tabletop. Along the same line where North America and its landmass neighbor once were joined, the two continents rifted apart. Ocean water lapped over the flattened margin of North America once again, and for the next few hundred million years, beds of Paleozoic limestone, sandstone, and mudstone were deposited over the remnants of the once-tall mountain range.

About 335 million years ago, the eastern and southern margins of North America began to collided with other continents to form the supercontinent Pangea. The mountain range formed by the collision rose across Texas along the same curved continental juncture that had formed nearly 1 billion years earlier. Geologists named this younger uplift the Ouachita Mountains. Paleozoic rocks that had been deposited on the old Proterozoic surface were tilted and eroded.

The familiar geologic cycle was repeated once more, as erosion reduced the mountains to a flat surface. After about 80 million years of quiet stability, the Earth rumbled again, and Pangea cracked and split apart. North America separated from the European continent and headed westward, while the southern margin sagged to form the early Gulf of Mexico in Jurassic time. As the Gulf continued to deepen, Cretaceous marine sediments were laid over the eroded surface, and the Proterozoic igneous rocks and tilted Paleozoic rocks together formed the underlying beveled surface.

After the Edwards Plateau was uplifted in the past 20 to 15 million years along the Balcones fault zone, erosion of the plateau began. The thick Cretaceous sequence of rocks was cut into valleys and ridges in the Hill Country and completely peeled off in the Llano Uplift to expose the igneous and metamorphic rocks.

In the road guides included in this section, you'll notice a pattern of the roads beginning or ending in the Cretaceous rocks, then proceeding across some Paleozoic limestones or sandstones before reaching the crystalline rocks at the heart of the Llano Uplift. The crystalline rocks form a bull's-eye at the center of the uplift and are flanked by Paleozoic rocks, which in turn are surrounded by a cover of Cretaceous rocks. The Cretaceous rocks, at the surface of the surrounding Edwards Plateau, form a resistant cap and thus are higher than the Llano Uplift, which has been eroded. So, the Llano Uplift is a modern lowland, even though it was once a region of lofty mountains.

US 87
COMFORT — FREDERICKSBURG — MASON — BRADY
93 miles

From Comfort, US 87 heads north toward Fredericksburg, passing low hills of Glen Rose Formation of the Trinity Group for several miles as the road climbs away from the Guadalupe River drainage. The road passes through these Cretaceous limestones for a few miles, climbing into the overlying Edwards Limestone, where the surrounding plateau countryside is noticeably flat. The road continues on the plateau for about 7 miles before again descending through Glen Rose beds as it encounters topography cut by the Pedernales River drainage. The road crosses the Pedernales River a few miles south of Fredericksburg, where the river has cut into the Hensell Sand. This unit, also of the Trinity Group, was deposited at the same time as the Glen Rose Formation but farther west on a shallow marine shelf. In the nineteenth century

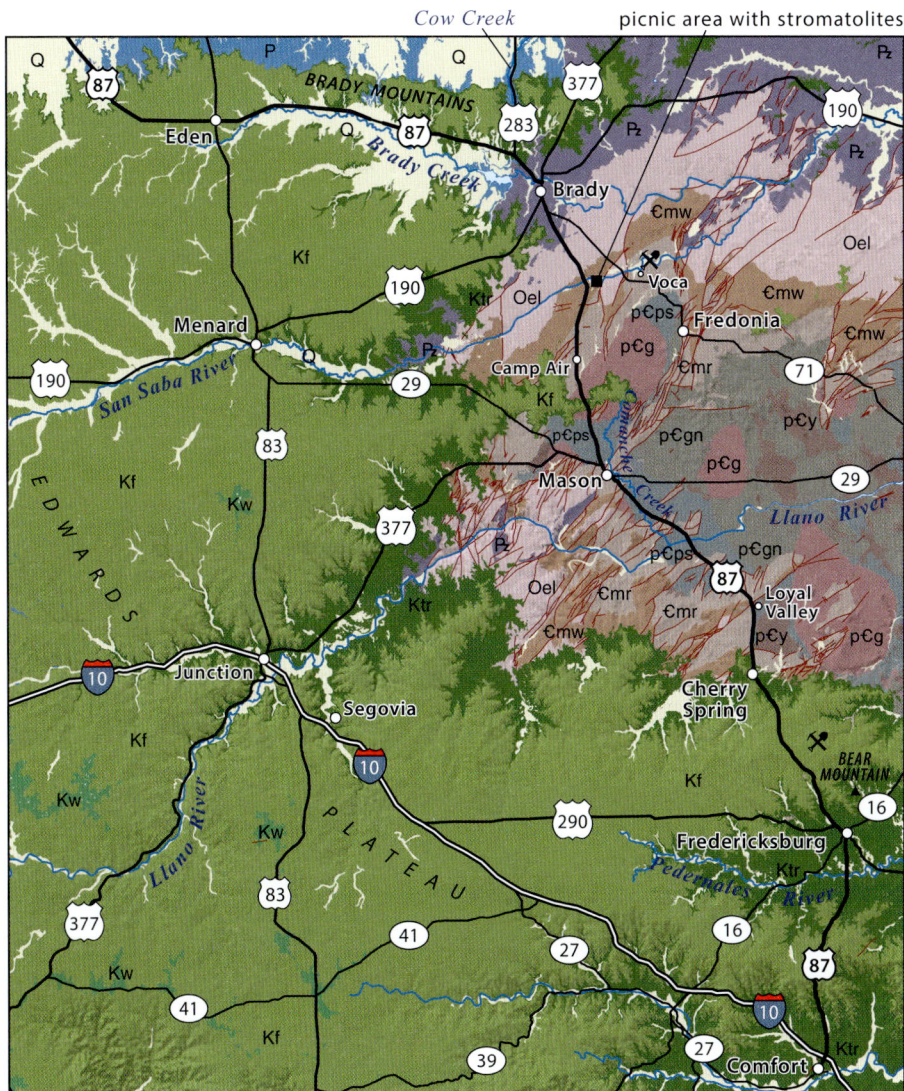

Geology along US 87 between Comfort and Brady.

CENOZOIC

| Q | sediments, undifferentiated (Holocene and Pleistocene) |

MESOZOIC (Cretaceous)

| Kw | Washita Group; includes Buda Formation and Del Rio Clay |

| Kf | Fredericksburg Group; includes Edwards Limestone and Segovia Member of the Edwards Limestone |

| Ktr | Trinity Group; includes Glen Rose Formation and Hensell Sand |

PALEOZOIC

| P | sedimentary rocks, undivided (Permian) |

| Pz | sedimentary rocks, undivided (Devonian–Pennsylvanian); includes Bend, Strawn, and Canyon Groups and Barnett Shale |

| Oel | Ellenburger Group; includes Gorman Limestone and Tanyard Formation (Ordovician) |

Moore Hollow Group (Cambrian)

| Єmw | Wilberns Formation; includes Morgan Creek Limestone, Point Peak Siltstone, and San Saba Dolomite |

| Єmr | Riley Formation; includes Hickory and Lion Mountain Sandstones |

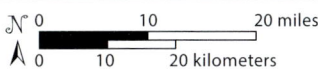

🗡 mine/quarry

— fault

PROTEROZOIC

pЄps	Packsaddle Schist
pЄgn	gneiss, undivided
pЄy	younger granitic rocks
pЄg	granitic rocks; includes Town Mountain Granite

in Fredericksburg, German immigrants built solid houses and shops of locally derived limestone blocks.

North of Fredericksburg, US 87 crosses Hensell Sand for 3 miles to ascend on to the Edwards Plateau, where creamy-colored rubble dots the plateau surface. Watch for several good roadcuts 10 to 15 miles north of Fredericksburg where the flat-bedded character of the limy, fossiliferous beds is on display. Gypsum was once mined in the area and is still mined from a gypsum-rich section of the Edwards Limestone 3 miles to the east of the highway on Cherry Mountain Loop. About 3 miles south of Cherry Spring, the road reaches the northern edge of the Edwards Plateau, and northbound travelers get a panoramic view northward across the low terrain of the Llano region. The road cuts through the limestone as it winds downward back onto whitish Hensell Sand.

BEAR MOUNTAIN

From Fredericksburg, Ranch Road 965 heads north across clayey, silty, sandy Cretaceous sediments to the limestone terrain of the Edwards Plateau. Early Cretaceous limestones, deposited in the shallow seas that once spread across Texas 100 million years ago, form hillside ledges all around. A few miles north of town, a vista to the north comes into view; to the left (west) is a ridge of flat-bedded limestone, but to the right (east) is a dark red, bouldery hill looking distinctly out of place. Known as Bear Mountain, this steep-sided hill is composed entirely of pink granite, which is mined here for architectural stone.

View looking north along Ranch Road 965 at Bear Mountain, a knob of pink granite among a sea of limestone. (30.3132, -98.8613)

The granite knob appears to be at about the same level as the limestone across the road. Did the granitic magma penetrate into the limestone or was Bear Mountain a huge knob on the Cretaceous seafloor against which the limestone was deposited? The ages of the two rocks provides the answer. The granite crystallized 1 billion years ago, while the limestone is "only" about 100 million years old. The granite, being much older, could not logically have intruded the limestone. The granite knob must have been a hill (or perhaps an island) on the Cretaceous seafloor, and the limestone beds lapped onto it, eventually covering it. Erosion has removed the relatively softer limestone from around the hard, crystalline granite, which now stands high as it continues to resist erosion.

Striking, red to tan, Cambrian-aged Lion Mountain Sandstone appears in roadcuts just south of the northern intersection with the road to Loyal Valley. (30.5822, -99.0144)

In the low country between Cherry Spring and Mason are several excellent road-cuts. Just south of the north intersection with RM 2242 (Loyal Valley Road) is a large roadcut in reddish-tan Lion Mountain Sandstone, a member of the Riley Formation (Moore Hollow Group). Note the thick and thin horizontal beds and how the grains of quartz sand stand out—if you look closely. This sand was deposited about 500 million years ago in Cambrian time, well after life had developed a wide spectrum of forms on Earth. North of the outcrop, skyline ridges on either side of the road are also composed of this Cambrian sandstone.

Bouldery outcrops along a westward curve between Simonsville Road and Ranch Road 152 are dark, platy Packsaddle Schist, some of the oldest rocks in Texas. On the north side of Comanche Creek (3 miles south of downtown Mason), schist and

Several light-brown pegmatitic dikes cut the dark, severely deformed Packsaddle Schist in an outcrop south of Mason. (30.7197, -99.1987)

Close-up of the intense folding in the Packsaddle Schist. (30.7197, -99.1987)

The Mason County Jail on Westmoreland Street is one of several historic buildings in Mason constructed of locally quarried Hickory Sandstone. (30.7478, -99.2320)

gneiss are exposed in low but spectacular roadcuts. Pegmatite dikes crisscrossing the dark metamorphic rocks were originally molten, watery liquids that were forced into fractures in the older metamorphic rocks. The liquid came from the huge, hot, molten bodies of granitic magma that rose through the metamorphic rocks 1.1 billion years ago.

Just south of Mason, roadcuts and outcrops of Cambrian sandstone appear again. This unit, the Riley Formation's Hickory Sandstone, grades laterally into the Lion Mountain Sandstone farther south. In Mason, note how many historical buildings were constructed of this sturdy, brown sandstone.

Between Mason and Brady, US 87 traverses a section of Cambrian, Ordovician, and Pennsylvanian rocks that flank the older Proterozoic rocks of the Llano Uplift. The Cambrian rocks are found near Mason, whereas the younger Pennsylvanian rocks occur near Brady. The sedimentary stack of these Paleozoic limestones and sandstones is quite regular. Arranged from south to north, the rocks are oldest to youngest, and bottom to top.

Small roadcuts of rather poorly exposed, brown Cambrian sandstone can be found in the northern part of Mason. Along the 4 miles of highway north of town, look for roadcuts where dark Packsaddle Schist is cut by dikes and for exposures of pink Town Mountain Granite. To the north along this stretch is an east-west-trending skyline ridge of yellowish-white limestone, which the road climbs and crosses about 6 miles north of Mason. This prong of the high-standing plateau of Cretaceous Edwards Limestone and Hensell Sand directly overlies the Proterozoic granite at this point. Excellent roadcuts at the ridgetop expose several limy beds, though few fossils are to be found here. In the low area immediately north of the ridge, more pink granite appears along the roadside, its crystalline minerals sparkling in the sun. Look for it at the junction with Kruse Road.

About 6 miles north of Mason, US 87 cuts through a ridge capped by Edwards Limestone, here with layers of differing thicknesses. (30.8377, -99.2657)

North and south of the hamlet of Camp Air, the countryside is fairly flat. The exposure-less area for 2 miles south of Camp Air is the lower part of the Hickory Sandstone, which is composed of well-rounded sand grains derived from river deposits, dune sand reworked by the wind, and the underlying, weathered Proterozoic rocks. Nine miles to the northeast near Voca, it is quarried for hydraulic fracturing sand that enhances oil production. Two miles north of Camp Air and just north of Katemcy Road, the highway crosses the dark-red or brownish-red upper part of the Hickory Sandstone. The sand is red because it contains about 12 percent iron and could be a potential source of iron in the future. In these low roadcuts on both sides of the highway, distinct sedimentary bedding in the sandstone is apparent. One of the oldest Paleozoic rocks in Texas, this sandstone was deposited directly on the old Proterozoic bedrock in Cambrian time.

A rest area where US 87 crosses the San Saba River about 8 miles south of Brady allows close inspection of some interesting outcrops. Gray mounds of rock in the riverbed are stromatolites, the remains of life that have been around on Earth's surface for more than 3.4 billion years. The stromatolites in the San Saba riverbed are from the Cambrian-aged Wilberns Formation, a group of carbonates of the Moore Hollow Group that were laid down in shallow seawater where cyanobacteria thrived and built a stromatolite reef. The colony of bacteria grew upward as sediment was trapped, creating mounds. Examination of the riverbed shows stromatolites of many different sizes, ranging from 4 inches to as much as 13 feet across.

Between the San Saba River and Brady, the highway crosses Ordovician rock units, though not much more than rock rubble in the fields can be seen. Brady lies on the southern tip of a band of Pennsylvanian-aged rocks, which form the bedrock over a wide area to the northeast.

Stromatolites, mounds formed where cyanobacteria trapped sediment, can be observed in the riverbed of the San Saba River south of Brady. (31.0042, -99.2687)

North of Brady, windmills take advantage of the higher elevation of the Brady Mountains, a ridge of resistant Cretaceous limestone. (31.2131, -99.3730)

North and west from Brady is a west-trending ridge known as the Brady Mountains. From the junction of US 87 near Brady, US 283 heads straight north through a gap in this high, steep-sloped ridge of Cretaceous rocks. The gap was carved by the now-intermittent Cow Creek, which is the namesake of the Cow Creek Limestone, an Early Cretaceous unit of the Trinity Group that is found in other parts of Texas.

US 281
BURNET — MARBLE FALLS —
JOHNSON CITY — SAN ANTONIO
99 miles

This section of US 281 begins in the charming little community of Burnet, which sits on the Early Cretaceous-aged Hensell Sand. South of town, US 281 crosses into late Cambrian-aged Wilberns Formation carbonates, a difference of about 375 million years. The highway has entered the far eastern extent of the Llano Uplift. Almost all the Paleozoic rocks that had been deposited over the top of the Proterozoic rocks of the Llano Uplift were eroded away when the region was uplifted during the Ouachita mountain building. The much younger Cretaceous rocks were deposited over the top of the eroded, tilted Paleozoic units that encircle the older granites and metamorphic rocks of the uplift.

No roadcuts are present between Burnet and Park Road 4, which leads west to Longhorn Cavern State Park. That cave, dissolved in an uplifted ridge of Ordovician limestones, is discussed in the road guide for TX 29 on page 198. Just south of Park Road 4, however, US 281 crosses Honey Creek, where Cambrian carbonates outcrop in the creek bed. Look on the right (west) side of the bridge as you cross it. The dolomitic rocks are part of the Wilberns Formation of the Moore Hollow Group. Another great set of outcrops, also Wilberns Formation, appears along the roadside about a quarter mile south of Honey Creek, but here they are limestone-dominant carbonates.

Between the roadcut and Marble Falls, the roadway continues south over mostly Cambrian carbonates, except for a couple slivers of Cambrian Hickory Sandstone, Ordovician Gorman Limestone, and Pennsylvanian-aged Marble Falls Limestone. None of these units are exposed along the highway.

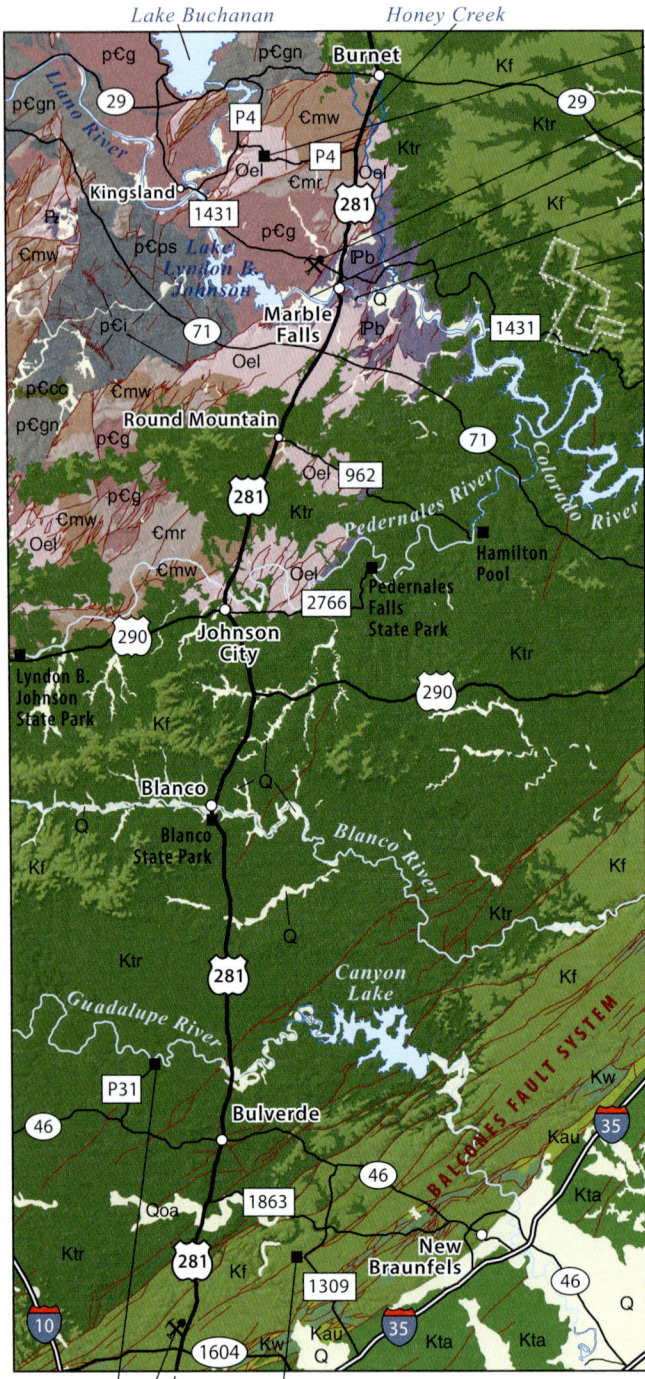

Geology along US 281 between Burnet and San Antonio.

Map labels (roads, places, and geologic features):

Lake Buchanan
Honey Creek
Burnet
Longhorn Cavern State Park
Granite Mountain
Lake Marble Falls
Max Starcke Dam
Balcones Canyonlands National Wildlife Refuge

pЄg
pЄgn
Llano River
29
Kf
Ktr
29
P4
Єmw
Ktr
Oel
Kingsland
Єmr
Oel
1431
281
pЄps
Lake Lyndon B. Johnson
pЄg
Q
71
pЄi
Marble Falls
ℙb
Oel
1431
pЄcc
Єmw
pЄgn
pЄg
Round Mountain
Єmw
Oel
Pedernales River
71
Єmr
962
Colorado River
Єmw
Oel
Ktr
281
Hamilton Pool
Oel
Pedernales Falls State Park
2766
Johnson City
290
Lyndon B. Johnson State Park
Ktr
290
Kf
Blanco
Q
Blanco State Park
Blanco River
Kf
Q
Kf
281
Q
Ktr
Canyon Lake
Kf
Guadalupe River
BALCONES FAULT SYSTEM
Kw
P31
Bulverde
Kau
35
46
46
Qoa
Ktr
1863
Kta
281
Kf
New Braunfels
Єmw
1309
46
10
Kw
Kau
35
Kta
Kta
1604
Q
Q

Guadalupe River State Park
San Pedro Quarry
to San Antonio
Natural Bridge Caverns

CENOZOIC

Q — recent sediments; includes alluvium and older terrace deposits (Holocene, Pleistocene, and some Pliocene)

MESOZOIC (Cretaceous)

Kta — Taylor Group

Kau — Austin Chalk

Kw — Washita Group; includes some Eagle Ford Group

Kf — Fredericksburg Group; includes Edwards Limestone

Ktr — Trinity Group; includes Cow Creek Limestone, Glen Rose Formation, and Hensell Sand

PALEOZOIC

ℙb — Marble Falls Limestone and Smithwick Shale of the Bend Group (Pennsylvanian); includes some Devonian and Mississippian rocks

Oel — Ellenburger Group; includes Honeycut Formation, Gorman Limestone, and Tanyard Formation (Ordovician)

Moore Hollow Group (Cambrian)

Єmw — Wilberns Formation; includes Morgan Creek Limestone, Point Peak Siltstone, and San Saba Dolomite

Єmr — Riley Formation; includes Hickory and Lion Mountain Sandstones

PROTEROZOIC

pЄps — Packsaddle Schist

pЄcc — Coal Creek Serpentinite

pЄgn — gneiss, undivided

pЄg — granitic rocks; includes Town Mountain Granite

0   5   10 miles
0   5   10 kilometers

—— fault

◆—◆ Precambrian intrusive dikes (pЄi)

⚒ quarry

Limestone beds of the Wilberns Formation appear in a roadcut a quarter mile south of Park Road 4. (30.6804, -98.2548)

GRANITE MOUNTAIN

FM 1431, the first major intersection on the north side of Marble Falls, leads west to a historic granite quarry operation at Granite Mountain, a granitic dome on the northwestern edge of town. FM 1431 crosses onto the granite about 1 mile west of the US 281 intersection. Heavily vegetated, upended mudstones of the Pennsylvanian-aged Smithwick Shale (Bend Group) are faulted against the sparsely vegetated Proterozoic Town Mountain Granite along a major fault having at least 3,000 feet of displacement. While we don't see that kind of displacement here, a noticeable change in elevation and rock type occur on opposite sides of the fault. Look for granite outcrops on the north side of FM 1431 where the vegetation thins.

About 1.7 miles west of US 281, a roadside park along FM 1431 provides excellent views of the quarry. The Town Mountain Granite has been quarried since the 1880s, when granite blocks were cut and shipped via a specially constructed railroad line to Austin to build the State Capitol. Marketed as Texas Pink Granite, the Town Mountain Granite graces many buildings throughout the United States and as far away as Iceland and Singapore. Blocks of granite from these quarries were also used to construct the stone groins that extend seaward from the seawall along the coast at Galveston. At the roadside park, look for tables constructed of the same granite you see across the highway.

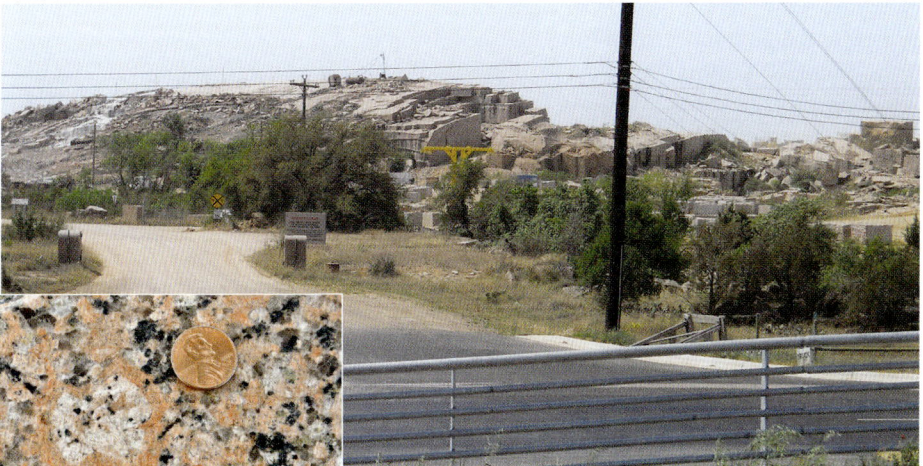

The quarry operations at Granite Mountain, viewed from the picnic area along FM 1431 west of Marble Falls. The close-up of the granite is from the top of one of the picnic tables. (30.5925, -98.2990)

East of Marble Falls, FM 1431 passes through arguably some of the most beautiful terrain of the Texas Hill Country. Just over 21 miles from the intersection in Marble Falls is the entrance to Balcones Canyonlands National Wildlife Refuge, which was established in 1992 to protect the nesting grounds of the black-capped vireo and the endangered golden-cheeked warbler, as well as to protect water quality for nearby Austin and central Texas.

The town of Marble Falls was named for the impressive waterfalls on the Colorado River before Max Starcke Dam (on the east side of town) was constructed and Lake Marble Falls inundated them in 1951. The dark-gray limestone around Marble Falls, named for the town, is Early Pennsylvanian age, around 320 million years old and in the Bend Group. The Marble Falls Limestone forms the canyon walls of the Colorado River beneath the US 281 bridge south of the town center. The best view is toward town from the south side of the bridge. These dark-gray limestones continue for 1 mile south of the river and form small roadcuts along US 281. Blocks of these Pennsylvanian rocks are caught and preserved between northeast-trending faults flanking the eastern edge of the Llano Uplift.

Just south of Marble Falls, the road to Max Starcke Dam on the left (east) exposes beautiful roadcuts of the Ordovician-aged Honeycut Formation carbonates of the Ellenburger Group. A closer look at these cuts reveals 6- to 10-inch-long chert nodules composed of hard silica that protrude from the softer limestone. Waters rich in silica flowed through the carbonate rock and replaced small areas with silica. The source of the silica is mainly from organisms such as the opaline silica of diatoms, radiolarians, and siliceous sponges. When these sponges and plankton died, they released their tiny silica spicules and shells that composed their skeletal framework. The dissolved silica percolated through the sediment and later precipitated out as the chert we see today.

Stromatolites can also be found in the Honeycut Formation, along with another interesting, albeit rare, snail fossil that grazed on the stromatolite colonies, much as cattle graze across a field of grass. Geologists think that in this area, an extensive sabkha (salt flats between a desert and ocean) had formed that was ultra-saline in nature. Only certain species of snails could live in such a hostile environmental of heat and high salinity, conditions that stromatolites of the time flourished in.

Along US 281 between Marble Falls and Johnson City, older Cambrian and Ordovician rocks lie directly beneath younger Cretaceous rocks. The older rocks outcrop where stream erosion has cut through the Cretaceous cover. Look for significant roadcuts of the Honeycut Formation just south of Rocky Road and another 1.3 miles farther south between Little Flatrock Creek and the TX 71 interchange.

South of the TX 71 junction, US 281 climbs a low drainage divide topped with soft, yellowish Cretaceous limestone and sandstone beds. These Cretaceous rocks are breached around the hamlet of Round Mountain. Look for Cambrian dolomite with interbeds of siltstone in the roadside rocks in the lower elevations for a few miles south of Round Mountain, including the creek bottom of South Cypress Creek.

At Round Mountain, you can head east 16.1 miles on RM 962E to reach Hamilton Pool, a charming, little swimming hole that requires reservations well in advance of visiting. Springs feed the blue-green pool with a constant supply of cool, clear water. Hamilton Creek spills into the pool over a 50-foot-high ledge of the Cow Creek Limestone, a Cretaceous unit of the Trinity Group.

A mile north of Johnson City, US 281 crosses the Pedernales River (locally pronounced "Perden-Alice"), where hard, gray Ordovician dolomites are well exposed by river erosion. Note how the Paleozoic rocks along this road are tipped at an angle so they tilt down to the southeast, reflecting their position on the flank of the Llano Uplift. Excellent exposures of tilted Paleozoic rocks can be seen along the river in nearby Pedernales Falls State Park, downstream from Johnson City.

Along Max Starcke Dam Road, curved, domed mounds of stromatolites can be found in the limestone walls of the Honeycut Formation. (30.5540, -98.2605)

Large roadcut of Honeycut Formation carbonates just south of Rocky Road. Behind this roadcut is an abandoned quarry where blocks of Honeycut were removed for various projects. (30.5378, -98.2798)

PEDERNALES FALLS STATE PARK

From Johnson City, head east 9 miles on Robinson Road (FM 2766), then turn north (left) on the park entrance road, which crosses the cover of yellowish-tan Early Cretaceous limestone all the way to the parking area. The easy hike down to Pedernales Falls reveals a whole new world, however, because the Marble Falls Limestone of Pennsylvanian age forms the lower bedrock in the cliffs and falls. The hard limestone contains abundant white calcite veins and large fossil crinoid stems, which are especially prevalent on surfaces in the lower part of the falls area. Note how the bedding is inclined, as opposed to the flat-lying Cretaceous limestones above. The Marble Falls Limestone was deformed in the Ouachita mountain building event near the end of Paleozoic time.

The falls, a series of rapids, are the product of this inclination. Note how the water slides down the surface of the inclined bedding planes and how the up-tipped ends of these beds form the edges of the pools. The erosive power of running water is evident everywhere on the limestone surfaces, with grooves, rills, and polishing. Look also for round potholes on the edges of the pools. These holes, 1 foot or larger across, have vertical sides and usually contain one or more rounded, hard rocks at the bottom. The potholes are ground out of the hard limestone by the swirling action of the passing water, which cause the round cobbles to act like grindstones, carving the pothole deeper and deeper. You can even see the grooved sides on some potholes, which record the past grinding action of the trapped cobbles.

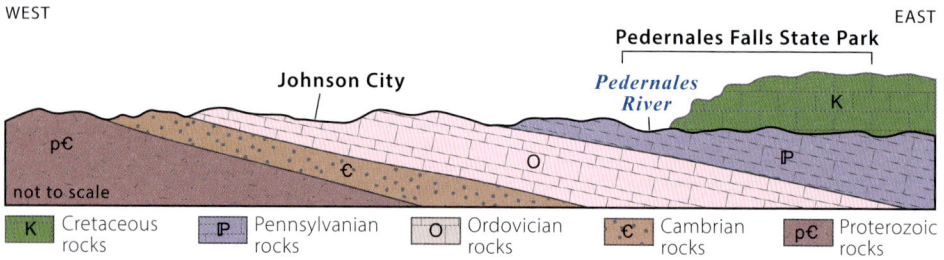

The cross section between Johnson City and Pedernales Falls State Park shows the flat-lying Cretaceous rocks overlying the tilted Pennsylvanian limestone.

Pedernales River falls and pools, viewed from the Hackenburg Loop Trail, formed on inclined beds of Marble Falls Limestone at Pedernales Falls State Park. (30.3373, -98.2518)

Fossil crinoid stem in Marble Falls Limestone at Pedernales Falls State Park. Swiss Army Knife for scale. (30.3383 -98.2522)

LYNDON B. JOHNSON STATE PARK

Lyndon B. Johnson State Park, along the Pedernales River 14 miles west of Johnson City on US 290, was a ranch owned by the former president. Though most rocks around the LBJ ranch are Early Cretaceous Hensell Sand, older rocks of Cambrian age crop out in the river bottom. The park was established in 1970 after enough money was raised to purchase land on the south side of the river, across from LBJ Ranch. In addition to being the home of the Texas State Bison Herd and the State of Texas Longhorn Herd, the park was also where Lyndon's wife Claudia "Lady Bird" Johnson dreamed of "beautifying" Texas by planting millions of wildflowers along Texas roads. Today, the roadsides are awash each spring with bluebonnets (a type of lupine), Indian blanket, and other colorful wildflowers thanks to Lady Bird's efforts.

The highway between Johnson City and Blanco passes low hills where yellow-tan limestone ledges peek out every now and then from the brush country. All the rocks are Early Cretaceous Glen Rose Formation, whose thin-bedded, alternating marl and limestone create a stair-stepped topography. The solid limestone is the steps with the softer marl forming the slopes between the steps. One hill a few miles north of Blanco shows the stair-step character particularly well. Also look along the roadway for erosion rills in the soft, yellowish marly layers. Fossil hash of clams and snails is quite common in these shallow marine beds. The Glen Rose Formation is also well exposed at the swimming area at Blanco State Park, on the Blanco River upstream from the US 281 bridge on the south side of the town of Blanco.

Between Blanco and Bulverde, US 281 crosses the Guadalupe River. To see the river and steep cliffs of Glen Rose Formation and Hensell Sandstone up close, head west 7 miles on TX 46 from Bulverde and then north 3 miles on Park Road 31 to Guadalupe River State Park.

Glen Rose Formation, the dominant bedrock of central Texas, creates a stair-step pattern in a large outcrop 2.7 miles north of Blanco. (30.1333, -98.4037)

Southeast of Bulverde are the Natural Bridge Caverns, where tours guide visitors past marvelous dripstone features, giant rooms, and cave pools. To reach it from US 281, take the exit for Bulverde (FR 1863) and continue east on FR 1863 for 8.6 miles. Turn right (south) on Natural Bridge Caverns Road and watch for signs to the entrance.

From Bulverde, US 281 continues south into San Antonio. Many of the roadcuts that used to be visible along the highway have been covered with grass or cement. The road passes some active limestone quarries, including the San Pedro Quarry just north of the Loop 1604 interchange, and Brackenridge Park, the site of a major quarry in the early 1900s. See page 85 in I-35 road guide for more on this park.

TX 16
FREDERICKSBURG—LLANO—SAN SABA
71 miles

North of Fredericksburg, TX 16 traverses Cretaceous rocks, crossing about 8 miles of poorly exposed, flat-lying, thin beds of the tan-colored Hensell Sand of the Trinity Group in hills on either side of the road. A view north into the valley of Palo Alto Creek comes up at about 2 miles north of town. Eight miles north of town, there is a high-standing prong of the Edwards Plateau, where the road crosses about 1 mile of Edwards Limestone of the Fredericksburg Group, then descends toward the Llano lowland proper, dropping through a short segment of Hensell Sand again. Watch carefully for flat areas of brown sandstone for about 1 mile on either side of the small hamlet of Eckert. The town lies on Cambrian-aged Hickory Sandstone of the Riley Formation, which is preserved in a faulted, down-dropped block on the edge of the Llano Uplift. In stretches without trees, look west to see the topographic edge of the Edwards Plateau, where the Cretaceous sandstone and limestone form a distinct cliff.

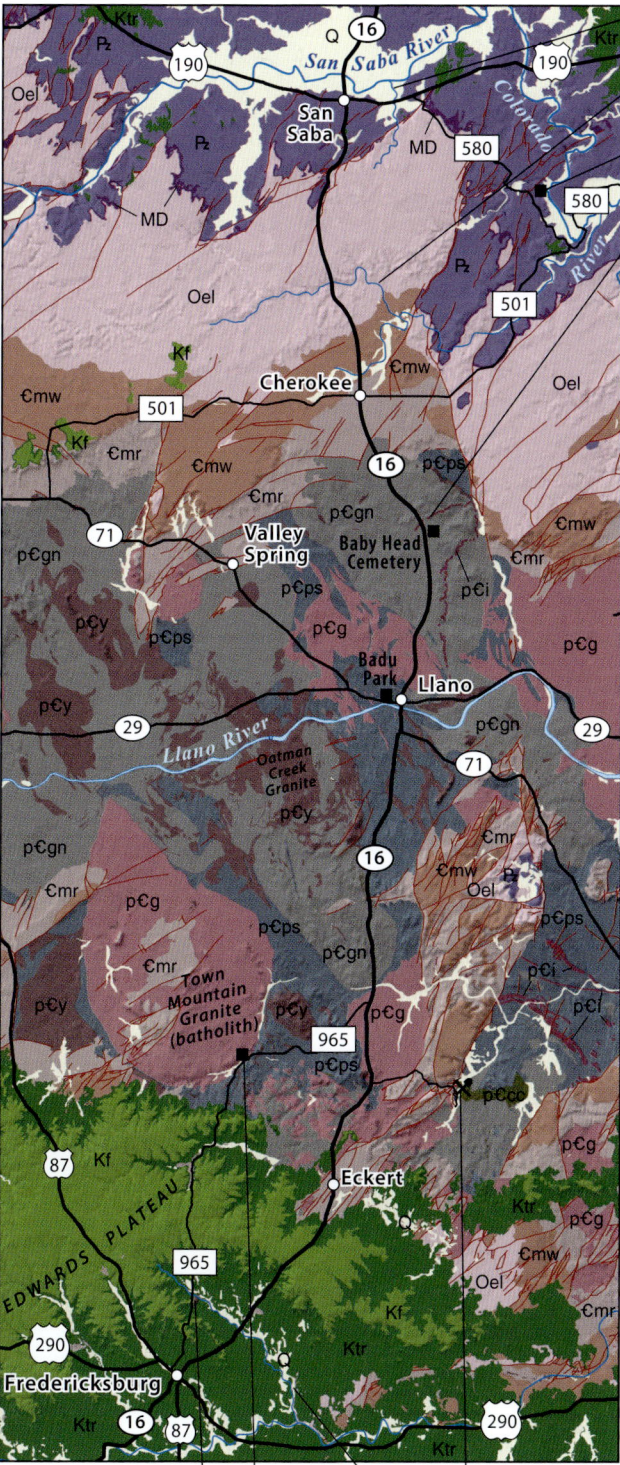

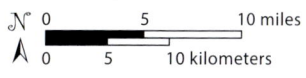

Geology along TX 16 between Fredericksburg and San Saba.

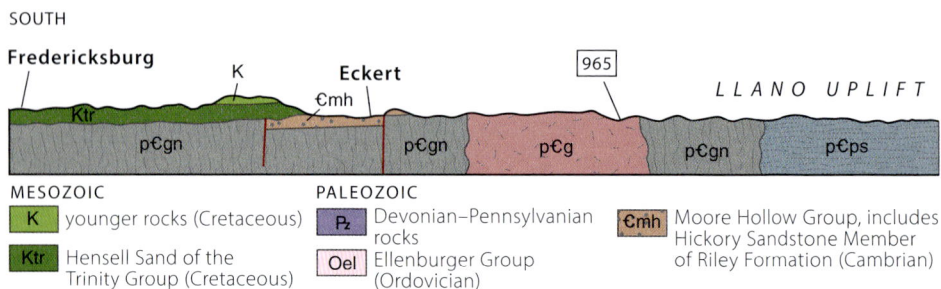

SOUTH

Fredericksburg   K   Eckert   965   *L L A N O   U P L I F T*

Ktr
pЄgn   Єmh   pЄgn   pЄg   pЄgn   pЄps

MESOZOIC

| K | younger rocks (Cretaceous) |
| Ktr | Hensell Sand of the Trinity Group (Cretaceous) |

PALEOZOIC

| Pz | Devonian–Pennsylvanian rocks |
| Oel | Ellenburger Group (Ordovician) |

| Єmh | Moore Hollow Group, includes Hickory Sandstone Member of Riley Formation (Cambrian) |

TX 16 cross section between Fredericksburg, Llano, and San Saba. Proterozoic gneiss, granite, and schist predominate in the center of the section.

COAL CREEK SERPENTINITE

In a remote corner of the Llano Uplift are rocks unlike any other in the state of Texas—bright-green serpentinite, a metamorphic rock that forms when rock of Earth's mantle is hydrated, or combined with water, under pressure. About 1.326 billion years ago, a chain of volcanic islands, called an island arc, began to form south of the continent of Laurentia. Island arcs are created when an oceanic plate (along with the uppermost mantle) subducts under another oceanic plate (this process occurs today along the Aleutian Islands of Alaska and in Japan). While much of the oceanic plate gets subducted, slices can get scraped off and added onto the crust of the island arc or continent during subduction. This particular island arc collided with Laurentia around 1.275 billion years ago during the Grenville mountain building event, which occurred along the eastern and southern margins of North America. In the process, the mantle portion of the oceanic plate that was added to the continent got metamorphosed into serpentinite. At the same time, intrusive igneous rocks from the magma chamber that fed the volcanoes of the island arcs were also metamorphosed into the nearby Big Branch Gneiss. The metamorphosed mantle rocks, called the Coal Creek Serpentinite, can be seen in a quarry on the north side of the Willow City Loop about 5 miles east of TX 16 near the junction with Willow City–Click Road.

Coal Creek Serpentinite, exposed in a quarry along the Willow City Loop, is a scraped off piece of ocean plate rock that was added to the continent in the Proterozoic Eon. (30.4794, -98.6310)

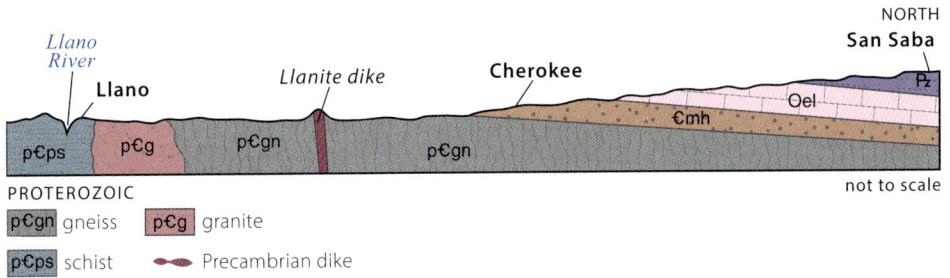

NORTH
San Saba

Llano River

Llano

Llanite dike

Cherokee

Pz

Oel

€mh

p€ps   p€g   p€gn   p€gn

PROTEROZOIC

not to scale

p€gn gneiss   p€g granite

p€ps schist   ◄●► Precambrian dike

North of Eckert, hills scattered with pink, blocky boulders of gneiss and granite are readily visible from the highway, signaling entry into the central Proterozoic portion of the Llano Uplift. Approximately 5.5 miles north of Eckert is the intersection with the north terminus of the Willow City Loop. While this scenic byway is famous for some of the most beautiful springtime flowers in the state, it is also known for a geologic oddity—the Coal Creek Serpentinite (see sidebar on facing page).

The prominent skyline hills east from the junction of TX 16 and Ranch Road 965 (the side road to Enchanted Rock State Natural Area) are a block of Paleozoic sedimentary rocks that occupy a fault block down-dropped into the Proterozoic rocks at the southeast corner of the Llano Uplift. The Paleozoic sedimentary rocks were more resistant to weathering than most of the surrounding Proterozoic crystalline rocks, so they stand elevated here as a ridge line.

North of the FM 965 junction, TX 16 crosses Valley Spring Gneiss and Packsaddle Schist, which are ocean floor sediments metamorphosed by colliding continents during the Grenville mountain building in Proterozoic time. Watch for low roadcuts where the light-colored gneiss looks like contorted, banded granite, whereas the schist is black and has platy minerals that sparkle in the sunlight. From the bridge over the Llano River in the town of Llano, look east (downstream) for views of sandbars

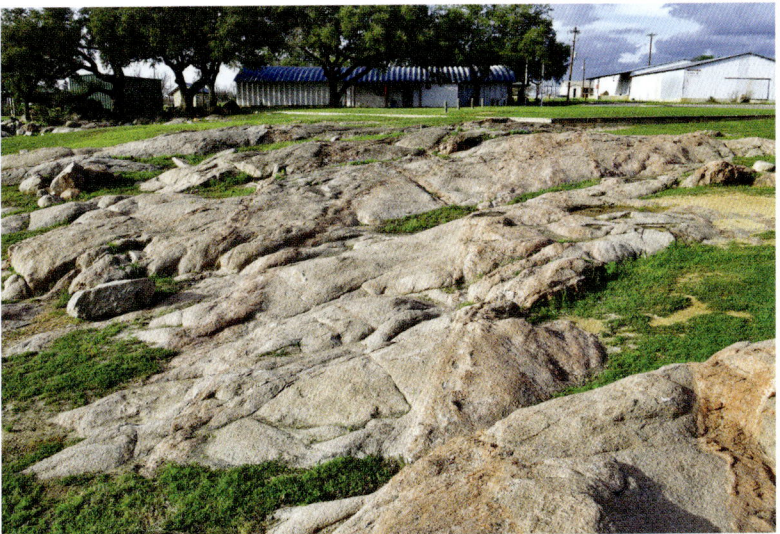

The Pack-saddle Schist in Badu Park in Llano. (30.7542, -98.6767)

ENCHANTED ROCK STATE NATURAL AREA

The granite domes at Enchanted Rock State Natural Area, 8 miles west of TX 16 on Ranch Road 965, loom ever larger as you approach the entrance to the park. Despite their size, however, the domes are but a small piece of a huge, round mass of granite that rose through the Packsaddle Schist like a giant, hot balloon about 1 billion years ago. The circular shape of the granite body, a batholith, looks like a bull's-eye on geologic maps. Geologic features of the domes are clearly laid out in raw-rock profusion, devoid of soil or vegetation—or concrete cover. The view from the summit of Enchanted Rock, elevation 1,825 feet and 445 feet above Sandy Creek, is a gorgeous 360-degree panorama.

The name Enchanted Rock comes from old Native American legends and pioneer observations of strange sounds and lights. Common creaking and groaning noises could easily be granite blocks grinding against one another as they expand and contract from heating and cooling between

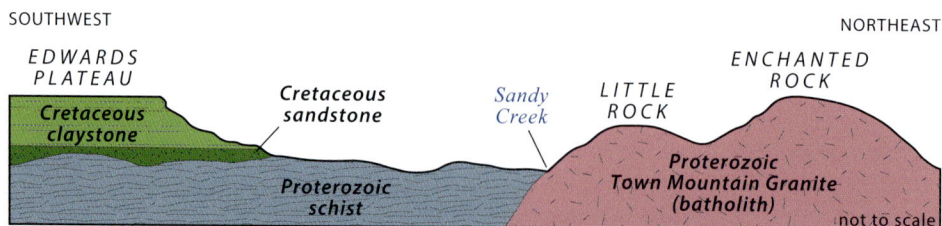

Cross section between the Edwards Plateau and Enchanted Rock.

The classic rounded shape of Enchanted Rock can be seen for miles around. (30.4988, -98.8188)

daytime and nighttime. Low light at dawn and sunset shimmers off crystals in the granite, and the sparkling reddish domes blend almost magically with the early morning or late evening sky.

The dome shape of the hills is the product of the interaction of erosion with zones of weakness in the granite. The domes are separated by linear zones of fractures that intersect at nearly right angles. Look for these blocky fractures in the low area between Enchanted Rock and Little Rock along the Echo Canyon Trail. Water and erosion attack these fractured areas with greater ease than the solid granite areas, creating valleys along the fractures, while leaving solid granite to form the high areas.

The rounded shape of the domes is caused by exfoliation. Granite forms deep within the Earth's crust. As thousands of feet of overlying rock are removed by erosion over geologic time, pressure caused by the weight of this pile of rock is reduced. The granite expands a little in response to the lessened pressure, which in turn causes the granite to split in curved sheets. Weathering cracks the sheets, creating blocks and slabs, which slide downslope. Exfoliation sheets and blocks are beautifully displayed on the north flank of Little Rock and can be seen from the flanks of Enchanted Rock.

As you climb around, look for odd-shaped rock pedestals created by the differential disintegration of the wetted, lower portion of boulders, where chemical weathering attacks the rock faster than the dry boulder top. Note also the round weathering pits, or pools, where standing water causes chemical weathering to disintegrate the granite. These pits, filled with water after rains, are important water sources for animals in many desert areas of the world.

The surface of Enchanted Rock is crossed by long, linear bumps and rills representing dikes, which filled cooling cracks in the early granite batholith. As the granite magma cooled and crystallized, it shrank, cracks developed, and hot liquid from the last phase of the magmatic mush rushed in to fill the cracks. If the filling material is finer-grained than the surrounding granite, it is slightly harder and resists erosion, so it stands up as a little ridge. But, if the granitic dike material is coarser grained, it weathers more easily, creating a linear depression in the surface rock. These dikes are particularly noticeable on the face of Enchanted Rock.

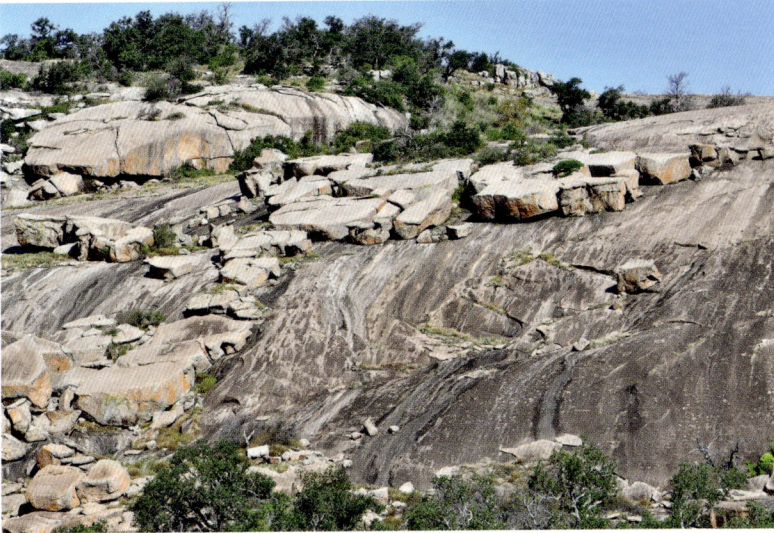

Weathering on Enchanted Rock causes long fractures to form, allowing large "sheets" and blocks of rock to slide downslope. (30.4979, -98.8206)

As water interacts with fractures in the rock, chemical reactions with minerals in the fractures create rounded surfaces, a process called spheroidal weathering. (30.4969, -98.8233)

A pegmatite dike cuts through the Packsaddle Schist in Badu Park in Llano. Note the coarse center and finer grained edges, indicating the dike cooled faster along its edges. (30.7546, -98.6774)

during low water periods. Boulders and outcrops of the Valley Spring Gneiss and Packsaddle Schist are also exposed in the river bottom. A small dam lies beneath the bridge, so the upstream view shows a lake. Badu Park, on the upstream side of the bridge (and dam), provides easy access to the rocky riverbank below the dam.

The road segment between Llano and San Saba offers the opportunity to view the igneous and metamorphic rocks, some quite unusual, of the Llano core, as well as the beds of hard Paleozoic sandstone and limestone that flank the uplift's crystalline center. Low, relatively featureless topography is encountered for 8 miles north of Llano, with pink granite and gneiss boulders here and there. Nine miles north of town, an east-west ridge held up by a hard igneous dike stands above the surrounding Valley Spring Gneiss. The dike can be seen in roadcuts on either side of the highway where the road cuts through the ridge between Baby Head Cemetery and the northern junction with County Road 226. The granitic dike contains reddish-pink feldspar crystals and blue quartz grains that float in a fine-grained, almost black groundmass. The rock, called llanite because it is unique to this area, is unusually hard and was quarried at one time for building stone. The quartz is blue because of chromium impurities. You really must get out of your car and examine this rock up close in the roadcut.

North of the llanite dike, TX 16 passes from Proterozoic igneous terrain into its fringe of Paleozoic sandstone and limestone. The Paleozoic rocks dip gently to the

This dike of llanite is 9 miles north of Llano near Baby Head Cemetery. The close-up of llanite shows its blue quartz and pink feldspar. (30.8905, -98.6585)

north and, from south to north, are stacked in proper sequence, oldest to youngest, Cambrian to Pennsylvanian. These sedimentary rocks were deposited as horizontal layers on the beveled surface of Proterozoic rocks. Watch for a roadcut on the east (right) side about 4 miles north of the llanite dike (or one-half mile north of County Road 445), where the oldest Cambrian unit, the Hickory Sandstone of the Riley Formation, is exposed. Here, the gray-tan, pebbly, and hard sandstone is shot through with thin calcite veins. This pebbly sand was laid down directly on the flat, eroded surface of the Proterozoic-aged Valley Spring Gneiss.

The landscape around the town of Cherokee is open and quite flat, but look for a quarry on the west side of the road a few miles north of town (near the intersection with County Road 407), where hard, glinty blocks of dark-gray Wilberns Group limestone, a unit of the Moore Hollow Group, are about 490 million years old from Cambrian time.

A few miles farther, in the stream valley of Buffalo Creek, the Tanyard Formation crops out, part of the Ordovician Ellenburger Group. This hard, dense limestone contains many chert nodules, formed from redistributed silica. Original layers composed of the remains of siliceous marine organisms, such as sponges and diatoms, contributed the silica to circulating water.

Between Buffalo and Simpson Creeks, abundant limestone boulders, weathered to a battleship gray, lie in fields on both sides of the highway. The boulders are derived from the underlying bed of Ordovician-aged Gorman Limestone. At the Simpson

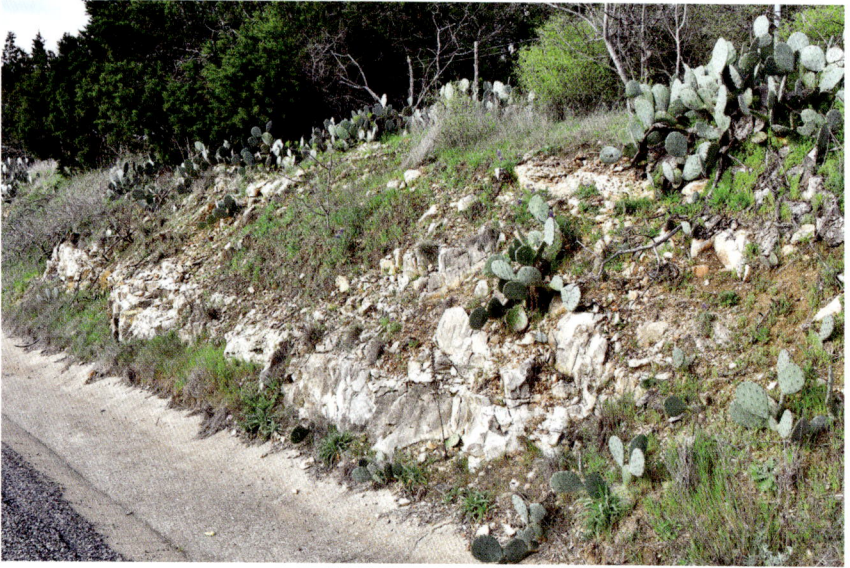

The Tanyard Formation, a hard, dense limestone of the Ellenburger Group, crops out in low roadcuts 5.7 miles north of Cherokee, just north of Buffalo Creek. (31.0613, -98.7317)

Creek crossing, just south of San Saba, thin beds of Pennsylvanian-aged Marble Falls Limestone are exposed in the roadcuts and stream beds.

COLORADO BEND STATE PARK

Located at the edge of the Llano Uplift and named for the sweeping bend of the Colorado River, Colorado Bend State Park features steep cliffs, numerous outcrops, and waterfalls. Much of the rock in the park is from the Ordovician-aged Ellenburger Group, including the Gorman, Honeycut, and Tanyard Formations. These rocks were deposited when the region was located at, or near, the equator, and a warm, shallow sea teeming with life covered much of North America. When organisms and their carbonate shells and skeletons died and settled to the seafloor, vast amounts of limestone formed and created what geologists call the Great American Carbonate Bank, extending from western Texas to eastern Canada. Some of the limestone within the Ellenburger Group was also altered to dolomite, which increased the porosity of the carbonates, allowing water to flow more freely through it. This water dissolved the carbonate rocks, creating a karst topography of caves, springs, and sinkholes at Colorado Bend. The caves are of interest to geologists because they were created by both surface water percolating downward and groundwater from deep underground migrating upward.

In addition to the karst topography, another type of porous limestone called tufa is also found in the park. Tufa mounds are created when cold water from springs discharge onto the surface, releasing some dissolved carbon dioxide into the atmosphere. The chemistry changes cause calcite to precipitate and build the tufa mounds. Carbon dioxide is also released through agitation of the water, such as spilling over a waterfall. This process has led to the formation of Gorman Falls, an enormous tufa mound that is also the tallest waterfall in Texas at 70 feet. A rocky, 1.5-mile trail leads to the falls. Along the way, looks for trace fossils, such as worm burrows, in the Gorman Limestone. To reach the park from San Saba, head east 3.5 miles on US 190, and then turn south on FM 580 and follow signs to the town of Bend and onto CR 436.

Gorman Falls flows over moss- and fern-draped tufa. (31.0585, -98.4824)

The Colorado River has cut a deep canyon into the Ellenburger Group carbonates in Colorado Bend State Park, as seen from a pullout along Park Road 446. (31.0267, -98.4553)

TX 29
MASON—LLANO—BURNET
63 miles

This road guide begins in the quaint Hill Country town of Mason where historical buildings, many of which are well-preserved, are constructed of locally quarried, brown Hickory Sandstone. This Cambrian-aged unit is found on the edge of and in the Llano Uplift in elongate, down-dropped blocks between linear faults.

The Mason County Courthouse in Mason is constructed of locally quarried Hickory Sandstone. (30.7478, -99.2320)

Known as the Gem of the Hill Country, Mason is also famous for the spectacular blue topaz that can be found in the area. While not as deep blue as its Brazilian cousins, the pastel blue shade of the topaz around Mason is pleasing to the eye and produces beautiful, faceted jewelry. The special Lone Star Cut features the five-pointed Star of Texas in the center. For several years after its discovery, area ranchers and other land-owners allowed the public to search for Texas topaz. Over the years, as the number of gems being found became less, fewer people came to search. In 2023, the last ranch to allow such hunts discontinued the practice. Occasionally, small stones are found in dry creek beds and other washes. It is much easier (and safer), however, to find them for sale at local shops.

Between Mason and Llano, TX 29 traverses the heart of the Llano Uplift, rolling along almost entirely on the billion-year-old granite, gneiss, and schist that characterize the Llano country. Here and there along the way, low, rounded knolls of pink granite and banded slabs of gneiss form the platform for a clump of trees, almost emulating a small scene from a Japanese garden. To the east, knobs and knolls coalesce in skyline ridges, reminding the observer that although erosion may have leveled once-tall mountains, it left a bit of topography on the modern land surface.

Inks Lake State Park; Devil's Waterhole

Longhorn Cavern State Park

Hoover Point

CENOZOIC

Q — sediments, undifferentiated (Holocene and Pleistocene)

MESOZOIC

Kf — Fredericksburg Group (Cretaceous)

Ktr — Trinity Group; includes Hensell Sand

— fault

⚒ quarry

Precambrian intrusive dikes (pЄi)

0 5 10 miles
0 5 10 kilometers

PALEOZOIC

P₂ — sedimentary rocks, undivided (Devonian to Pennsylvanian); includes Bend Group

Oel — Ellenburger Group; includes Gorman Limestone and Tanyard Formation (Ordovician)

Moore Hollow Group (Cambrian)

Єmw — Wilberns Formation; includes Morgan Creek Limestone, Welge Sandstone, Point Peak Siltstone, and San Saba Dolomite

Єmr — Riley Formation; includes Hickory and Lion Mountain Sandstones

PROTEROZOIC

pЄps — Packsaddle Schist

pЄcs — Coal Creek Serpentinite

pЄgn — gneiss, undivided; includes Valley Spring and Big Branch Gneisses

pЄy — younger granitic rocks; includes Oatman Creek Granite

pЄg — granitic rocks; includes Town Mountain Granite

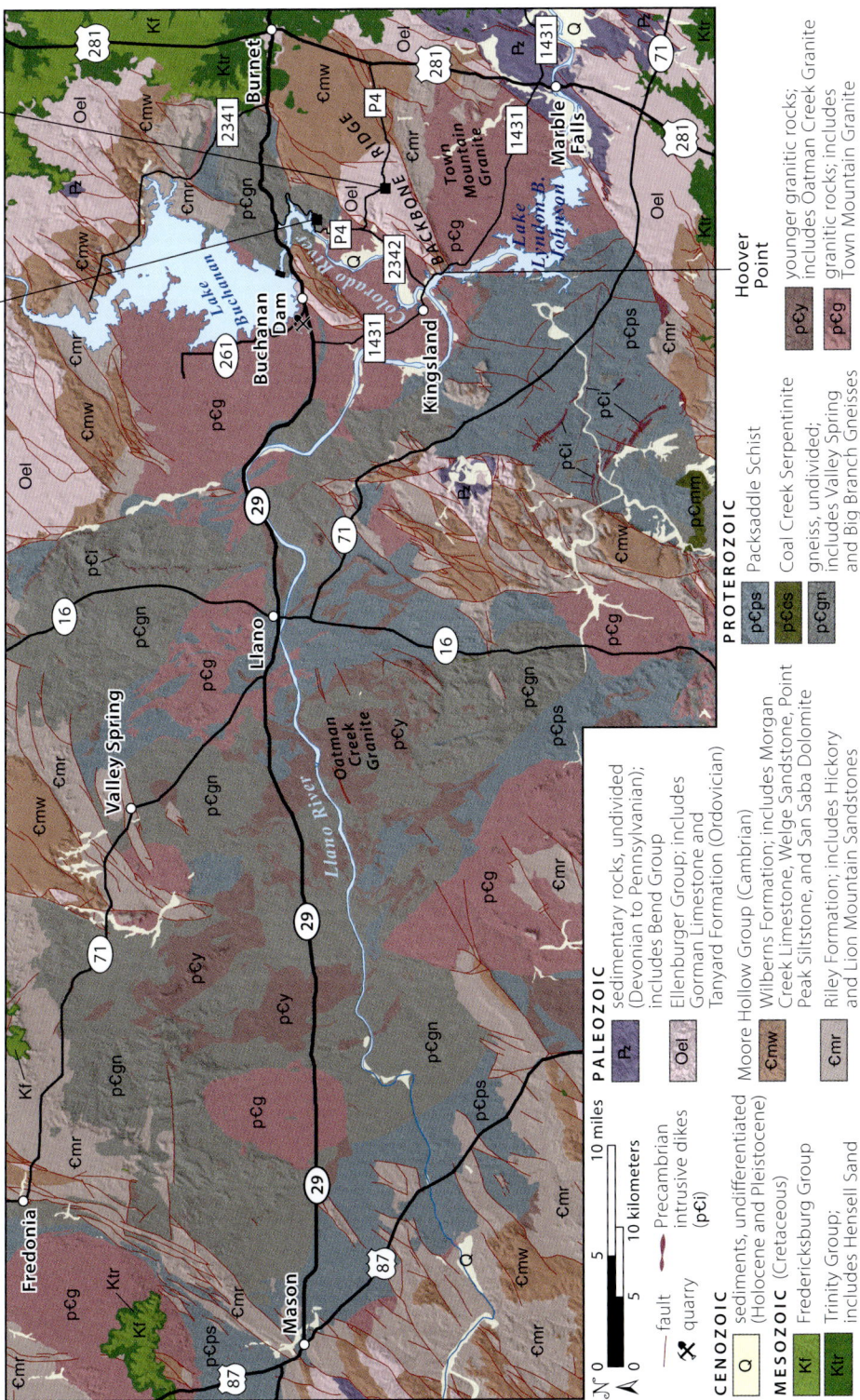

Geology along TX 29 between Mason and Burnet.

Blue topaz, the state gem of Texas, weathers out of the local pegmatites and collects in dry stream-beds. This stone, in the author's collection, measures 0.75 inch across and was found near Mason.

About 2 miles east of Mason on the crest of a hill, watch for a sizeable roadcut on the north side of the highway, where grayish bands of gneiss and schist (Packsaddle Schist) are laced by lighter-colored granitic dikes.

In Llano, well-exposed outcrops and boulders of Packsaddle Schist cut by pegmatite dikes can be seen in Badu Park in the Llano River bed. See the photos on page 187 in the road guide for TX 16, which intersects TX 29 at Llano. The road between Llano and the town of Buchanan Dam follows a large bend in the Llano River for part of the route, traversing Proterozoic crystalline rocks the entire distance. A few miles east of Llano, granite and gneiss outcrops near the highway appear as low platforms, protruding above the grasses in surrounding fields. Note especially how jointed, or fractured, the rocks are here. Look for exfoliation sheets where the granite is weathering in concentric layers, like an onion. Coarse-grained, crystalline dikes of pegmatite snake through these rocks as well.

As the highway nears the town of Buchanan Dam, watch for the exit north to FM 261 on the west side of Lake Buchanan. About one-quarter mile north of TX 29 on FM 261 is a roadside quarry where large, zoned feldspar crystals stand out of the pink

View to the north from TX 29 of irregular knobs of Valley Spring Gneiss east of Buchanan Dam. (30.7600, -98.3606)

granite. The feldspar zoning represents a change in chemistry of the cooling magma, such that the composition and color of the feldspar crystals changed as they grew in the hot, liquid melt.

East of Buchanan Dam, TX 29 crosses the Valley Spring Gneiss, a rock more than 1 billion years old. It formed originally from nearly remelted rhyolitic volcanic rocks, ash-flow tuffs, and tuffaceous sediments that were metamorphosed during an episode of Proterozoic mountain building.

HOOVER POINT AT KINGSLAND

About 14 miles east of Llano, turn south on FM 1431 and drive 7.7 miles to see a world-famous roadcut south of Kingsland. The Colorado River cuts through Backbone Ridge, and FM 1431 curves past road-cuts at a high point known as both Lookout Mountain and Hoover Point. A large pullout on the west (river) side provides parking and a spectacular view. What makes this particular cliff so special is the green sand composed of the mineral glauconite, an iron-potassium phyllosilicate similar to the mica family of minerals. While not a rare mineral, glauconite in the quantity it is found here is unique.

The famous roadcut at Hoover Point displays a dark-green, 6- to 8-foot-thick glauconite layer in the lower third of the cut. The diagonal feature (center left) is a fault with about 2 feet of offset. (30.6453, -98.4157)

The rock layers in this roadcut come from the latter part of the Cambrian Period when a shallow sea covered much of North America during the Sauk transgression, a global rise in sea level. The copious amounts of glauconite probably formed as a result of the alteration of fecal pellets within a nearby lagoon. Those pellets then got washed into a poorly oxygenated tidal inlet between barrier islands where they mixed with sand and then later lithified into sandstone. White pods of carbonate lenses, the result of later deposition around trilobite fossils, are also found within the sandstone, as well as broken shells of the brachiopod *Lingula*. The glauconitic sandstone is found at the base of the cliff in the Lion Mountain Sandstone Member of the Riley Formation. Overlying the Lion Mountain is the Welge Sandstone and Morgan Creek Limestone members of the Wilberns Formation, both deposited in a shallow marine environment.

In addition to the glauconitic sandstone, numerous geologic structures can also be seen in the roadcut. Well-defined normal faults offset bedding in several places. At the south end of the cut, you can see tilted limestone abruptly faulted down against sandstone. Small fault blocks have been dropped down relative to blocks on either side. These faults are related to tectonic activity during the Pennsylvanian Period when the Ouachita mountain building created tensional stress in the region.

East of the Park Road 4 intersection, TX 29 continues over the Proterozoic rocks of the Llano Uplift. A marked change occurs, however, before reaching US 281 at Burnet. Around the intersection of Ranch Road 2341, the rocks change from intrusive igneous rocks of the Llano Uplift to the Paleozoic sandstone and limestone that flank the uplift. These younger Paleozoic rocks include Cambrian Wilberns dolostone and the Ordovician Tanyard Limestone. While the contact between this change in rock types is buried and cannot be seen along the highway, the difference is significant. East of the intersection with Ranch Road 2341, the hill north of TX 29 is capped with Cretaceous limestones: units of the Fredericksburg Group overlying the Glen Rose Formation of the Trinity Group. These much younger rocks were deposited over everything in Cretaceous time but have since been eroded away from the Llano Uplift.

PARK ROAD 4 TO INKS LAKE AND LONGHORN CAVERN STATE PARKS

To see several great geologic locales, turn south onto Park Road 4 from TX 29 east of Buchanan Dam. The scenic drive through Inks Lake State Park weaves past many colorful, pink boulders and natural outcrops in the knobby, irregular terrain of the Valley Spring Gneiss. Among the oldest rocks in Texas, and the oldest on publicly accessible land, the Valley Spring Gneiss formed between 1.28 and 1.23 billion years ago when rhyolitic lava flows, ash-flow tuffs, and sedimentary rocks were metamorphosed into gneiss during the Grenville mountain building. The gneiss surrounds Inks Lake, a reservoir that was created in 1938 with a dam on the Colorado River. One of the best places to see the Valley Spring Gneiss up close is at the Devil's Waterhole within Inks Lake State Park.

The gneiss is thought to have been more than 12 miles deep in Earth's crust when the Grenville Mountains formed during the assembly of the supercontinent Rodinia. At these depths, rock is subjected to tremendous pressure and extremely high temperatures to the point that rock can partially melt, producing a metamorphic rock called migmatite, which is found at the park. In addition to migmatite, several other indications of intense pressures have been found by geologists, including at least five foliation trends where minerals aligned to combat the pressure, various folds in the rock, and a unique structure called boudinage where weak and strong layers of rock deform together to form a sausage-like structure. In addition to the deformation features, dikes associated with the younger Town Mountain Granite cut the gneiss. The Town Mountain Granite, also exposed in the area, was magma that formed in the roots of the Grenville Mountains as they towered over this region of North America some 1.08 billion years ago.

At the south end of Inks Lake State Park, Park Road 4 crosses onto granite rocks in Hoover Valley. Backbone Ridge forms the skyline hills to the east. Watch for a sign to Longhorn Cavern State Park and a left turn off Park Road 4. Immediately after making the turn, notice that the geology changes abruptly from pink granite to hard, bedded, gray limestone. You have just crossed a fault that marks the edge of a large, wedge-shaped block of early Paleozoic sedimentary rocks preserved in the surrounding igneous terrain. The road climbs the flank of Backbone Ridge, and limestones in the exposures along the ascending roadway are part of the Ellenburger Group of Ordovician rocks, approximately 485 million years old.

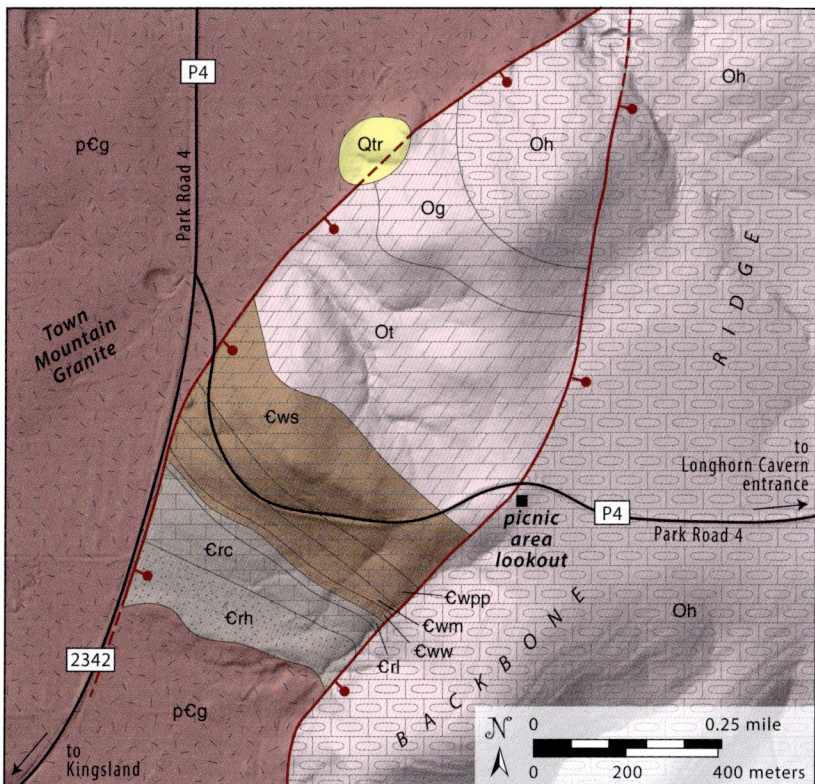

CENOZOIC

| Qtr | recent travertine deposits |

PALEOZOIC

| Oel | Ellenburger Group: Honeycut Formation (Oh) Gorman Limestone (Og) Tanyard Formation (Ot) (Ordovician) |

Moore Hollow Group (Cambrian)

| Єmw | Wilberns Formation: San Saba Dolomite (Єws) Point Peak Formation(Єwpp) Morgan Creek Limestone (Єwm) Welge Sandstone (Єww) |

| Єmr | Riley Formation: Lion Mountain Sandstone (Єrl) Cap Mountain Limestone (Єrc) Hickory Sandstone (Єrh) |

PROTEROZOIC

| pЄg | Town Mountain Granite |

━━•━ fault; dashed where concealed; ball and bar on downthrown side

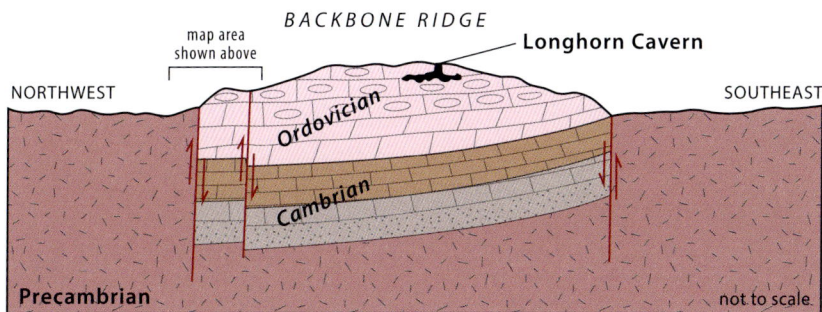

granite   limestone   cherty limestone   fault

sandstone   dolomite

Map and cross-section of the fault wedge along Backbone Ridge on the road to Longhorn Cavern.

Devil's Waterhole, a popular swimming area of Inks Lake, features Valley Spring Gneiss cut by numerous light-colored pegmatitic dikes, some several feet across. (30.7479, -98.3592)

The entrance to Longhorn Cavern is cut into the Gorman Limestone. (30.6843, -98.3500)

At the top of Backbone Ridge is the side road to Longhorn Cavern. Around the stone administration building are surface exposures of the dark-gray limestone of the Ellenburger Group. Longhorn Cavern is within the Gorman Limestone of the Ellen-burger Group. The limestone was deposited in a warm, shallow sea approximately 485 million years ago as part of the Great American Carbonate Bank, which is found from Texas to eastern Canada. Some of the limestone was altered into dolomite when magnesium replaced some of the calcium in the limestone.

The exact origin of the cavern is still debated by geologists. Early studies suggested surface waters dissolved the limestone along fractures, creating a passageway that allowed an underground river to widen out the cavern. More recent research suggests deep groundwater rose through the rocks and dissolved the cavern. After the cavern formed, many beautiful speleothems, such as stalactites, stalagmites, flowstones, dripstones, and large calcite crystals, were deposited on its surfaces.

Native Americans knew of the cave system long before Europeans arrived, and fossils of extinct horses and camels found inside suggest there was entrance at least back to the Pleistocene Epoch. During the Civil War, Confederate soldiers used the cave as a place to make gunpowder. During the 1920s and 30s, a church with services on Sunday and a nightclub with a dancefloor and live band used the cavern. It was eventually sold to the State of Texas, who then enlisted the Civilian Conservation Corps (CCC) to assist in clearing mud from the cavern, as well as constructing the original Park Road 4 and building the first visitor center. Today, the park provides a guided 90-minute walking tour of the system. Wild cave tours can also be booked in advance to see deeper parts.

WEST TEXAS

The formidable wall at the southern end of Sierra Diablo is capped with the younger Permian Hueco Limestone that overlies the older Cambrian Van Horn Sandstone and Proterozoic Hazel Formation. (31.2117, -104.8513)

A Goldilocks curl of sand whisks across the distance as you quietly watch the sun retreat below the black, saw-blade edge of an unnamed mountain range. The moment is interrupted by a coyote's quick yelp, and something small rustles behind you in the graying shadows of a skeletal yucca. Welcome to West Texas.

This land of sand, distance, and mountains fosters contemplation, and you can't help thinking about rocks because they're overhead, underfoot, everywhere. If contemplation leads anywhere, it is to the inescapable conclusion that to understand this raw, rough-hewn part of Texas, you must know something of its geology.

Behind the scenery is a vibrant tale of crunching mountains, ripping crustal plates, and violent volcanoes. In quieter intervals, reefs grew silently at the edge of blue oceans, sediments poured into deep, water-filled basins, and erosion chewed at uplifted landscapes. Public tracts have been set aside to preserve much of the story in its natural state—Guadalupe Mountains National Park, Davis Mountains State Park, Balmorhea State Park, Big Bend Ranch State Park, and Big Bend National Park. Very old rocks are found in this western corner of Texas, where 1-billion-year-old Protero-zoic metamorphic, igneous, and sedimentary rocks appear at the surface.

The second oldest suite of rocks in West Texas, Paleozoic in age, are grandly laid out north of Big Bend National Park in a tortured string of low mountains around Marathon. Known as the Marathon Uplift, it is part of the ancient Ouachita Moun-tains that towered over Texas but are mostly in the subsurface today. The roots of the old mountains are exposed in the Llano Uplift in central Texas, coming to air again only at Marathon and in El Solitario Dome in Big Bend Ranch State Park.

NORTHWEST

SOUTHEAST

M A R A T H O N   U P L I F T

| K | Cretaceous sedimentary rocks | | P | Permian sedimentary rocks | | Pz | folded and faulted Paleozoic rocks | | pϵ | Proterozoic rocks |

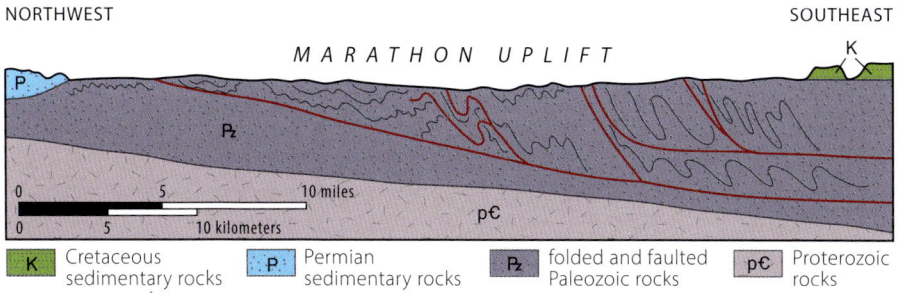

Cross section of the Marathon region, showing the intensely folded and faulted Paleozoic rocks. The Marathon area is a structural uplift because old rocks are elevated to the surface, but topographically, the Marathon area is low, surrounded by higher, younger rocks. Erosion has removed the younger rocks to expose the older section.

A tremendous pile of these Paleozoic sedimentary rocks, more than 14,000 feet thick, was laid down, mostly in deep marine water, from the Cambrian to the Permian Period, a span of nearly 300 million years. These deep-water sediments are intriguing because of the way they were deposited. Many of the sandstone and boulder beds in the Mississippian-to-Pennsylvanian Tesnus and Haymond Formations, for example, were deposited by turbidity currents. These instantaneous bursts of water and sediment, triggered perhaps by earthquakes, flow at breakneck speeds down the steep, submerged edges of continents, only to finally deposit their sediment load in the deep sea at the base of the slope.

Equally intriguing are the white, hard, pure silica rocks of the Caballos Novaculite of Silurian to Mississippian age. Novaculite, a term borrowed from the Ouachita Mountains of Arkansas, is used there to describe similar hard, dense, light-colored, silica-rich sedimentary rock. Because the Marathon region is also part of the Ouachita Mountain belt, it is appropriate to use the term here. This rock type extends from West Texas all the way to Arkansas, although it is mostly in the subsurface. The hard novaculite resists erosion and is easily recognizable as white ridges across much of the Marathon Uplift.

The Ouachita and Marathon Mountains grew upward during the collision of the North and South American crustal plates about 300 million years ago when the supercontinent Pangea was assembled. The trough that had been receiving sediment for millions of years was arched upward, and the pile of sediment was pushed laterally for at least 125 miles to the northwest. In the process, the sedimentary rocks were contorted into long, northeast-trending folds, while fault after fault cracked through the rocks. What goes up must come down in the Earth, and the forces of erosion began wearing down the new range. In the short time of only a few million years, the range was worn low enough that seas again lapped over the surface. Stacks of Permian limestones were laid over the area, burying the deformed Paleozoic rocks.

As the Ouachita and Marathon Mountains rose, the crust to the north (on the inland side of the collision) buckled downward to form deep depressions—the Delaware and Midland Basins—in which thick sections of sediment accumulated and around which magnificent reefs grew in profusion. In between the two basins was

ERA	GEOLOGIC AGE	NOTABLE ROCK UNITS IN WEST TEXAS		
CENOZOIC	PALEOGENE — OLIGOCENE / EOCENE / PALEOCENE	Big Bend Park Group: South Rim Formation (rift basin sediments, lava, flow breccia) / Chisos Formation (lava, ash, tuff, conglomerate, sandstone)	Buck Hill Group	Tascotal Formation (volcanic sediments) / Mitchell Mesa Formation (ignimbrite, rhyolite, welded tuff) / Duff Formation (rhyolitic tuff, breccia, conglomerate)
MESOZOIC	CRETACEOUS	Terlingua Group	Pen Formation	dark marl, mudstone (weathers to yellow) concretious
			Boquillas Formation	flaggy limestone, chalk, marl; shoreline
		Washita and Fredericksburg Groups	Santa Elena Limestone / Boracho Limestone / Del Carmen Limestone / Edwards Limestone	thick and thin limestone beds; carbonate, shelf
		Trinity Group	Glen Rose Formation	thick limestone beds, marl, dolomite, sandstone, mudstone, conglomerate
	PERMIAN	*unconformity*		
		Ochoa Group	Castile Formation	gypsum
		Guadalupe Group	Capitan Limestone	massive white limestone
		Delaware Mountain Group	Cherry Canyon Formation	mostly sandstone and limestone, some shale
			Brushy Canyon Formation	
		Bone Spring Group	Bone Spring Formation	dark limestone, interbedded
		Hueco Limestone		massive gray fossiliferous limestone with some shale and sandstone
PALEOZOIC	PENNSYLVANIAN	Strawn Group	Haymond Formation	mostly shale, some sandstone and conglomerate
		Bend Group	Dimple Limestone	limestone
	MISSISSIPPIAN	Morrow Group	Tesnus Formation	sandstone, shale, chert, and some conglomerate
		Barnett Shale		yellow-gray to black shale
	DEVONIAN / SILURIAN	Caballos Novaculite		chert and siliceous mudstone
	ORDOVICIAN	Maravillas Chert		limestone and chert
	CAMBRIAN	Van Horn Sandstone		massive coarse red sandstone and conglomerate
PROTEROZOIC		Hazel Formation		red sandstone
		Carrizo Mountain Group		metamorphic rocks, including schist, slate, and quartzite
		Castner Marble		limestone; locally altered and metamorphosed to marble

Stratigraphic column of important rock units in West Texas.

a limestone bank called the Central Basin platform. Permian rocks deposited in the basins are exposed in the Delaware, Apache, and Guadalupe Mountains, and collectively, the entire basin area is known as the Permian Basin. The water contained in the basins got progressively shallower as sediments from the adjacent, eroding mountains filled in the lowlands and the climate dried. The Midland Basin filled in by the middle of Permian time, but the Delaware Basin, with its circling Capitan Reef, continued as a mostly inland sea with a small connection to the greater ocean system.

In Triassic time, the supercontinent Pangea wrenched apart, and the Gulf of Mexico opened as North America separated from Eurasia, Africa, and South America. As North America moved westward with this new pulse of seafloor spreading, the North American plate overrode an oceanic plate along a subduction zone at the western edge of the continent. Collision after collision impacted the western coastline, and thrusting far inland began forming the Rocky Mountains. During the same time, shallow seas filled the foreland basin east of the rising mountains, and Cretaceous limestones were deposited at what is known as the Diablo platform.

The mountain building lasted almost 100 million years, and the final punch from 75 to 40 million years ago is known as the Laramide mountain-building event. Folds

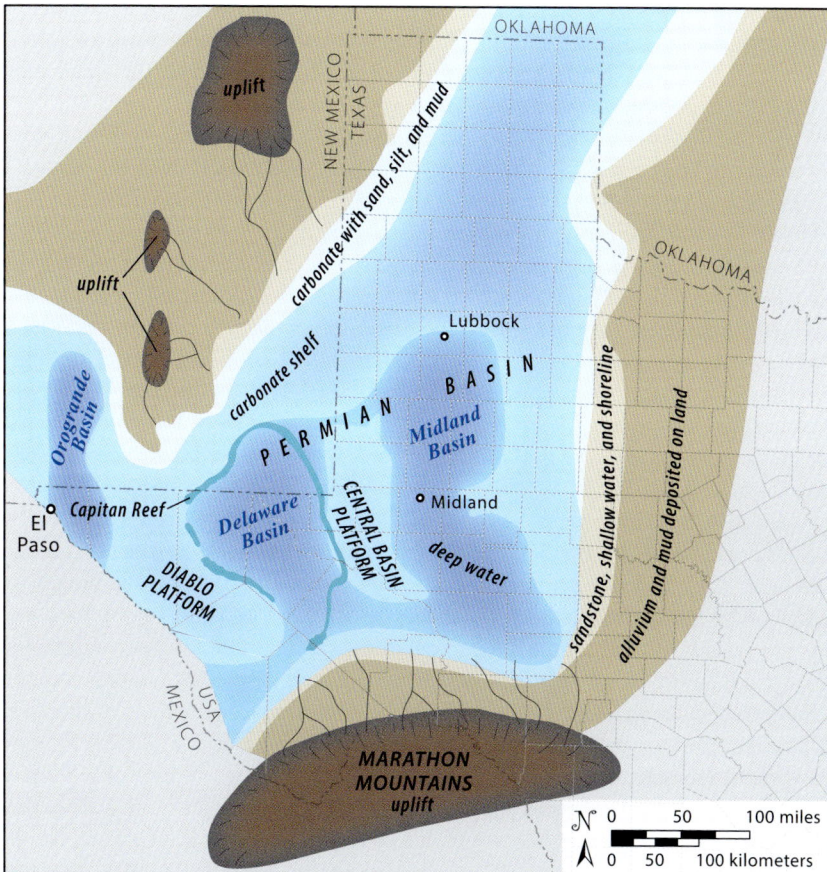

Geography of West Texas during the Permian Period. —Modified from Ewing, 2016

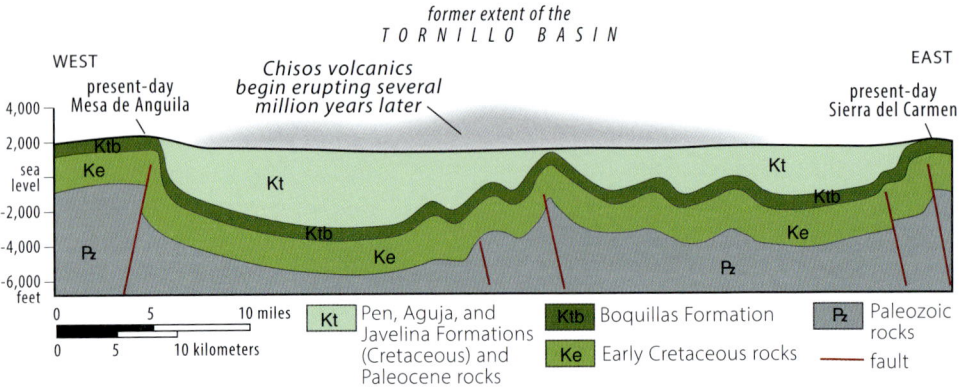

Cross section across the Tornillo Basin about 48 million years ago. The Cretaceous rocks of West Texas were folded during the Laramide mountain building, before Chisos volcanism and before Basin and Range normal faulting. Today, the remnants of the Tornillo Basin extend from the Rio Grande in Big Bend National Park northwest to near Van Horn and Kent. —Modified from Ewing, 2016

and faults of Laramide age are seen in the mountains of West Texas, such as folds of the Marathon Uplift and the Mesa de Anguila and Sierra del Carmen monoclines, which are giant draping warps in the sedimentary layers in the Big Bend area. In the Big Bend of the Rio Grande, the older Ouachita Mountains meet the trend of the younger Rocky Mountains to create a fascinating but complex landscape.

While the Rocky Mountains grew, erosion acted upon the landscape and moved the mountains to the sea, grain by grain. In places, erosion removed the blanket of Permian and Cretaceous rocks to expose the old, underlying Paleozoic rocks.

About 46 to 27 million years ago, in the Eocene and Oligocene Epochs, western North America stretched, or relaxed, following the intense compression that built the Rocky Mountains. Hot magma rose through fractures from the mantle below. Volcanoes erupted violently, spreading lava and ash over a wide area of West Texas. Some bubbles of lava never made it to the surface but stalled as blisters between layers in the sedimentary rock cover, only later to be exposed by deep erosion. The eruptive volcanic episodes are recorded in the Davis Mountains and Paisano volcano system west of Alpine. Round knobs and hills of volcanic intrusions add their peculiar forms to the scenery of Big Bend.

Beginning 25 million years ago in Texas, Basin and Range extension created a series of north-south normal fault zones that uplifted blocks of Earth's crust relative to the intervening, down-dropped blocks. The basins dropped and the ranges rose incrementally with each earthquake. The largest historic earthquakes in Texas have occurred within the Basin and Range, including a magnitude 5.8 near Valentine in 1931 and a magnitude 5.7 near Alpine in 1995. We have this recent faulting and uplift to thank for exposing much of the geology in West Texas. Thick beds of limestone, once buried, now form cliff bands across steep mountain sides.

Sediment eroded from the rising ranges filled the basins. It wasn't until Pliocene time, from 5 to 2 million years ago, that the basins in West Texas, as well as those farther north in New Mexico, became integrated with the young Rio Grande. Terraces of stream gravel along the river, stranded as the river cut downward, were deposited during the higher water flow of the Pleistocene ice ages.

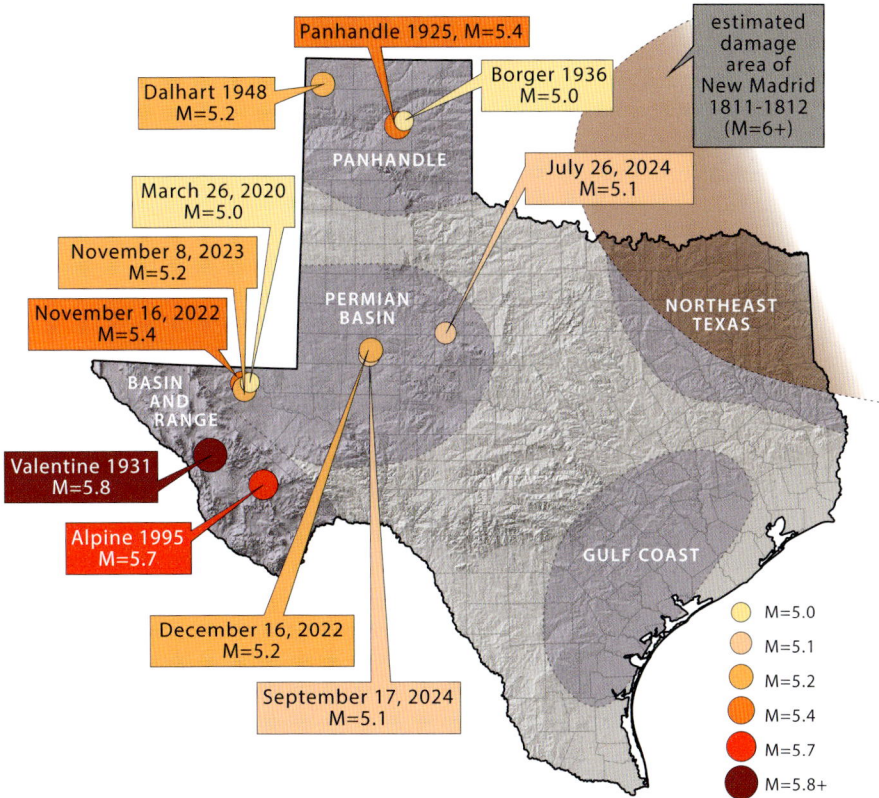

Historic earthquakes in Texas with a magnitude of 5.0 or greater.

In the map:

- Panhandle 1925, M=5.4
- Dalhart 1948 M=5.2
- Borger 1936 M=5.0
- July 26, 2024 M=5.1
- estimated damage area of New Madrid 1811-1812 (M=6+)
- March 26, 2020 M=5.0
- November 8, 2023 M=5.2
- November 16, 2022 M=5.4
- Valentine 1931 M=5.8
- Alpine 1995 M=5.7
- December 16, 2022 M=5.2
- September 17, 2024 M=5.1

Map labels: PANHANDLE, PERMIAN BASIN, NORTHEAST TEXAS, BASIN AND RANGE, GULF COAST

Legend:
- M=5.0
- M=5.1
- M=5.2
- M=5.4
- M=5.7
- M=5.8+

The impressive Guadalupe Mountains form the backdrop of the Salt Basin Dunes, both part of Guadalupe Mountains National Park. (31.9243, -104.9871)

Erosion continues to be the dominant shaper of the landscape. Look up at Santiago Peak, a volcanic intrusion south of Marathon, to realize how much rock has been weathered, eroded, and removed from this country. The intrusion, which fed an overlying volcano, formed entirely underground but now stands several thousand feet above the surrounding region. Alluvial fans and talus piles at the edge of every West Texas mountain slope also give some notion, even in today's dry climate, of the power of erosion. Quiet, relentless, daily erosion, operating over millions of years, is perhaps the most dynamic geologic process of all.

West Texas, much like the High Plains region, is also facing a serious challenge when it comes to the availability of water. In addition, an aging water infrastructure is quickly deteriorating, contributing to massive losses from leaks and complete pipe failures. Contamination of available water is also a major concern. Increased use by a growing population, combined with oil and gas industry use, have caused wells to either become unusable or go dry, or entire aquifers to become too polluted to use. Efforts are underway to promote water conservation and develop alternative water sources. Funding is being sought to address the state's aging water infrastructure. Legislative efforts are ongoing to better regulate and oversee oil and gas wastewater disposal and other water-related issues. In 2024, the State of Texas set aside more than one billion dollars to develop the Texas Water Fund, a program dedicated to water conservation projects across the state.

I-20
I-10 — Monahans Sandhills State Park
85 miles

I-20 between I-10 and Pecos crosses wide, flat, and lonesome desert, a surface of sand and gravel deposits laid down during the Pleistocene ice ages by streams flowing out of New Mexico highlands to the north. At exit 3, about 4 miles east of the I-20/I-10 junction, are outcrops of Boracho Limestone on either side of the road. These Cretaceous rocks of the Sixshooter Group are part of a ridge uplifted along northwest-trending faults.

Between Pecos and Monahans is vast, flat country punctuated by sand dunes. I-20 crosses the wide, dry Pecos River bottom near milepost 48 east of the town of Pecos. Occasionally, low, varicolored outcrops of Triassic-aged rocks can be seen on the east bank when they are not completely covered by the sand dunes. The sand in the Pecos River valley, exposed and dry most of the year, is the source for the sand dunes east of the river. Look for white caliche beds in the area around the Triassic outcrops east of Pecos.

Five miles west of Monahans, the pumps of the North Ward Estes and South Sealy oil fields spread far out from the highway both north and south of the road. The primary producing formations are the Permian-aged Yates Sand and Queen Sand of the Artesia Group. Both large oil fields are part of the much more expansive Permian Basin oil field of West Texas.

The area east of Monahans is a gigantic sand dune field. Tan sand is seen for miles on either side of the highway, piled in dunes and mostly trapped by vegetation. On windy days, sand blows across your windshield in pelting streams of abrasive

CENOZOIC

Q — recent sediments; includes alluvium, playa deposits (small dots), and landslides (Holocene and Pleistocene)

Qaf — alluvial fan deposits (Holocene)

Qcc — caliche (Quaternary)

Qw — windblown sand, silt, and sand sheet deposits (Holocene)

Qd — windblown dunes (Holocene)

QTs — older terrace, high gravel, and alluvial fan deposits (Miocene to Pleistocene)

MESOZOIC

K — sedimentary rocks; includes Boracho Limestone (Cretaceous)

TR — Dockum Group (Triassic)

VOLCANIC and IGNEOUS ROCKS (Neogene and Paleogene)

Tv — volcanic rocks, undivided; includes Davis Mountains volcanic complex

Ti — igneous rocks

PALEOZOIC

P — sedimentary rocks, undivided; includes Ochoa Group in northwest (Permian)

⬥ oil field

— normal fault

· · · · · county boundaries

Red Bluff Reservoir

NEW MEXICO

LOVING

South Sealey oil field

North Ward Estes oil field

Monahans Sandhills State Park

Kermit

WINKLER

WARD

Monahans

Pecos

Grandfalls

REEVES

Pecos River

PECOS

Balmorhea
Balmorhea State Park

BARRILLA MOUNTAINS

STAR MOUNTAIN

DAVIS MOUNTAINS

JEFF DAVIS

Ft. Davis

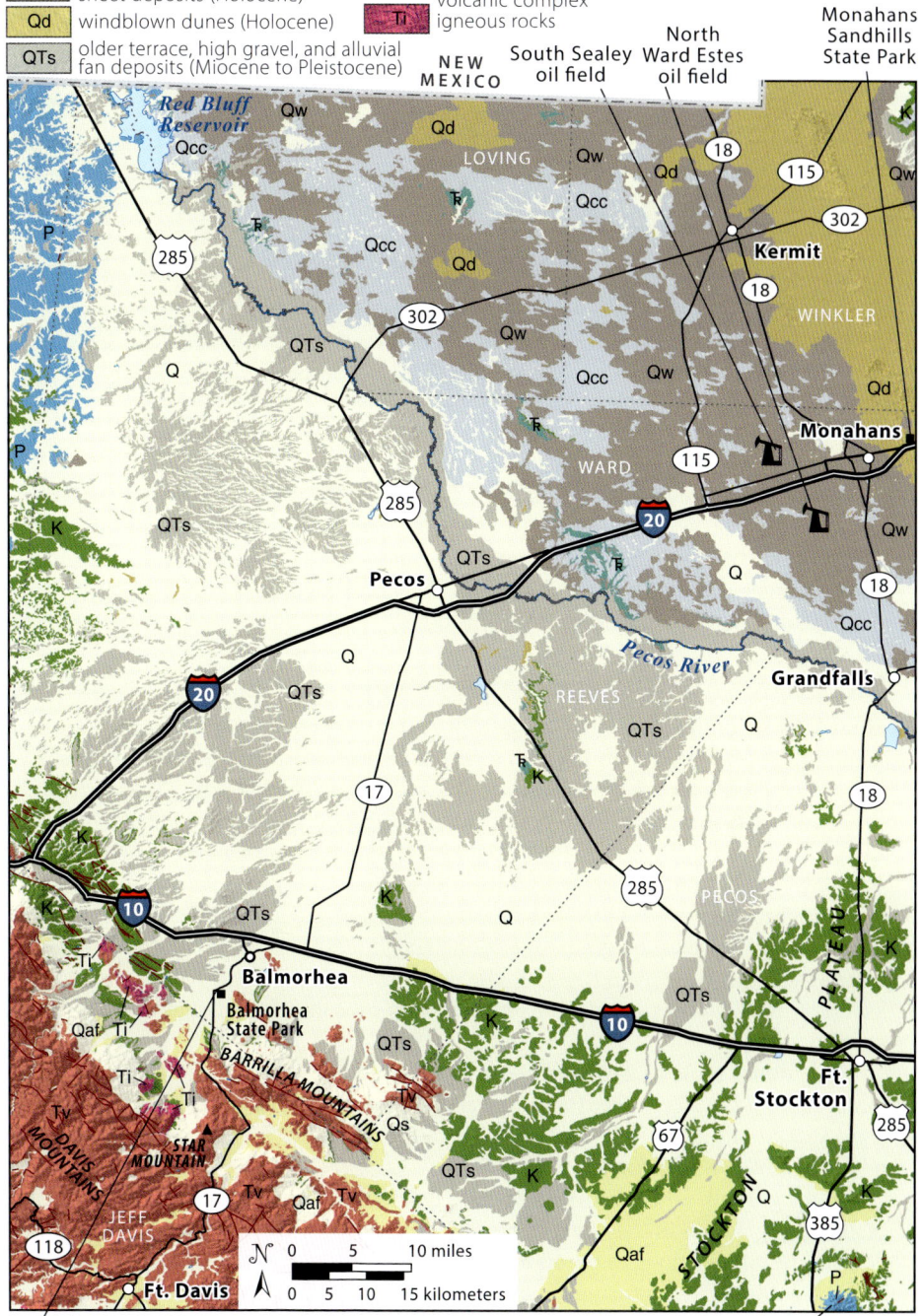

San Solomon Spring

Ft. Stockton

STOCKTON PLATEAU

Sierra Madera astrobleme

N

0   5   10 miles

0   5   10   15 kilometers

Geology along I-20 between I-10 and Monahans.

Small, vertical faults that look like fractures cut the Boracho Limestone in roadcuts on both sides of the road near exit 3. (The evenly spaced lines are drill holes.) (31.1244, -104.0312)

particles. About 5 miles east of Monahans at exit 86 is Monahans Sandhills State Park, where the State of Texas has preserved a beautiful segment of dunes for public use and appreciation. The entry and headquarters are right next to the interstate. Take a break from driving and spend a few minutes to a few hours enjoying the dunes, including climbing up one side and sliding down the other.

MONAHANS SANDHILLS STATE PARK

Monahans Sandhills State Park, along I-20 at exit 86, preserves a delightful tract of sand dunes with classic features such as ripples, cross-beds, and interdune deposits. The Monahans dunes are representative of a wide swath of sand dunes that virtually cover the Pecos River area from Ft. Stockton northward to the New Mexico border and eastward through Kermit and Monahans. A ready supply of sand has been available since the Pleistocene ice ages in the eroded, dry flats of the Pecos River valley. Prevailing westerly winds blow the sand steadily eastward out of the Pecos valley, only to drop the sand grain by grain. The Monahans dunes lie in a topographic depression, whose east (downwind) side is the escarpment of the High Plains, the Llano Estacado. As westerly ground winds rise to climb over the plateau, they lose velocity and drop sand.

Large dunes grow as sand marches up the dune's windward side, only to fall as sandy rain on the steep side. Inclined wedges of sand are thus added to this downwind side, resulting in large-scale cross-bedding, a distinctive feature preserved in many ancient dune deposits.

On the ground, sand is moved by wind in two ways. The wind agitates the surface of dry sand grains, and they start to bounce around. A grain bounced up into the air flow will be pushed along in the air for a few inches until it settles back down, striking another grain, which in turn is bounced up. This process is called saltation, and the length of the bounce depends on wind speed and grain size. The low, straight-crested ripples formed by this process can be seen everywhere in the park. The agitation of the surface also causes grains to continuously edge forward in a second motion aptly called creep. While you're looking down at the ripples, search for the tracks and trails of insects and small animals among the rippled dune surfaces.

Dunes form and move as sand travels along the surface and up the windward side of a dune to accumulate at the crest. Once enough sand accumulates, it falls down the slipface. As more sand moves and gathers, the dune migrates in the direction of the wind.

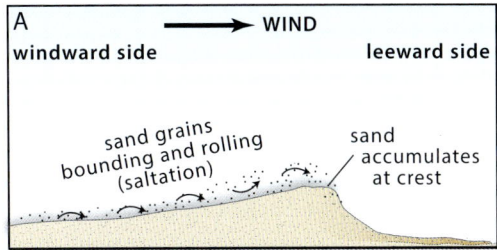

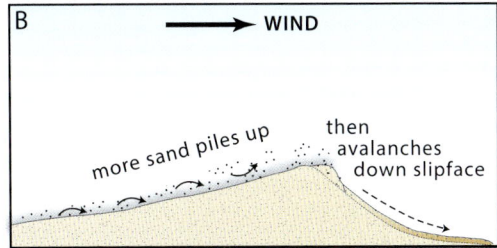

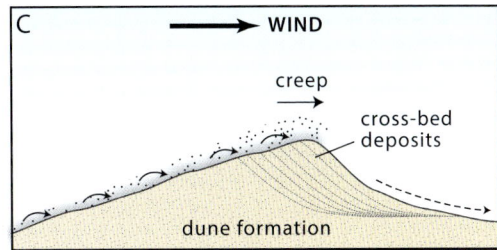

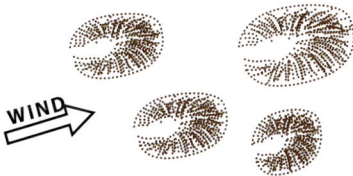

Dome dunes are small, circular, and do not have a slip face. These are fast moving dunes.

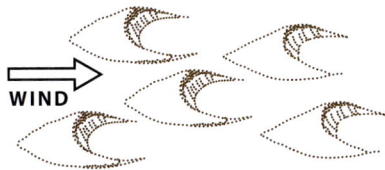

Barchan dunes are thickest around their middle. They form when wind blows sparse sand in one direction across a flat surface.

Transverse dunes, with long gentle windward slopes, form when sand is abundant, wind blows from one direction, and there are few plants.

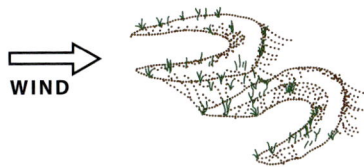

Parabolic dunes develop when vegetation anchors a dune's curved arms while the wind blows away sand at its center. These are slow-moving dunes.

Several types of dunes that you can see when walking through the dune field at Monahans Sandhills.

Dunes come in different shapes depending on wind conditions and sand supply. Two main types are seen at Monahans Park. Transverse dunes are long ridges oriented perpendicular to a single prevailing wind direction where sand supply is high. Barchan dunes are C-shaped sand piles formed in a single prevailing wind where sand supply is low. Note that the horns of barchan dunes point downwind. Barchans appear commonly on the outer edges of larger dune fields.

Green patches of vegetation get established between the dunes because these depressions receive a bit more shade than other nearby areas and are often cooler and tend to retain more moisture—just enough to allow plants to grow. Depending on how wet these depressions get, sometimes a small pond forms where interdune deposits collect. The ponds, however, don't last long in the West Texas heat.

Wind produces ripples on dunes at Monahans Sandhills State Park. Note the irregular ripple forming at the rock in the lower left. (31.6401, -102.8188)

Vegetation has become established in one of several interdune areas within Monahans Sandhills State Park. (31.6240, -102.8114)

I-10
Ft. Stockton—El Paso—New Mexico
259 miles
See map on page 209 for Ft. Stockton to I-20 junction

This long segment of I-10 traverses a splendid variety of geology, from windswept flats near Ft. Stockton past volcanic spires and fault block mountains to windblown sand dunes that parallel the verdant Rio Grande before reaching El Paso and heading north toward the New Mexico border. Fill up the tank, grab some water, and get ready for a geologic odyssey.

Around Ft. Stockton, I-10 sweeps across flat, gravelly open land of the Stockton Plateau, where alluvial fans and streams have filled up low areas with rubbly debris eroded and transported from surrounding mesas and mountains. Approximately 20 miles south of Ft. Stockton along US 385, some hills break the flat terrain at the southern edge of the Stockton Plateau. These hills are the rebound structure of the Sierra Madera astrobleme, or impact crater, one of three such meteor craters located in Texas. A rebound structure forms as the impact crater floor rebounds after the initial impact, creating a central peak or ring within the crater. The peak rises 793 feet above the desert floor with the crater itself about 8 miles in diameter. The age of the meteor impact is uncertain, but it probably formed less than 100 million years ago. In 1972, NASA astronauts Gene Cernan and Jack Schmitt used Sierra Madera as training for geological studies during the Apollo 17 mission. The crater itself is located on the private La Escalera Ranch, so access is limited, but US 385 crosses onto the crater's edge as it passes next to the hills.

West of Ft. Stockton, mesas with limestone ledges appear off in the distance to the south. The road cuts through Early Cretaceous limestone beds that are dipping at a very gentle angle to the west near milepost 256—evidence of faulting and movement of the rock section here. A few miles farther west at milepost 234, the same rocks are dipping in the opposite direction, toward the east. You have just driven across a syncline, a downwarp caused by the Laramide mountain building to the west.

The Barrilla Mountains form a distinctive profile on the southwest skyline between mileposts 236 and 225. This volcanic range is composed mainly of extrusive lava flows resting on gently tilted Cretaceous sedimentary rocks. The Barrilla Mountains heaved up about 35 million years ago as part of the Davis Mountains volcanic complex.

More mesas of Early Cretaceous limestone of the Washita Group are seen around milepost 229. West of the mesas, around exit 222, the high peaks of the Davis Mountains form a dark profile to the west. Flat alluvial fan deposits spread out northward from the highway.

Near exit 209 to TX 17 south toward Balmorhea and Ft. Davis, low rolling hills of black volcanic rocks are visible to the south. These hills are unorganized and rubbly, typical of landforms composed of volcanic rocks.

San Solomon Spring in Balmorhea State Park, only 7 miles south of I-10 on TX 17, has drawn campers for thousands of years, as indicated by Native American artifacts in the area. As early as 1851, canals were built from the springs to irrigate crops, and in 1935, the Civilian Conservation Corps began to construct the park's swimming pool, concession buildings, and park residence. The Balmorhea area lies on gravel beds and alluvial fans built northward from the Davis Mountains. Early and Late Cretaceous limestone, sandstone, and shale deposits form the bedrock around the park.

0 10 20 miles
0 10 20 kilometers
NEW MEXICO
MEXICO
USA
NM
Transmountain Drive
El Paso
Rio Grande
FRANKLIN MOUNTAINS
HUECO BOLSON
HUECO MOUNTAINS
RIO GRANDE RIFT
EL PASO
HUDSPETH
CULBERSON
JEFF DAVIS
GUADALUPE MOUNTAINS NATIONAL PARK
GUADALUPE MOUNTAINS
SALT BASIN
DELAWARE MOUNTAINS
APACHE MOUNTAINS
SIERRA DIABLO
BAYLOR MOUNTAINS
WYLIE MOUNTAINS
DAVIS MOUNTAINS
Van Horn
Kent
Plateau
Allamoore
Hazel Mine
Sierra Blanca
Little Blanca Mountain
Round Top
Sierra Blanca
Bonanza and Alice Ray Mines
QUITMAN MOUNTAINS
BEACH MOUNTAINS
CARRIZO MOUNTAINS
THREEMILE PEAK; Red Rock Ranch
CENOZOIC
Q recent sediments; includes alluvium, landslides, alluvial fan, and lake deposits (Holocene and Pleistocene)
Qw windblown sand (Holocene)
QTs older gravel and bolson deposits (Miocene–Pleistocene)
MESOZOIC
CRETACEOUS
Kw Washita Group (includes Boracho Limestone), Taylor, Austin, and Terlingua Groups
Kf Fredericksburg Group and other Cretaceous rocks
Ktr Trinity Group
Jm Malone Formation (Jurassic)
PALEOZOIC
PERMIAN
Po Ochoa Group
Pwa Wichita Albany Group
Pdm Delaware Mountain, Guadalupe, and Whitehorse Groups, undivided
Pbs Bone Spring Group
Ph Hueco Limestone; includes Mingus Shale of the Strawn Group (Pennsylvanian)
Pz sedimentary rocks, undivided; includes Magdalena Formation (Pennsylvanian) and Cutoff Shale (Permian)
MS Silurian–Mississippian rocks, undivided; includes Fusselman Dolomite (Silurian) and Helms Shales (Mississippian)
O€ Ordovician rocks, undivided; includes Montoya Dolomite, El Paso Group and Bliss Sandstone (late Cambrian–Ordovician)
PROTEROZOIC
p€ms metasedimentary rocks; includes Carrizo Mountain Group, Castner Marble, and Thunderbird Group and Van Horn Sandstone (Cambrian)
fault
county boundary
national park boundary
mine/quarry
VOLCANIC and IGNEOUS ROCKS
Tv volcanic rocks, undivided; (Eocene–Oligocene)
Tdm Davis Mountains volcanic complex (Eocene–Oligocene)
Ti igneous rocks, undivided (Tertiary)
p€g granitic rocks; includes Red Bluff Granite (Proterozoic)
p€m meta-igneous rocks; includes Hackett Peak Formation (Proterozoic)

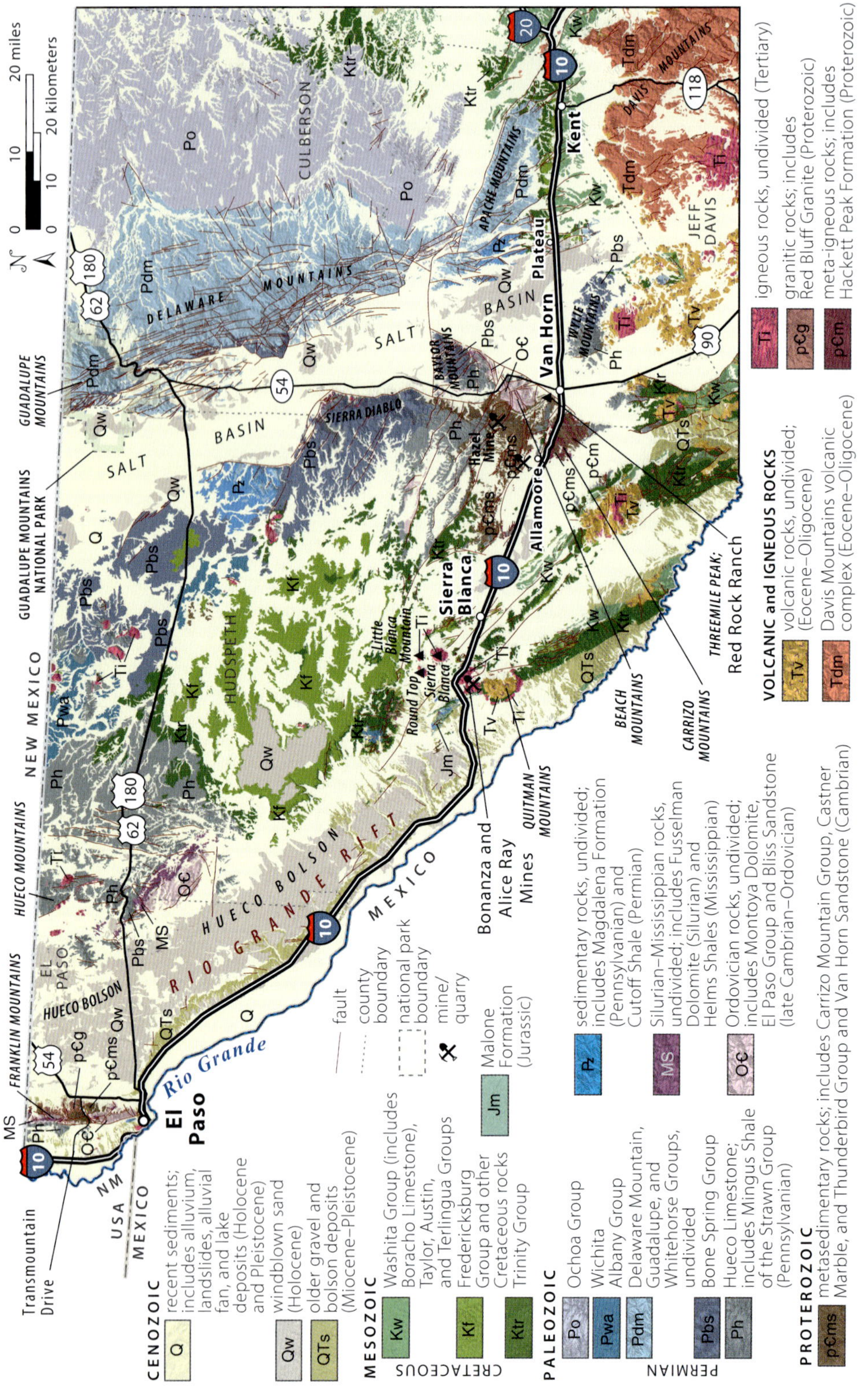

Geology along I-10 between the I-20 junction and El Paso. For the section of I-10 west of Ft. Stockton, see the map on page 209.

San Solomon Spring at Balmorhea State Park forms wetlands that support wildlife in this dry landscape. (30.9449, -103.7844)

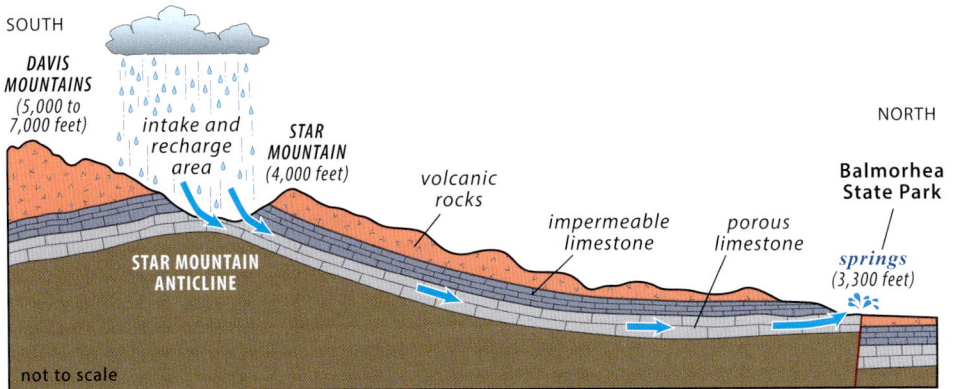

Cross section of intake and recharge area and springs at Balmorhea State Park.

Rainfall in the Davis Mountains flows downslope, making its way into the subsurface through cracks, holes, and small caverns etched in the limestone beds. It emerges at the surface by spewing up along fractures and faults in the limestone, nearly 700 feet lower than where it entered the aquifer at Star Mountain.

The panorama of the volcanic Davis Mountains parallels the highway for many miles, indicating the vastness of the volcanic field. Around 35 million years ago, a string of volcanoes erupted, forming the pile. Where I-10 crosses dry stream washes, note the mix of limestone fragments and dark volcanic cobbles, just what you would expect to be eroded from the volcanic Davis Mountains and surrounding skirts of Cretaceous limestone beds.

West of the intersection of I-20 (exits 187 and 186), I-10 enters the Basin and Range, where it snakes between north-south-trending mountain ranges. A long, flat plateau on the horizon north of I-10 west of Kent (exit 176) is the southern flank of the Apache Mountains, part of a string of mountains that extends northward to the New Mexico border and includes the Apache, Delaware, and Guadalupe Mountains. These ranges expose huge carbonate reefs and deeper water deposits of sandstone, limestone, and shale, deposited during Permian time in the Delaware Basin. The best place to see these rocks is in Guadalupe Mountains National Park, featured in pages 229–232.

Low mesas near the highway between Kent and Plateau (exit 159) are eroded remnants of flat-lying beds of Early Cretaceous limestone that once covered this area. The roadcut at Plateau is Early Cretaceous–aged conglomerate and sandstone that fill the valley to the west known as Salt Basin. Between Plateau and Van Horn, I-10 crosses this topographically low desert area, complete with blowing sand dunes and salt deposits left from shallow lakes that existed in wetter times during the Pleistocene ice ages. The Delaware-Apache Mountains are the uplifted, eastern border of the Salt Basin, whereas the Baylor and Beach Mountains and Sierra Diablo form the steep west flank. To the south are the Wylie Mountains, another uplifted block of Permian carbonate rocks, which rise like a sentinel from the floor of the Salt Basin.

Red Rock Ranch, a private area northwest of Van Horn, boasts spectacular outcrops of the Van Horn Sandstone, deposited in Cambrian time. Van Horn is the portal to a beautiful pass where I-10 crosses between the Carrizo Mountains south of the highway and Threemile Peak to the north. Cambrian sedimentary rocks, laid down at the dawn of Paleozoic time, form the lower slopes of the distinctive Threemile Peak, whereas hard Permian limestones form the resistant cap.

Proterozoic-aged rocks are rarely seen at the surface in Texas, but you can see some deformed 1.1-billion-year-old varieties next to the picnic tables at the westbound rest

The distinctive silhouette of Threemile Peak northwest of Van Horn is composed of the Permian-aged Hueco Limestone. (31.0374, -104.8597)

area a few miles west of Van Horn. Notice the green-black color and dense packing of small, parallel grains in these very hard metamorphic rocks. They once were mudstones but were metamorphosed into phyllite when they were heated and pressured at great depths within the Earth's crust. Veins of white silica shoot through the metamorphic rock in many places. A highway roadcut at the Hudspeth/Culberson county line (or the Mountain/Central time zone change) just west of the rest area shows similar dark phyllites at the west end. They lie in sharp contact with pink metamorphosed rhyolite, all part of the Proterozoic Carrizo Mountain Group. Take exit 133 to see a similar roadcut along the frontage road. A few miles to the north is the Hazel Mine, which targeted a rich silver vein in the Hazel Formation, a thick series of Proterozoic-aged red shale and sandstone.

Dark phyllites (left) rest against pink metamorphosed rhyolite (right) in a roadcut just west of the rest area at the Hudspeth/Culberson county line. Both units are part of the Proterozoic-aged Carrizo Mountain Group. (31.0434, -104.9108)

The Carrizo Mountain Group exposed in a roadcut near exit 133. Here, the rocks are composed of amphibolite and granulite, metamorphic rocks formed at high temperatures and pressures deep within the Earth's crust. Amphibolite, a dark rock, contains the minerals amphibole and plagioclase but very little quartz. Granulite is a lighter-colored rock containing the minerals quartz, feldspar, and pyroxene. (31.0551, -104.9535)

A talc quarry north of Allamoore and exit 129. (31.1043, -105.0027)

Between the Carrizo Mountain pass and the town of Sierra Blanca, the highway rides across flatter desert country, built on the surface of alluvial fans that extend outward from the surrounding mountains. At exit 129 is the former mining town of Allamoore. In the mountains to the north, look for open-pit mines where talc, a soft metamorphic mineral rich in magnesium, was extracted between 1952 and the early 1980s, when the last processing plants shutdown. The district was the second-largest producer of talc in the United States at the time. Talc forms during the metamorphic alteration of dolomite, a carbonate rock.

West of Allamoore, ranges on the distant skyline to the south are the Quitman Mountains, uplifted during the Laramide mountain building about 60 million years ago, the same event that built the Rocky Mountains. Within these mountains, the Bonanza and Alice Ray Mines were major producers of lead and zinc and smaller occurrences of molybdenum and tungsten. The mineralization occurred where the Laramide intrusions interacted with Cretaceous-aged limestones, sandstones, and shale units.

West of Sierra Blanca, rubbly, dark-brown hills and rounded mountain topography indicate volcanic rocks. Three distinctive, conical volcanic peaks of Oligocene age north of the highway are Round Top (the primary focus of rare earth elements and critical mineral exploration in Texas), Little Blanca Mountain, and Sierra Blanca, best seen from exit 99 (Lasca Road). The rest areas near exit 99 are where the interstate curves to the north around a large quartz-monzonite stock, a granitic rock that intruded during the Laramide mountain building. The highway then descends southward into a broad valley, where colorful, pink-white-tan badlands are etched into the formerly smooth-sloping top of the alluvial fans. Here, the drainages erode deeply as they approach their confluence with the Rio Grande, now only a few miles away to the south.

In the 60 miles between exit 85 and El Paso, I-10 closely follows the course of the Rio Grande. Serrated Laramide ranges on the Mexico side of the river are outlined against the sky, while the green fields on the river floodplain nestle at their base. This

The quartz monzonite stock in the Quitman Mountains weathers into rounded blocks so typical of granitic rock. Viewed from the westbound rest area at exit 99. (31.2184, -105.4849)

area sits along the Rio Grande rift, a tensional feature that originates in central Colorado and extends south through New Mexico and into Mexico, roughly following the Rio Grande. These tensional forces pulled the crust apart, forming fault-block mountains. Normal faults along the east side of the rift rotated, uplifted, and exposed mountain blocks. One example of these blocks is the Franklin Mountains in the distance to the north, which were tilted about 40 degrees to the west toward the center of the rift.

The road runs on a high terrace above the river, and much of the surrounding landscape is windblown sand and dune deposits. Abundant sand in the dry flat of the river bottom is readily available for the wind to distribute. In some places, the dunes are stabilized by the sparse vegetation, but look for rounded sand dunes and wind ripple marks on free sand faces. They are especially obvious in the low sun angles of early morning or late afternoon.

Near El Paso, low hills far in the distance to the northeast are the Hueco Mountains. The Franklin Mountains provide a rugged backdrop to El Paso. The long stretch of flat desert and dry sand between the Quitman Mountains and El Paso is the southern end of the Hueco Bolson, a low, down-dropped fault block between the uplifted Franklin and Hueco Mountains. Sedimentary fill in the Hueco Bolson is nearly 9,000 feet thick, in part because as a structural depression, or basin, it has no drainage outlet.

Between El Paso and the New Mexico border, I-10 passes along the west side of the Franklin Mountains. A fault at the western base of the mountains has uplifted a thick section of Paleozoic sedimentary rocks, which overlie a section of Proterozoic rocks, exposed on the steep east flank. If you look carefully from El Paso, you can see the westward dip of the rocks in the Franklin Mountains because the block is tilted away from the Rio Grande. As the highway heads north toward the New Mexico border, it transects alluvial fans that extend toward the Rio Grande from their heads in the Franklin Mountains. Look for sand and gravel of these alluvial fans exposed along the roadway.

Transmountain Drive
through the Franklin Mountains

The Woodrow Bean Transmountain Drive (TX 375) is a beautiful, dramatic drive through the Franklin Mountains that exposes some of Texas's rarely seen Proterozoic rocks. The old rocks are on the east side of the mountain, so we'll head west on Transmountain Drive from US 54. See the geology map for US 62/US 180 on page 223 for a closer look at this area.

Approximately 1.5 miles west of the US 54 intersection is a set of outcrops on the right (north), opposite the rest area. This colorful, but puzzling, array of rocks known as Confusion Hill reveals geologic events that likely occurred just before and during the Grenville mountain-building event as the supercontinent Rodinia began

A panoramic view of Confusion Hill from the rest area on Transmountain Drive. (31.8955, -106.4567)

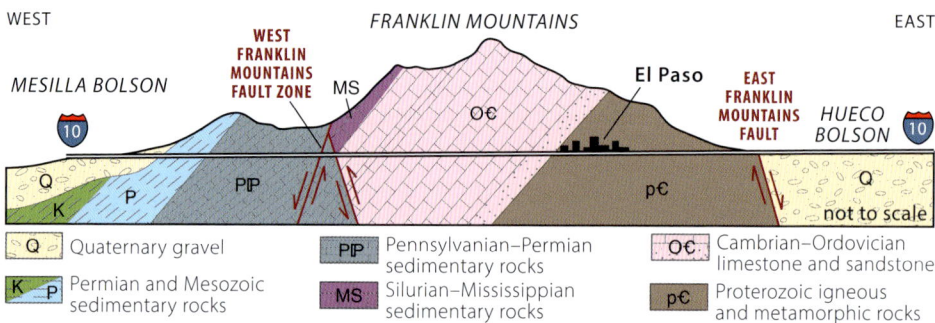

Cross section of Franklin Mountains near El Paso.

to assemble. The Castner Marble, the oldest rock in this outcrop (and in the Franklin Mountains), formed as layers of limestone and mudstone in a shallow sea or tidal flat about 1.25 billion years ago. The next youngest unit, the Mundy Breccia, consists of large blocks of basalt. Both of these units were metamorphosed when several pulses of magma intruded into the older sedimentary and volcanic rocks during the Grenville mountain-building event. The magma crystallized as the 1.12-billion-year-old Red Bluff Granite Suite.

About 1 mile farther west is a very large roadcut of Castner Marble on the right (north) side of the highway. Close examination of this rock reveals fossils of stromatolites and algal mats that indicate the Castner originally formed in a shallow sea or tidal flat. In addition, black igneous sills intruded the marble, and small red garnets formed

Within the Castner Marble are small, red garnets that developed as the carbonates were metamorphosed. (31.8941, -106.4713)

Large roadcuts of the Castner Marble about 1 mile west of Confusion Hill. (31.8941, -106.4713)

as the limestone was metamorphosed when the Red Bluff granites intruded these sedimentary rocks.

At Smugglers Pass, 2 miles west of the Castner roadcut, large roadcuts dominate both sides of the highway. Here, the youngest of the Precambrian-aged rocks in the Franklin Mountains, the Thunderbird Group, includes ignimbrites deposited by pyroclastic flows, ash-fall deposits, and conglomerates composed of volcanic debris. Geochemical analysis of this package of rocks suggests that the Thunderbird Group erupted from volcanoes fed by the Red Bluff granite magma chamber.

Research by geologists in Antarctica found that granites located in Coats Land have the same age, appearance, and, most importantly, geochemical signatures as the Red Bluff Granite. The western edge of Precambrian North America was likely connected to Antarctica during the reign of Rodinia. The granite bodies were separated during the breakup of the supercontinent.

US 62/US 180
El Paso—New Mexico
130 miles

As the combined US 62/US 180 heads eastward from El Paso, you get a good view to the north of the Franklin Mountains, which rise more than 7,000 feet above sea level. This north-south range is a tilted fault block uplifted in Miocene time as part of the Basin and Range extension. Faults lie along both sides and uplifted old Proterozoic granite and metamorphic rocks to their highest elevation in the state. Because of this resulting tilt, more than 10,000 feet of sedimentary, igneous, and metamorphic rocks are exposed in the Franklin Mountains.

The road crosses desert country for about 20 miles before reaching the Hueco Mountains. The depression between the Franklin and Hueco Mountains is called the Hueco Bolson. *Bolson*, meaning "pocket" or "purse" in Spanish, is a basin with no drainage outlet. Usually, bolsons are filled with great thicknesses of sediment derived from the erosion of the adjacent mountains. The Hueco Bolson is no exception—9,000 feet of clay, silt, sand, and gypsum are between your tires and solid rock

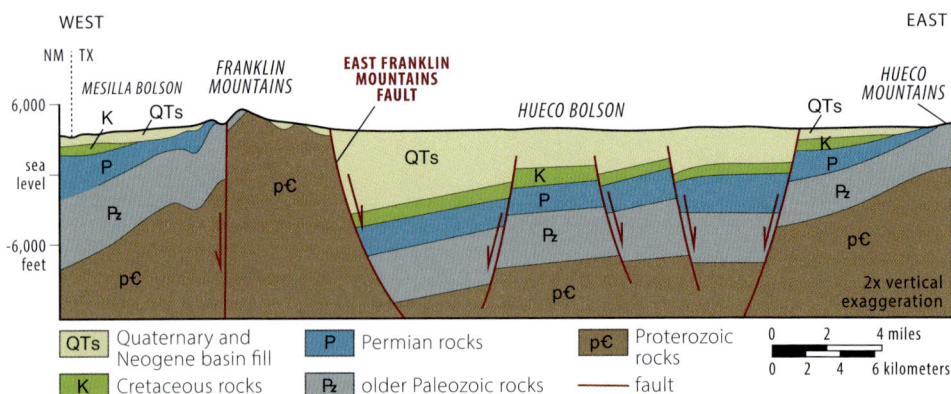

Schematic cross section of the Hueco Bolson and the Basin and Range fault blocks crossed by US 62.
—Modified from Ewing, 2016

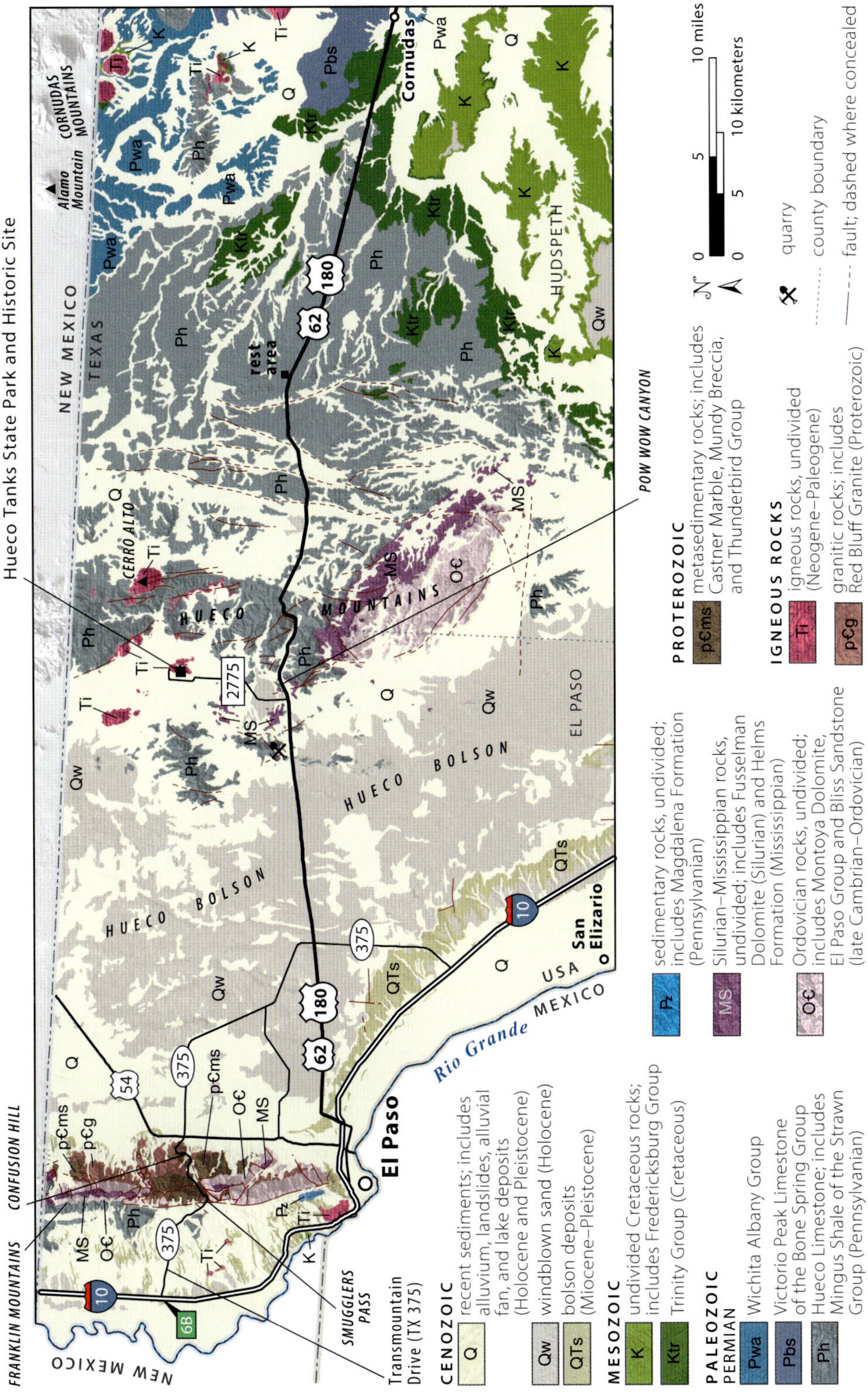

Hueco Tanks State Park and Historic Site
NEW MEXICO
TEXAS
Alamo Mountain
CORNUDAS MOUNTAINS
Cornudas
rest area
CERRO ALTO
HUECO
MOUNTAINS
POW WOW CANYON
HUECO BOLSON
HUDSPETH
EL PASO
San Elizario
USA
MEXICO
Rio Grande
El Paso
CONFUSION HILL
FRANKLIN MOUNTAINS
SMUGGLERS PASS
Transmountain Drive (TX 375)
NEW MEXICO
62
180
375
54
10
2775
6B
CENOZOIC
Q recent sediments; includes alluvium, landslides, alluvial fan, and lake deposits (Holocene and Pleistocene)
Qw windblown sand (Holocene)
QTs bolson deposits (Miocene–Pleistocene)
MESOZOIC
K undivided Cretaceous rocks; includes Fredericksburg Group
Ktr Trinity Group (Cretaceous)
PALEOZOIC
PERMIAN
Pwa Wichita Albany Group
Pbs Victorio Peak Limestone of the Bone Spring Group
Ph Hueco Limestone; includes Mingus Shale of the Strawn Group (Pennsylvanian)
Pz sedimentary rocks, undivided; includes Magdalena Formation (Pennsylvanian)
MS Silurian–Mississippian rocks, undivided; includes Fusselman Dolomite (Silurian) and Helms Formation (Mississippian)
O€ Ordovician rocks, undivided; includes Montoya Dolomite, El Paso Group and Bliss Sandstone (late Cambrian–Ordovician)
PROTEROZOIC
p€ms metasedimentary rocks; includes Castner Marble, Mundy Breccia, and Thunderbird Group
IGNEOUS ROCKS
Ti igneous rocks, undivided (Neogene–Paleogene)
p€g granitic rocks; includes Red Bluff Granite (Proterozoic)
quarry
county boundary
fault; dashed where concealed
0 5 10 miles
0 5 10 kilometers

Geology along US 62/US180 between El Paso and Cornudas. For the eastern end of the road near Guadalupe Mountains National Park, see map on page 226.

below! The Hueco Bolson formed during the great extensional Basin and Range period that occurred throughout the west, continuously filling with sediment as its bedrock base lowered because of faulting.

The surface of the Hueco Bolson is fairly flat, covered by alluvial fan deposits and gravelly stream deposits along with sand hills and dunes. Some sand can be seen trapped in piles behind the sparse sage and creosote bushes, but most of the sand moves west to east across the bolson where it is piled on the western flank of the Hueco Mountains. The prevailing westerly winds hug the ground and move sand eastward, but where the Hueco Mountains are encountered, the winds rise, lose their velocity, and drop sand.

East of Loop 375 on the outskirts of El Paso, you begin to see panoramic views to the northeast of the Hueco Mountains, where Cerro Alto is the highest peak at 6,787 feet above sea level. The bolson surface is about 4,000 feet in elevation. As the road approaches the foothills of the Hueco Mountains, the hills you see north of the highway are composed of the Permian-aged Hueco Limestone.

Good roadcuts of steeply dipping, westward-tilted Paleozoic limestones and shales on the western flank of the Hueco Mountains are the Pennsylvanian-aged limestones of the Magdalena Formation and the underlying Mississippian-aged shales of the Helms Formation. The Magdalena is particularly interesting because it shows a time when global sea levels were rising at the beginning of the Absaroka transgression. Because of this, fossils such as petrified wood, algae mounds, chert nodules, and crinoid fossils are all found within the Magdalena. A quarry near State Park Road (FM 2775) is in limestone of the Magdalena Formation, mined for its pure calcium carbonate.

The roadcuts get bigger and the surrounding hills taller as US 62 passes through the Hueco Mountains via Pow Wow Canyon. Limestone of the Magdalena Formation is predominant in these cuts, but as you continue eastward through the canyon into younger rocks, Permian Hueco Limestone exposures prevail. Notice that the Hueco

Approximately 8 miles east of the Loop 375 interchange, sage, yucca, and creosote bush have trapped sand along the roadside. (31.8249, -106.1316)

Limestone beds tilt down toward the east. Within the Early Permian section is an angular unconformity, where relatively flat-lying beds of the Hueco Group lie directly over tilted layers of Magdalena Limestone. This relationship shows that the Magdalena was tilted before the Hueco was deposited. Near Hueco Ranch Road on the east flank of the Hueco Mountains, Hueco Limestone composes the undulating topography, created by the irregular solution of the underlying limestone by the relentless dissolving action of rainwater.

HUECO TANKS STATE PARK AND HISTORIC SITE

One of the larger huecos, a low depression in the syenite rocks that collects water, is seen on one of the many hiking trails within Hueco Tanks State Park and Historic Site. (31.9249, -106.0463)

On the eastern side of the barren Hueco Bolson is Hueco Tanks State Park and Historic Site, where a relatively small, rugged outcrop of reddish rocks protrudes above the desert floor. The rocks are syenite, a coarse-grained, intrusive igneous rock similar in composition to granite, but with significantly less quartz, and characterized by the presence of the mineral alkali feldspar (predominantly orthoclase). The syenite magma intruded into Permian-aged carbonate rocks, then solidified approximately 34 million years ago. Erosion over millions of years removed the overlying rocks, exposing the syenites at the surface. Once exposed to the ravages of weathering, the rocks fractured and weathered into the pinnacles and boulders we see today. Within the cracks and shaded overhangs are numerous hollows, or huecos, that hold water. This water is a precious commodity in a region that only gets around 8 inches of rain a year. During infrequent rainstorms, water gets funneled through the fractures and into small depressions on the rocky surfaces, providing a brief life for tiny fairy shrimp before the water evaporates. In addition, plants not native to this part of Texas, such as the mountain mulberry, can be found among the rocks. Water has drawn people to this site for at least 10,000 years. The earliest permanent settlement were members of the Jornada Mogollon peoples who constructed a village of subterranean dwellings and farmed the surrounding soils around 1075 CE. Native American tribes such as the Kiowa, Mescalero Apache, and Tigua visited the area and painted more than four thousand pictographs.

Geology along US 62/US 180 between Cornudas and the New Mexico border.

From the rest area about 40 miles east of El Paso, look north to see Alamo Mountain, a flat-topped landmark just over the border in New Mexico. Alamo Mountain gets its flat-topped shape primarily due to horizontal layers of relatively resistant rock that are capped by an even harder layer, which resists erosion and creates a flat plateau-like surface. To its east are the rounded hills of Sierra Tinaja Pinta, a set of 35-million-year-old igneous intrusions—small outliers of the same huge volcanic event that created the Davis Mountains near Big Bend National Park. On the far northern horizon, round hills on the Texas–New Mexico border are the Cornudas Mountains, also cored by Cenozoic intrusive bodies.

Telephoto view of the flat-topped silhouette of Alamo Mountain to the north in New Mexico, as seen from the rest area about 40 miles east of El Paso. (31.8162, -105.6822)

Beyond the rest area, watch for small mesas of the white Cretaceous-aged Campagrande Limestone on either side of the road. These marine limestones once extended across the region, but uplift followed by erosion has severely reduced their footprint, leaving only these remnant mesas as hints of their former extent.

The road continues through low outcrops of Permian-aged limestone, shale, and siltstone of the Bone Spring Formation before dropping into the white salt flats of Salt Basin. The flats are the dry salt remnants of lakes that once shimmered in the bolson, fed by runoff and springs from the Guadalupe and Delaware Mountains during wetter times in the Pleistocene ice ages of the last 2.6 million years. Lest you think these are worthless salt pans of no consequence, disputes between Mexican and American interests over the rights to mine, transport, and market this salt led to the El Paso Salt War of 1877. The conflict culminated in the battle of San Elizaro (then the county seat of El Paso County).

The Salt Basin Dunes, sometimes referred to as the White Sands of Texas, are a group of gypsum dunes that occupy a basin adjacent to the Guadalupe Mountains. During the Pleistocene ice ages, the climate was much cooler and wetter, and a lake (called Lake King) covered the basin. At its maximum extent, this lake covered an area of approximately 350 square miles and had an average depth of about 35 feet. Toward the end of the last ice age, the climate warmed and became much drier, causing the lake to dry up. Today, many white playas dot the ancient lake floor. Gypsum and halite currently form in these playas, and prevailing winds pick up loose material and deposit it into a dune field covering approximately 2,000 acres at the eastern edge of

An immense salt pan, the floor of a lake that occupied the valley during the Pleistocene, lies west of the Guadalupe Mountains. (31.7539, -104.9916)

A beautiful roadcut a half mile south of the Pine Springs Visitor Center exposes a major fault separating gray and tan limestone members of the Permian-aged Delaware Mountain Group on the right side of the photo. (31.8854, -104.8226)

the basin. The dunes vary in height up to 60 feet and are composed almost entirely of white gypsum, much like their New Mexico cousins to the north.

East of the salt flats is the magnificent west cliff-face of the Guadalupe Mountains to the north and the lower Delaware Mountains stretching to the south. The Sierra Diablo can also be seen stretching to the south, forming the west flank of Salt Basin. Salt Basin is a down-dropped fault block between the Sierra Diablo and the Guadalupe and Delaware Mountains.

A few miles east of the TX 54 junction, the roadway climbs Guadalupe Pass into Guadalupe Mountains National Park, where the geology of a giant Permian reef is magnificent, unique, and well exposed in roadcuts, cliff faces, and outcrops.

US 62/US 180 continues into New Mexico and to Carlsbad Caverns, which are located at the east end of the Guadalupe escarpment. The reef rocks were lifted out of their deep burial place, and in the process, sulfuric acid–rich groundwater etched its way into fractures in the limestone to create the caverns.

Guadalupe Mountains National Park

Inaugurated as a national park in 1972 largely through the conservation efforts and land donations of Wallace Pratt, a renowned petroleum geologist, Guadalupe Mountains National Park preserves the largest fossil reef in the world. The Capitan Reef is so well known in the geologic world that the Permian epoch in which it existed, from 273 to 259 million years ago, is known as the Guadalupian.

The horseshoe-shaped reef, nearly 400 miles long, formed around the edge of the Delaware Basin, which was an arm of the ocean that covered West Texas and extended into southeast New Mexico. Portions of this reef are also exposed in the Apache Mountains near Van Horn and in the Glass Mountains near Marathon.

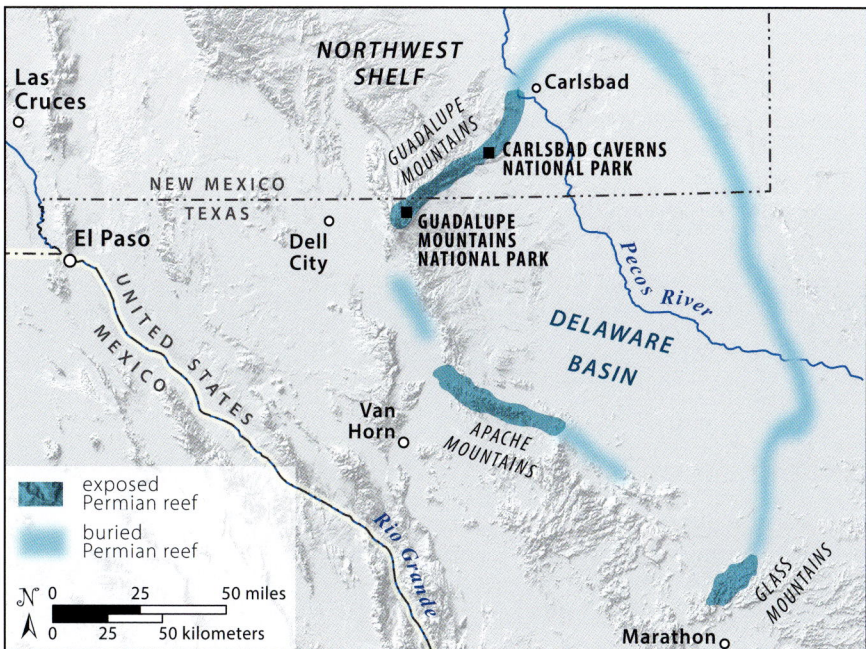

During the Permian, the largest-known reef complex in the world formed around the Delaware Basin.

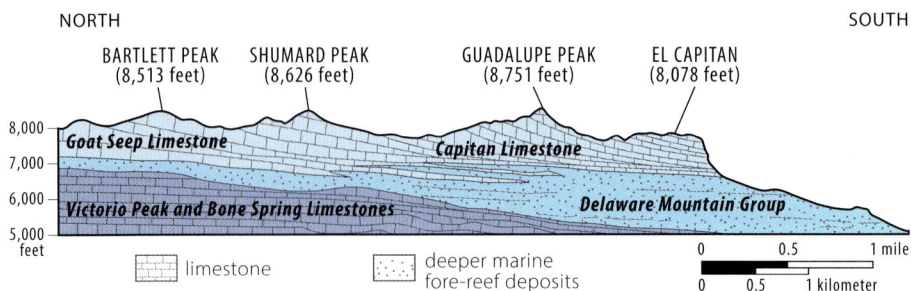

NORTH   SOUTH

BARTLETT PEAK   SHUMARD PEAK   GUADALUPE PEAK   EL CAPITAN
(8,513 feet)   (8,626 feet)   (8,751 feet)   (8,078 feet)

Goat Seep Limestone   Capitan Limestone

Victorio Peak and Bone Spring Limestones   Delaware Mountain Group

limestone   deeper marine
                   fore-reef deposits

Geologic cross section of Guadalupe Mountains. Note how the inclined limestone beds,
which formed the reef, tongue into the deeper marine deposits.

Modern reefs are found in shallow, warm, clear ocean water, built by the growth of lime-secreting marine organisms, mainly corals. Similar conditions in the sea around the Delaware Basin probably existed, but the principal reef-builders were calcareous sponges, lime-secreting algae, some solitary corals, and an enigmatic encrusting creature named *Tubiphytes*. Bryozoans (moss-animals) and clam-like brachiopods were the other contributors to the Capitan Reef. The limy skeletons of these organisms piled up through millions of years, creating a formidable carbonate wall in the Permian Sea. The reef reached a height of 1,300 feet but also migrated seaward for miles. As waves pounded the reef's seaward face, loosened pieces of reef rock fell in front of the reef, creating a wide talus debris apron. The organisms grew out over the talus, extending the realm of the reef seaward.

In this region, sea level changes were greatly influenced by several factors, the most important of these were glaciation and basin subsidence. At times when sea level fell, the reef was left high and dry, killing the organisms. When the sea level rose

Reef rocks of the Capitan Limestone form the top of El Capitan. Below are sandstone, claystone, and limestone beds of the Delaware Mountain Group. (31.8539, -104.8445)

again, the animals repopulated the reef to start the cycle all over again. These repeated cycles of sea level change, as well as a record of life and death on the reef, can be seen in the different colored portions of the high walls on the Guadalupe Escarpment.

On the backside of the reef, a shallow lagoon trapped sediment from streams, and limestone, mudstone, and sandstone built up the floor of the lagoon. The lagoon waters commonly evaporated, leaving a soupy brine too salty for most marine organisms, but it was ideal for the formation of vast expanses of evaporative minerals such as anhydrite and gypsum, along with algae-laden limestone.

Eventually, the connection between the Delaware Basin and the open sea was permanently cut off, and the basin water evaporated, leaving behind thick deposits of salt and potash. Over millions of years, the low areas filled up with sand and silt brought in by streams, and the old reef, lagoon, and basin were buried under thousands of feet of sediment.

Regional uplift that began with the Laramide building of the Rocky Mountains about 70 million years ago was enhanced with regional extension that began about 25 million years ago. As uplift occurred, erosion relentlessly removed the sediment that was deposited over the deeply buried Permian-aged reef. The hard limestone resisted further erosion, while the softer rocks surrounding it continued to be removed. Today, the reef stands tall above the deep-water rocks of the Delaware Basin in a near mirror-image of the Permian reef-talus-basin scene of 250 million years ago. It looks almost as if someone had simply pulled the plug on the sea!

The upper cliff faces of El Capitan are the limestone reef rock, while the lower shaly and thin-bedded limestone slopes are the deep-water sediments deposited in the basin in front of the reef. Large blocks of limestone debris, representing talus deposits, can be seen in lower slopes if you look carefully. Look for the steeply inclined beds in the upper limestone cliffs that represent various stages of the steep reef front.

A fine view of El Capitan can be had from the scenic turnout along US 62/US 180. Outcrops along the road cut through flat-bedded limestone, sandstone, and shale beds of the Delaware Mountain Group that were deposited in deep water in front of the reef.

Flat-lying beds of limestone (whitish layers), sandstone, and shale compose most of the Delaware Mountain Group rocks seen along the roadway. (31.8539, -104.8445)

The Pine Springs and McKittrick Canyon Visitor Centers have fantastic geological displays explaining the area. For a closer look at the geology of the Delaware Mountains, the Permian Reef Trail leaves from the McKittrick Canyon Visitor Center. This strenuous, 8-mile-round-trip trail takes you through the heart of the Permian reef complex.

From the park to the New Mexico border, the roadway follows the high scarp created by the Permian reef rocks, seen north of the road. Roadcuts, beginning with the Permian-aged Cherry Canyon Formation, followed by the Permian-aged Bell Canyon Formation, are seen for several miles east of the park. Both formations consist of cyclic marine deposits of sandstone and siltstone with several limestone interfingers. These roadcuts are cut from flat-bedded rocks originally deposited in the deep Permian seawaters at the front of the reef.

US 67
Marfa — Presidio
60 miles

South of Marfa, US 67 heads generally southwest through a broad valley with occasional low hills. In the distance to the south and west, remnants of Texas's volcanic past, including the Chinati Mountains, loom large on the horizon. About 1.5 miles south of Marfa at a pull-off, a roadcut of Oligocene-aged Tascotal Formation of the Buck Hill Group provides the first evidence of volcanism in this area. The Tascotal is composed of a mixture of tuffaceous sediment, sandstone, and conglomerate—material eroded from the volcanoes and deposited in a widespread blanket across the area and as an apron around the base of the Chinati Mountains. In many places, the

Tascotal Formation sediments south of Marfa show the jumbled nature of these volcanic deposits. (30.2900, -104.0252)

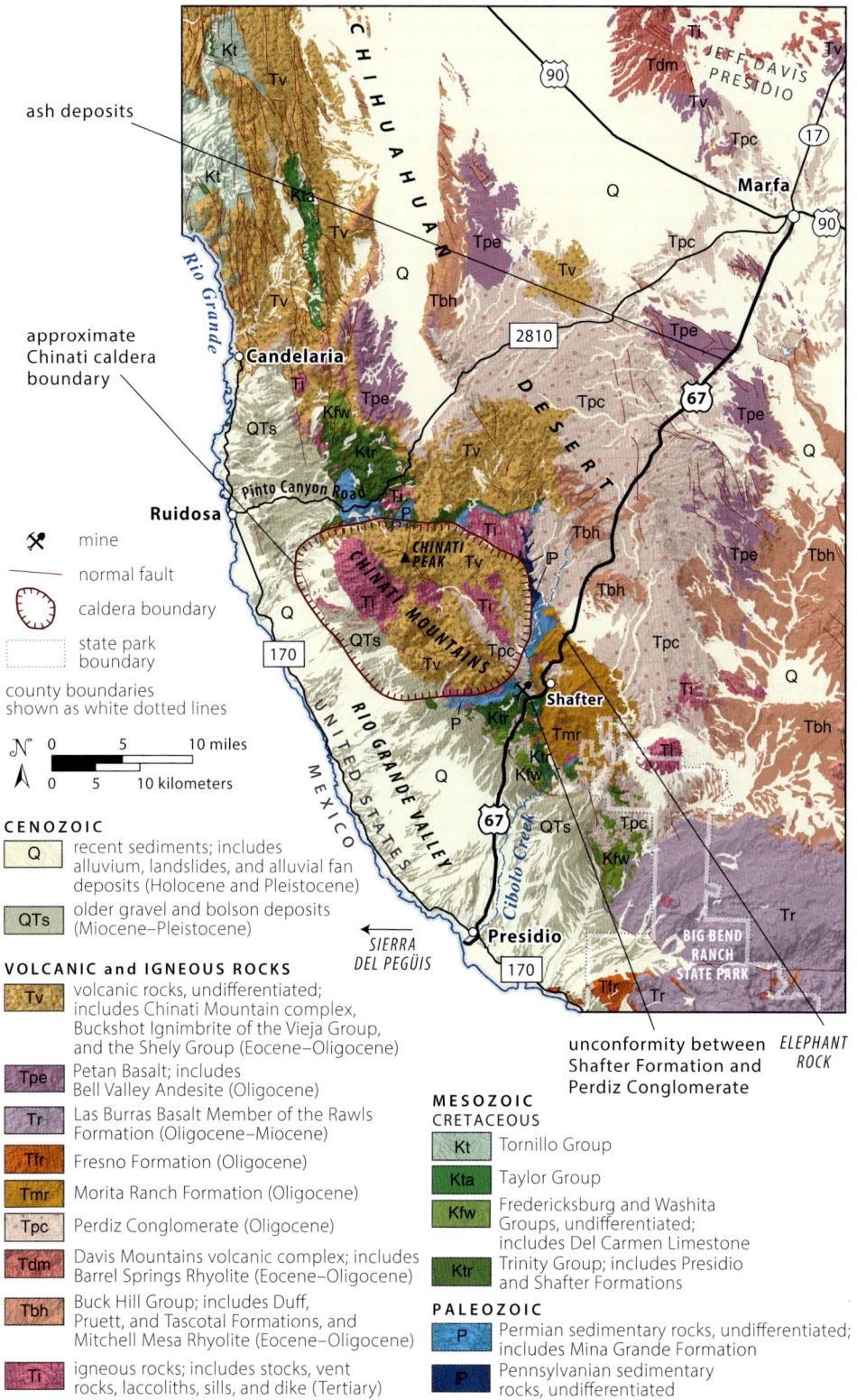

ash deposits

approximate
Chinati caldera
boundary

mine
normal fault
caldera boundary
state park
boundary
county boundaries
shown as white dotted lines

N
0   5   10 miles
0   5   10 kilometers

CENOZOIC

Q recent sediments; includes
alluvium, landslides, and alluvial fan
deposits (Holocene and Pleistocene)

QTs older gravel and bolson deposits
(Miocene–Pleistocene)

VOLCANIC and IGNEOUS ROCKS

Tv volcanic rocks, undifferentiated;
includes Chinati Mountain complex,
Buckshot Ignimbrite of the Vieja Group,
and the Shely Group (Eocene–Oligocene)

Tpe Petan Basalt; includes
Bell Valley Andesite (Oligocene)

Tr Las Burras Basalt Member of the Rawls
Formation (Oligocene–Miocene)

Tfr Fresno Formation (Oligocene)

Tmr Morita Ranch Formation (Oligocene)

Tpc Perdiz Conglomerate (Oligocene)

Tdm Davis Mountains volcanic complex; includes
Barrel Springs Rhyolite (Eocene–Oligocene)

Tbh Buck Hill Group; includes Duff,
Pruett, and Tascotal Formations, and
Mitchell Mesa Rhyolite (Eocene–Oligocene)

Ti igneous rocks; includes stocks, vent
rocks, laccoliths, sills, and dike (Tertiary)

unconformity between
Shafter Formation and
Perdiz Conglomerate

ELEPHANT
ROCK

MESOZOIC
CRETACEOUS

Kt Tornillo Group

Kta Taylor Group

Kfw Fredericksburg and Washita
Groups, undifferentiated;
includes Del Carmen Limestone

Ktr Trinity Group; includes Presidio
and Shafter Formations

PALEOZOIC

P Permian sedimentary rocks, undifferentiated;
includes Mina Grande Formation

P Pennsylvanian sedimentary
rocks, undifferentiated

Geology along US 67 between Marfa and Presidio

Tascotal overlies the more explosively formed Mitchell Mesa Rhyolite deposits, also of the Buck Hill Group. The Mitchell Mesa is not exposed along US 67 but can be observed from the scenic Pinto Canyon Road that connects Marfa with Ruidosa on the Mexican border.

Continuing south, the road passes over various alluvium and terrace deposits, as well as Tascotal Formation sediments and its contemporaries, the Perdiz Conglomerate of the South Rim Formation and tuffaceous sediments of the Fresno Formation. At just over 12 miles south of Marfa, a new rock type is encountered in large roadcuts on both sides of the highway. The mostly barren landscape gives way to stark, white ash deposits of the Oligocene-aged Bell Valley Andesite in the Petan Formation. While the Petan is primarily a vesicular basalt, at this location, ash covers the basalt. At the southern end of this roadcut, Petan basalts are exposed, providing evidence that vast lava flows covered this area. In addition, look for orangish sand dikes that intruded the andesite ash sometime after it was deposited. Sand dikes (also called clastic dikes) are seams of sedimentary material that fill open fractures cutting across preexisting rock layers. Sand dikes form when fluid pressure in saturated sand layers beneath an overlying, impermeable layer of mud or other material becomes high enough to force the sand upward into fractures or cracks.

South of the ash roadcut, US 67 continues to pass over more of the Perdiz Conglomerate and Tascotal Formations. Along the way, look to the west for striking views of the Chinati Mountains and Chinati Peak. The Chinati are primarily composed of igneous and metamorphic rocks that are the remains of a number of explosive, caldera-building eruptions during the Eocene. Many of the rocks seen along this route are the result of those eruptions, including Elephant Rock, a distinctive

Large white ash deposits of the Bell Valley Andesite. The close-up shows the Bell Valley's rounded blocky texture and small, orangish sand dikes. Swiss Army Knife for scale. (30.1510, -104.1058)

Elephant Rock, a weathered block of Morita Ranch Group basalt. (29.8768, -104.2809)

landform about 5 miles north of Shafter. Standing guard over the Chihuahuan Desert, Elephant Rock is the remnants of a large lava dam composed of Morita Ranch Group basalts that, along with the Mitchell Mesa Rhyolite, formed during the cataclysmic eruption and collapse of the Chinati caldera 32.3 million years ago. Elephant Rock gets its shape from preferential erosion of the basalt along weaker, more susceptible cracks and joints, thus creating the profile of an elephant.

South of Elephant Rock, the roadway passes numerous roadcuts as it winds its way through the eastern edge of the Chinati Mountains. The road passes through basalts of the Morita Ranch Formation, an Oligocene-aged porphyritic basalt derived from the last eruptive phases of the Chinati after the caldera collapse. Just over 2.5 miles south of Elephant Rock, as US 67 crests a hill, large roadcuts of Perdiz Conglomerate line the road. This weakly cemented fanglomerate, a conglomerate formed in an alluvial fan, is composed of rocks from the Chinati caldera that were shed eastward from the Chinati Mountain complex. Clasts as large as 3 feet in diameter can be observed in the roadcut, indicating a high-energy depositional environment.

After dropping down the south side of the hill for about 1 mile, a change in rocks occurs again. At a roadside pullout, dark rocks of the Buckshot Ignimbrite showcase the more explosive nature of the Chinati caldera. Ignimbrites consist of hardened tuff that solidifies from a pyroclastic flow, the superheated clouds of ash, rock particles, and gases that flow rapidly from a volcano. Imagine the modern images of hot, gray ash clouds flowing over the landscape after a major volcanic eruption. Studies from the early 1990s suggest that the total thickness of this deposit could be as much as 5,900 feet, indicating a massive eruption took place here when the Chinati caldera collapsed in the Eocene.

Various outcrops and roadcuts of the Buckshot continue for another 1.5 miles until the road crosses Cibolo Creek upon entering the mining town of Shafter. The creek has cut down into much older rocks, so a profound and very noticeable change in the rocks occurs at Shafter. The light-gray to white carbonate rocks are Early Cretaceous in age, part of the Shafter Formation of the Trinity Group. Beneath the Shafter lies the Permian-aged Mina Grande Formation that forms the host rock for one of the most important mining areas in Texas. The Shafter deposits formed from rising fluids heated by local igneous intrusions from the Chinati Mountains. The fluids were rich in metals from either older rock in the area or from the intrusions themselves. Once these fluids reached and interacted with the limestone, a flat-lying mineral deposit called a manto formed by the replacement of country rock during hydrothermal activity.

Perdiz Conglomerate showing large boulders and cobbles eroded from the Chinati caldera and deposited in an alluvial fan. (29.8470, -104.2928)

Buckshot Ignimbrite deposits, formed during the collapse of a caldera, show the jumbled nature of rock particles and superheated ash that rained down in this area around 32 million years ago during massive eruptions associated with the Chinati Mountains. (29.8395, -104.3060)

Like many mining towns of the American West, Shafter sprang up after the discovery of silver ores on the southeast edge of the Chinati Mountains. After initial attempts in the early 1880s to mine the ore failed, a second, more successful operation began in 1882 when a group of miners from California established the Presidio Mining Company. In less than a year, the Presidio was turning a profit, and the town of Shafter grew from less than one hundred people to as many as four thousand. In 1931, the mine closed, but it reopened in 1933 and experienced its greatest years of production before its closure in 1942 due to declining ore grades and flooding of the mine's lower levels. Over its 59 years of operation, the Presidio produced 35.15 million ounces of silver and 8,400 ounces of gold. Multiple attempts to reopen the mines have

This specimen of vanadinite and descloizite was recovered from the mines at Shafter.
—Photo courtesy Jonathan Woolley

Shafter Formation limestone just south of Shafter near the junction of Upper Shafter Road. (29.8109, -104.3149)

been made, with one of the more ambitious erecting a new steel headframe in the late 1970s that can be seen to the west of Shafter. However, this attempt, like many before it, failed when silver prices crashed in the early 1980s. Recent renewed interest in the mines has once again instilled hope that the silver mines of Shafter can become profitable.

Just south of Shafter, US 67 climbs out of a valley and to a hilltop with a very interesting roadcut. At first glance, the observer might think they are looking at a fault. Upon closer inspection, however, it becomes obvious there is no movement along the contact. Instead, the roadcut is a large angular unconformity between Cretaceous-aged limestones of the Shafter Formation and the overlying, Oligocene-aged alluvium deposits of the Perdiz Conglomerate shed from the Chinati Mountains. An angular unconformity occurs where younger sedimentary rock (in this case, the alluvium) is deposited on older, tilted and eroded layers of rock.

A major angular unconformity south of Shafter between the Shafter Formation and Perdiz Conglomerate. (29.8105, -104.3209)

Continuing south from this location, the roadway passes various roadcuts in Cretaceous rocks including the Shafter and Presidio Formations of the Trinity Group and Del Carmen Limestone of the Fredericksburg Group before leaving the Chinati Mountains and beginning a slow, gradual descent over alluvium deposits shed from the Chinati Mountains into the Rio Grande Valley. While the land here is relatively flat, the large Sierra del Pegüis appear in the distance to the southwest across the Rio Grande in Mexico. Their distinctive, slotted pattern is due to differential weathering, where softer, less resistant rocks are being eroded, perhaps along vertical fractures, faster than their harder counterparts surrounding them. US 67 continues south to the border town of Presidio, made famous in 1959 by the western movie *Rio Bravo* starring John Wayne. Presidio also serves as the gateway to Big Bend Ranch State Park and FM 170 (River Road), which follows the Rio Grande for several miles to the east. See the road guide for FM 170 on page 269.

The vertical slots in the Sierra del Pegüis in Mexico are the result of differential weathering. (29.6867, -104.3611)

US 90
SANDERSON — MARATHON — VAN HORN
184 miles

For about 30 miles west of Sanderson, US 90 winds its way through canyons and mesas built of flat-lying sedimentary rocks of Early Cretaceous age. Much of the layering is carbonate rock from the Del Carmen, Santa Elena, and Edwards Limestones, depending on which side of US 90 you are looking. The same limestone found around Sanderson is also found far to the southwest at Santa Elena Canyon in Big Bend National Park. The conditions for the sedimentation of these rocks were spread over a wide area. Roadcuts a few miles west of Sanderson, where the canyon widens to a broad valley, give a close-up view of these Cretaceous limestones.

At the west edge of Sanderson, south of US 90, is a large quarry into a sliver of Del Carmen Limestone. Rock from this quarry was removed to build the earthen dam opposite it, north of US 90 and adjacent to US 285. The tan color of the fresh limestone contrasts with the weathered, gray, natural outcrops of the same rock. North of the highway, Edwards Limestone dominates the landscape.

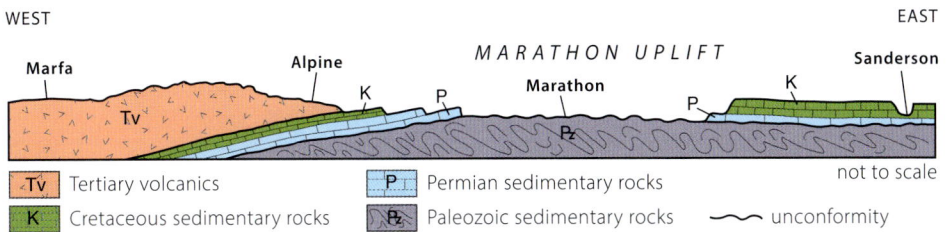

Cross section from Sanderson to Marfa on US 90.

Geology along US 90 between Sanderson and Marfa. For the section of US 90 west of Marfa, see the map on page .

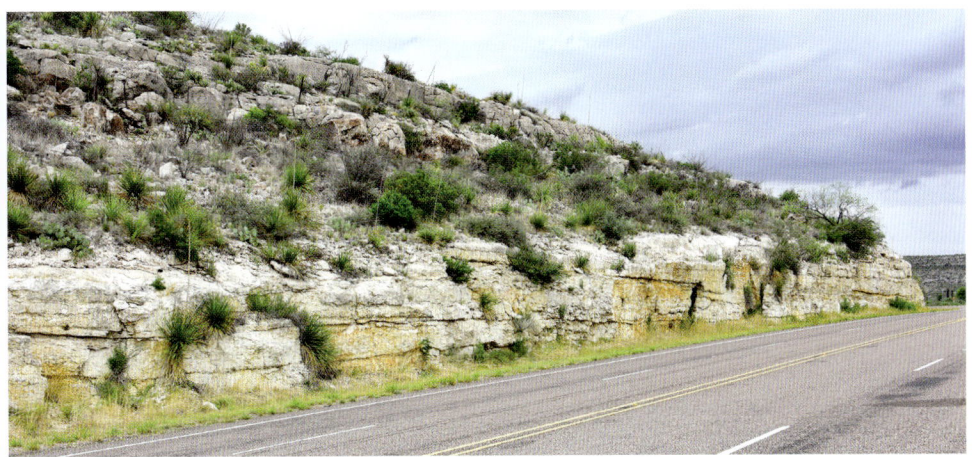

A large roadcut of Edwards Limestone on the right (north) side of the highway, 2.6 miles west of the US 285 intersection, shows its layered, blocky features. (30.1498, -102.4510)

In the wide valley west of Sanderson, alluvial fans spread outward from mesas. Roadcuts are cross sections through these fans and display their layers of sand and gravels. Streams, which are dry most of the year, carry immense amounts of water in short bursts after rainy thunderstorms, which carve the landscape and create desert landforms. Evidence of wind erosion can be observed on signs and displays along the roadway, but the wind mostly polishes rocks and builds small sand dunes in this area.

Talus slopes below the flat-lying mesas are typical in the desert. The mesas of hard caprock, whether limestone, sandstone, or even lava, are usually fractured and jointed. Blocks periodically fall off the mesas' edges to form talus slopes below.

About 35 miles west of Sanderson (or 18 miles east of Marathon), watch for the first blasted roadcut because here the geology changes dramatically. To the south are the Housetop Mountains, a large, flat mesa of Cretaceous limestone. In the roadcut, however, are dark, gray, hard limestone beds that stand up at a high angle. You are seeing the exposed edge of the old Ouachita Mountains, where beds of the Dimple Limestone of Early Pennsylvanian age (about 315 million years old) are tilted in a northeast-trending fold.

From this roadcut and extending for almost 3 miles to the historical marker is a series of beautiful roadcuts where Late Mississippian to Pennsylvanian rocks stand literally on end. These vertical beds of alternating sandstone and shale, the Tesnus and Haymond Formations, were deposited in the deep sea by turbidity currents, slurries of mud and sand that rapidly flow down continental slopes and canyons and deposit their load on the seafloor. Modern turbidity currents have been clocked at more than 50 miles per hour! Earthquakes probably trigger most turbidity currents by jarring loose the soft sediment lying on the outer continental shelf and upper continental slope. This sediment then flows downslope as a turbulent mass, driven by the force of gravity. The oldest turbidity currents began depositing the Tesnus in Mississippian time. After an interlude when the Dimple Limestone accumulated, deposition by turbidity currents resumed with the younger Haymond Formation and continued into the middle of the Pennsylvanian Period.

The exposed edge of the Ouachita Mountains is clearly seen in this roadcut of Pennsylvanian-aged Dimple Limestone. (30.2021, -102.9461)

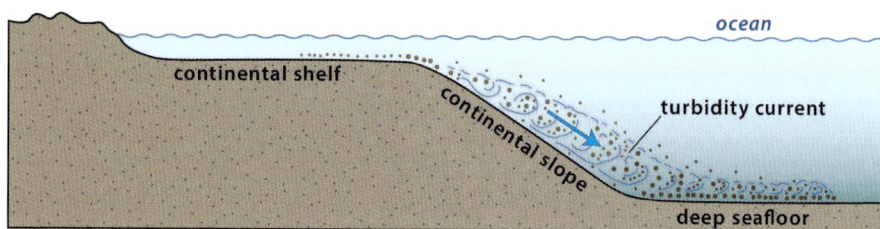

A turbidity current, a fast-moving slurry of sediment and water set loose by an earthquake, flows down the continental slope and is deposited on the deep seafloor. Much of the Haymond and Tesnus deposition was by turbidity currents. Each sand bed seen in roadcuts represents a separate turbidity current.

Each sand bed you see in the roadcut represents the deposit from a single turbidity current. The shale in between is the product of the gentler settling of very fine mud stirred up by the turbidity current, as well as the fine rain of mud continuously dumped on the sea bottom after being brought to the sea by rivers. An amazing 14,000 feet of sediments built up on the seafloor in this area during Paleozoic time.

Originally, of course, all these sand and shale beds were laid down horizontally. So, the fact they are now standing upright means they must have been subjected to epic mountain-building forces. The uplift, faulting, and folding of these rocks took place between 290 and 275 million years ago in the Permian Period as part of the Ouachita mountain-building episode. Ridges of the folded remains of these mountains can be seen running across the countryside to the south of US 90.

Near the town of Marathon, a prominent ridge north of the highway is formed by an east-west fold. Most of the fold axes in the Haymond and Tesnus are north-south, so this shows how abruptly the direction can change in the mountain folds. Capping the ridge is the white Caballos Novaculite, a hard, dense chert from Silurian to Mississippian time.

The roadcut opposite the historical marker shows vertical sandstone beds of the Tesnus Formation that are curved on top possibly due to the drag forces exerted by recent soil creep. (30.2125, -102.9877)

About 2.5 miles east of Marathon is a ridge to the north (right) capped by the Caballos Novaculite, a very resistant layer of chert. (30.2066, -103.2046)

Iron Mountain northwest of Marathon is a distinctive Eocene to Oligocene intrusive body. (30.2105, -103.2678)

Prominent above the alluvial fan and gravel surface northwest of Marathon is Iron Mountain, a distinctive, dark-colored, rough-looking peak composed of Eocene to Oligocene-aged intrusive rock. The dark, misshapen blob of a mountain stands out among the bedded, light-gray sedimentary rocks of the surrounding ranges.

A veritable wall of limestone faces you in the first string of hills west of Marathon. North of the highway, the westward-tilted ridges of Permian-aged rocks of the Glass Mountains attest to another major unconformity. These Permian rocks, about 200 million years old, rest directly on the deformed older Paleozoic rocks, more than 300 million years old, of the Marathon Uplift. The Permian rocks are the southeastern-most exposure of the Permian reef complex that forms the Guadalupe Mountains to the north. At the southwestern end of the Glass Mountains is the aptly named Cathedral Mountain, one of several mountains in this region with this name.

Permian limestones are well exposed in an impressive cliff on the face of the mountain south of the highway about 4 miles east of the US 67 junction (or 13 miles east of Alpine). At the west end of the mountain face, see if you can spot the small volcanic intrusion at the mouth of Ramsey Draw. In Ramsey Draw and to the west of it are younger rocks of Cretaceous age, lying on the back side of the westward-tipped Permian section.

Two miles east of Alpine, igneous dikes cut Eocene to Oligocene volcanic rhyolite beds. By now, you have again crossed a major geologic boundary, having stepped from the territory of Paleozoic rocks in the Marathon Uplift to much younger volcanic rocks of the Davis Mountains and Paisano volcano.

In Alpine, the Museum of the Big Bend on the campus of Sul Ross State University has geology displays as well as history and other science exhibits. Looking to the west from Alpine, you can see the profile of the Paisano volcano on the skyline.

A large cliff of the Permian Capitan Limestone and a small volcanic intrusion at the far right viewed from 13 miles east of Alpine. (30.3408, -103.4783)

West of Alpine at the edge of the Paisano volcanic field is a ridge of quartz-trachyte lava flows of the Decie Member (Duff Formation) from the eruptions that took place here during the Eocene and Oligocene. (30.3402, -103.7234)

PAISANO VOLCANO

Have you ever driven through a volcano? No? Then get ready because you are about to do that as you travel US 90 west of Alpine. Marvelous roadcuts expose the internal workings of the Paisano volcano, which erupted in Eocene time about 36 million years ago. If the magma feeding a volcano has a lot of water or gas in it, the eruption will generally blast forth explosively, spreading ash and various sizes of particles and debris over a wide area. If, however, the magma contains very little gas or water, it will ooze to the surface and flow laterally from the vent as a hot liquid, or lava flow. Lava may also move upward along fractures and cracks to produce walls of hardened rock called dikes. In other instances, the lava never makes it to the surface, instead creating a plug that penetrates some of the surrounding rock but never completely cuts through the overlying section. It is not uncommon for the caldera of a volcano

Geologic map of Paisano volcano west of Alpine.

Map labels:

roadcut through dike
roadside pullout

Tb
Tdm
JEFF DAVIS
Q
Tbh

MITRE PEAK

PUERTACITAS MOUNTAINS
Tdm
Qf
BREWSTER

Tpc
Tdd
Tb
Ti
Tdd
118
Tbh

Q
Tb
Tdd
Qf
lava and ash deposits
Tbh
Tbh

QTs
rhyolite
Qf
Alpine
Tbh

Qf
Ti
Qf
Tbh

26 miles from Alpine to Marfa
LIZARD MOUNTAIN
TWIN SISTERS
PAISANO PEAK
RANGER PEAK
Ti

90
collapse breccia
Tdd
Ti
Tl

PRESIDIO
Tb
Ti
Tdd
Tbh
Ti
Tbh

Q
Tdd
Qf
Ti
Qf
CATHEDRAL MOUNTAIN
Tbh

Tbh

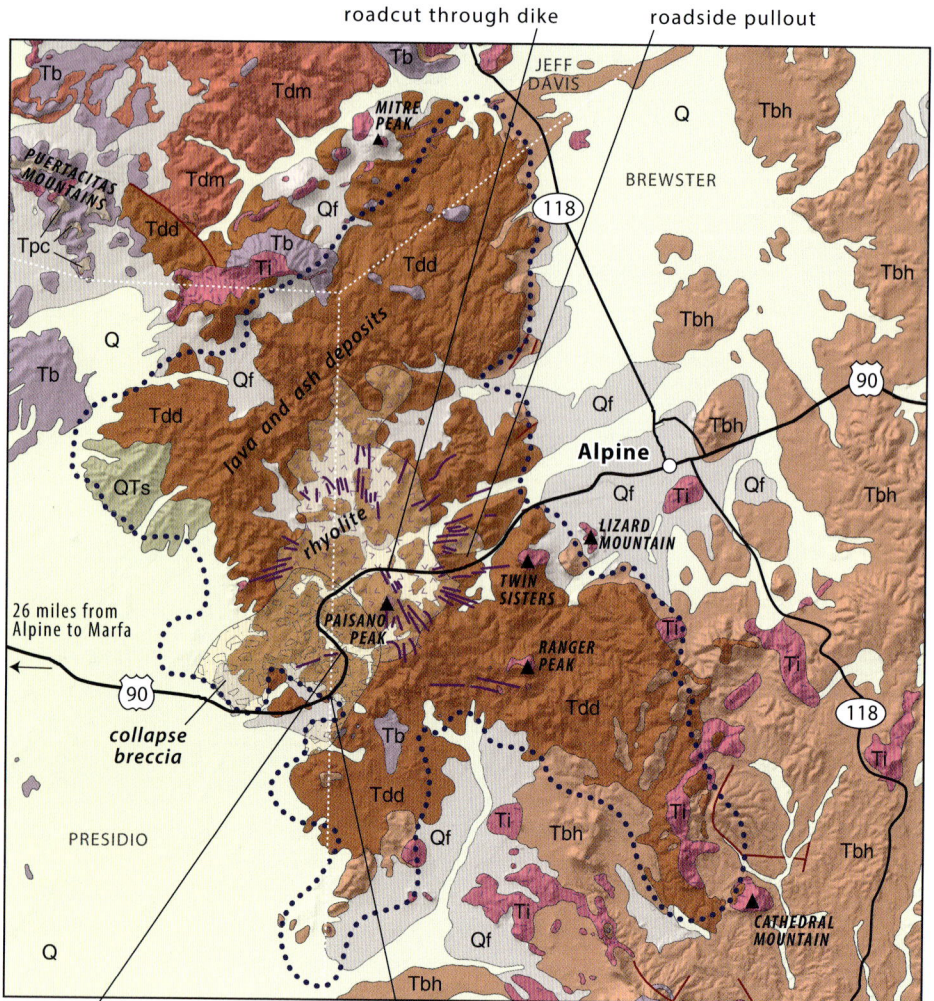

Baptist Encampment Dike at Paisano
ash-flow tuffs and rhyolites

N   0   5 miles
      0   5 kilometers

Legend:

—— fault
—— dike
•••• outline of Paisano volcano
county boundaries shown as white dotted lines

CENOZOIC

Qf alluvial fan deposits (Quaternary)
Q   alluvium (Holocene and Pleistocene)
QTs older gravel deposits (Miocene–Pleistocene)

VOLCANIC and IGNEOUS ROCKS

Tb Petan basalt (Oligocene)
Tpc Perdiz Conglomerate (Oligocene)
Tdm Davis Mountains volcanic complex; includes Sleeping Lion Formation (Eocene–Oligocene)
Tbh Buck Hill Group; includes Duff Formation, Crossen Trachyte, and Tascotal Formation (Eocene-Oligocene)
Tdd Decie Member of the Duff Formation (Oligocene)
Ti igneous rocks; includes stocks, laccoliths, sills, and dikes (Neogene and Paleogene)

to collapse after magma has been removed from below. It forms a broken jumble of angular blocks, called breccia, that collect in the volcanic neck. All these aspects of a volcano are recorded in the US 90 roadcuts west of Alpine.

West of Alpine and just east of a roadside pullout, views of the collapsed caldera rim and a zone of lava composed of very light-colored rhyolite, which is a very fine-grained version of granite, can be seen. Rhyolite is fine-grained because it came from molten rock that was extruded to the surface where cooling was so rapid that big mineral crystals had no time to form. Granite, on the other hand, comes from the same type of molten rock soup as rhyolite, but having solidified below the surface, it cools much more slowly and consequently grows large mineral crystals. The remainder of the Paisano volcano is composed of dark-colored lava flows and ash deposits.

The roadside pullout 5.5 miles west of Alpine offers beautiful views of Paisano Peak to the west, a prominent dike on the hillside to the south, and rocks of the Duff Formation formed during the caldera collapse to the north. Paisano Peak is an intrusion of nepheline syenite (a quartz-free intrusive igneous rock) within the collapsed crater of the volcano that erupted during the later stages of the volcanism associated with the Davis Mountains. The hard volcanic rocks of the peak cooled in the neck of the volcano to form a plug. Erosion and removal of the softer volcanic rock surrounding the plug have exposed the inner workings of this volcano. Several dikes extend across the caldera, forming black "walls" in areas where they have been exposed at the surface. At this roadside park, one of these dikes is spectacularly exposed in the hills to the south. Like Paisano Peak, the dikes are associated with later volcanism of the Paisano caldera where mafic (basaltic) rock intruded older volcanic rocks from earlier eruptions. An informative sign explaining the geology of the Paisano volcano can also be seen at the roadside pullout.

Paisano Peak, one of the sources for the volcanic rocks we see in this area, can be observed to the west from the roadside pullout 5.5 miles west of Alpine. (30.3230, -103.7430)

The roadcuts along US 90 west of Paisano Peak are mainly in the collapsed caldera of the Paisano volcano. Much broken and fragmented rock is seen along here. At the roadcut 0.2 mile west of the Paisano Baptist Encampment turnoff is a large dike that cuts broken rocks of the collapsed zone. This 3-mile-long dike, the largest in the Paisano volcano, formed when magma moved upward through a long crack. Striations on the dike faces record flow motion along the edge of the molten rock as it intruded the rhyolite. At the signed Presidio County line, south of the encampment, beautiful roadcuts of ash-flow tuff and rhyolites of the Decie Member (Duff Formation) can be seen.

From the roadside park 5.5 miles west of Alpine, an excellent view of one of the many dikes in the area can be seen across the hilltop on the south side of the roadway. (30.3230, -103.7430)

A younger, approximately 35-foot-wide grayish quartz-trachyte dike (center third of photo) cuts a bed of slightly older, baked Paisano rhyolite 6.8 miles west of Alpine. (30.3179, -103.7686)

Just west of the Paisano Baptist Encampment, the road cuts through a large dike. Collapse breccia is visible at the far-left side of the photograph. (30.2901, -103.7948)

MARFA TO VAN HORN

See map on page 262.

West of the Paisano volcano, the road crosses the flat, alluvial plain of the Marfa Basin. US 90 crosses Alamito Creek at the east edge of Marfa, where an outcrop of alluvial fan conglomerates and stream gravels is exposed on the south side of the highway east of the creek. Note all the different kinds of rounded cobbles and pebbles, debris from the surrounding mountains that was deposited here. The Puertacitas Mountains form the skyline to the north and the volcanic mountains of the Paisano volcano are on the skyline to the east. The distinctive profile of Cathedral Mountain (6,800 feet) is very clear to the east.

Marfa was named by the wife of the chief engineer of the Southern Pacific Railway in 1882. She was reading Dostoevsky's 1880 book *The Brothers Karamazov* and gave the Karamazov household servant's name, Marfa, to this 1880s railroad watering stop and freight headquarters. Two miles north of Marfa on TX 17 are beautiful, small-scale terraces along Alamito Creek. Take a good look at their shape and form to better understand how streams meander back and forth as they cut downward, leaving behind these flat, ledge-like terraces.

Just west of Marfa, low, grayish-white roadcuts of the Oligocene-aged Tascotal Formation come into view on the north side of the highway. The Tascotal is composed of tuffaceous sediments that cover the Mitchell Mesa Rhyolite and form a widespread blanket around the Chinati Mountains. Within the Tascotal, fine-grained, white ash layers can also be seen. Heading west, US 90 follows a flat valley basin between two mountain ranges, the Davis Mountains 15 miles to the east (right) and the Sierra Vieja 11 miles to the west (left). The sediments in this valley were shed from these mountains during millions of years of erosion.

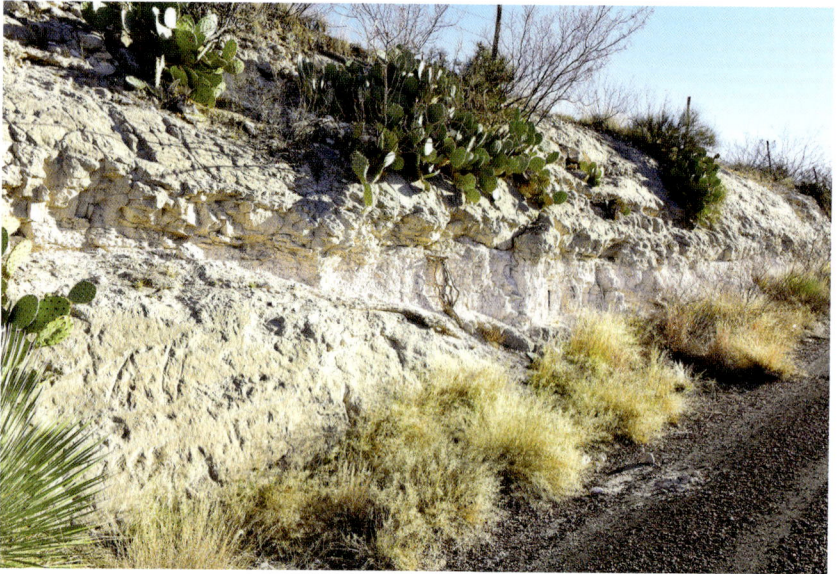

A white ash layer stands out from the tuffaceous sediments of the Oligocene-aged Tascotal Formation just west of Marfa. (30.3054, -104.0438)

Telephoto view of the Davis Mountains standing above the desert floor west of Marfa. (30.3404, -104.1386)

The small, unassuming town of Valentine, 35 miles northwest of Marfa, invokes the cliché "blink and you'll miss it." On August 16, 1931, however, Valentine was the site of the largest earthquake to ever strike Texas. At 5:40 a.m. Central Standard Time, a magnitude 5.8 earthquake struck approximately 7 miles southwest of town at a depth of just over 6 miles. While Texas does not see many earthquakes, this area of West Texas is the most seismically active part of the state due to the many faults that exist at the confluence of several geologic features including the Central Basin platform, the Ouachita Mountains, and the Rio Grande rift. The earthquake was felt over a large region that included most of Texas and portions of Kansas, New Mexico, Oklahoma, and northern Mexico. While no deaths were reported, several people in Valentine sustained minor injuries. In addition, almost every structure in Valentine and the surrounding area was either damaged or destroyed.

In the lonely stretch of road between Valentine and Van Horn is the ghost town of Van Horn Wells, noted by a state historic marker about 10 miles south of Van Horn. For centuries, Native Americans used the wells as a source of water in this dry, arid terrain. These natural springs originate from the base of the Van Horn Mountains to the west. The Wells are thought to have been named by Major Jefferson Van Horne, who passed through this area in 1849 on his way to establish a fort at El Paso.

North of Van Horn Wells, US 90 follows the valley toward Van Horn. Look west to see beautiful views of the Carrizo Mountains, and Threemile Peak and the Beach Mountains appear to the north as you reach town.

US 385
MARATHON — BIG BEND NATIONAL PARK
40 miles

In this remote, dry corner of Texas, an almost magical aura surrounds a gnarled, contorted tract of geology known as the Marathon Uplift. Here, rocks stand on end, tipped vertically to point to the sky, or they twist around hillsides imitating a snake. Others glisten like white battlements in the rising heat waves above the stony desert. Nowhere else in the entire state of Texas are rocks so deformed, so bent, so crushed, folded, and faulted as in the Marathon region. Travelers have the unique opportunity to peer into the deformed flanks of an old mountain range, which has been brought to the surface by intense crustal squeezing, then exposed by equally intense erosion. The Marathon Uplift is one of only three exposed pieces of the buried Ouachita Mountains that arc across Texas. The others are in the Llano Uplift and the nearby El Solitario Dome northwest of Big Bend National Park.

US 385 south of Marathon illustrates the folded and faulted character of the Paleozoic rocks within the Marathon Uplift and provides up close views of the strange Caballos Novaculite, a siliceous rock deposited on the seafloor in Silurian to Mississippian time. Just south of Marathon, small roadcuts on US 385 expose the Ordovician

A beautiful roadcut of the Caballos Novaculite is 3.3 miles south of Marathon. Swiss Army Knife for scale in the inset. (30.1588, -103.2367)

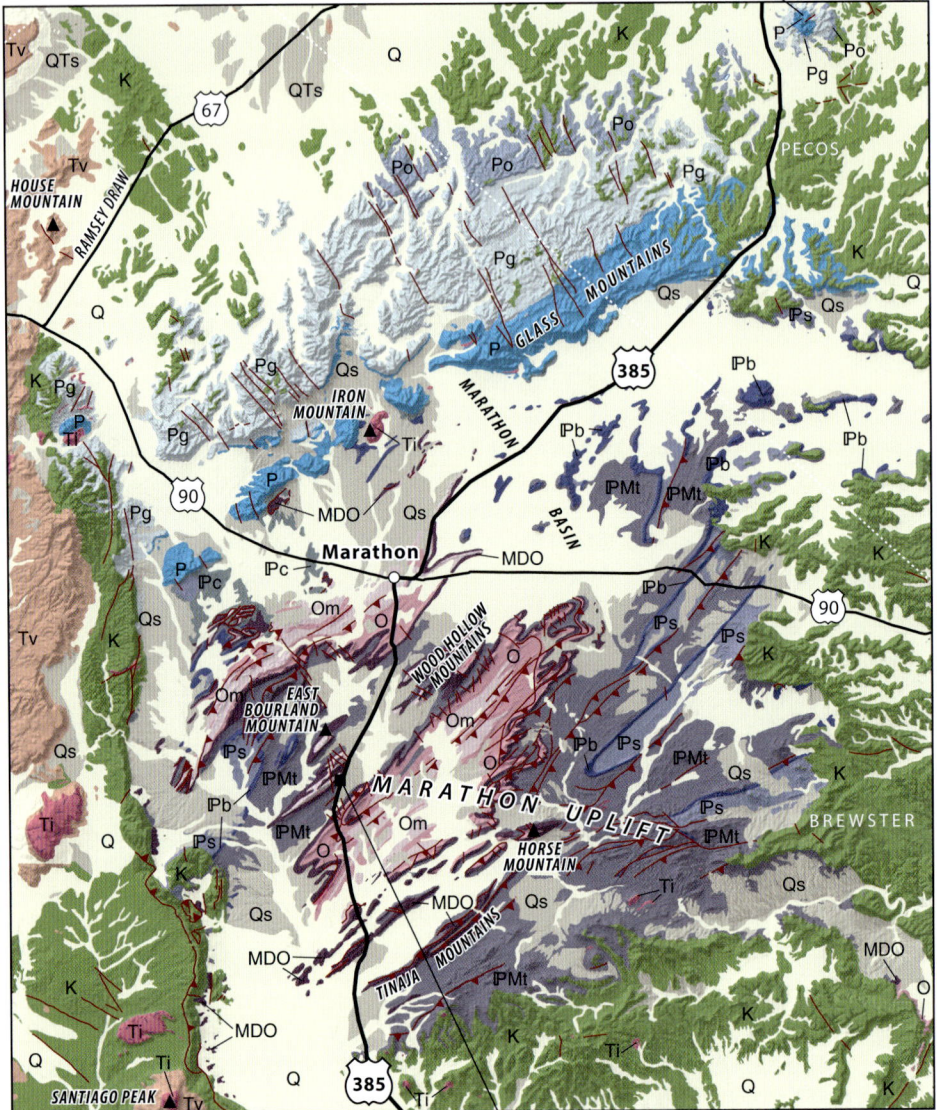

Geology along US 385 through the Marathon Uplift.

CENOZOIC

Q — recent sediments; includes alluvium and alluvial fan deposits (Quaternary)

Qs — older sedimentary deposits (Qs–Pleistocene); includes older gravel deposits (QTs–Miocene to Pleistocene)

VOLCANIC and IGNEOUS ROCKS

Tv — volcanic rocks; includes mostly Buck Hill Group (Neogene and Paleogene)

Ti — igneous rocks, undifferentiated (Tertiary)

MESOZOIC

K — sedimentary rocks; includes Terlingua, Fredericksburg, Washita, Trinity, and Coahuila Groups (Cretaceous)

41.2 miles from US 90 to Big Bend National Park

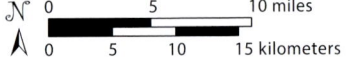

picnic area

PALEOZOIC

PERMIAN

P — sedimentary rocks, undivided

Pg — Guadalupe Group

Po — Ochoa Group

PENNSYLVANIAN

℔c — Cisco Group

℔s — Haymond Formation of the Strawn Group

℔b — Dimple Limestone of the Bend Group

thrust fault

normal fault

county boundaries shown as white dotted lines

MISSISSIPPIAN–PENNSYLVANIAN

℔Mt — Tesnus Formation of the Morrow Group

ORDOVICIAN–MISSISSIPPIAN

MDO — Caballos Novaculite and Maravillas Chert

ORDOVICIAN

O — sedimentary rocks, undivided

Om — Marathon Limestone

Marathon Limestone, which is lower in the Paleozoic section. On hills and ridgetops south of town, the white gleam of the Caballos Novaculite is striking, particularly to the east in the Wood Hollow Mountains and on the long ridges to the west of the highway. It is mapped with the older, underlying Maravillas Chert, also composed of silica. These rocks form ridges because they are hard and resistant to erosion.

Three miles south of Marathon, the road curves through a ridge of Maravillas Chert/Caballos Novaculite. Note how the beds are tipped steeply to the south, and the white, siliceous novaculite is hard, dense, fractured, and sharp (be careful!). It is likely that the bodies of marine sponges were significant contributors of the silica in the novaculite because tiny, distinctive sponge spicules can be seen in it under a microscope. Silica needles are the only skeletal strengtheners in the body of sponges.

Watch for a picnic area about 10.3 miles south of Marathon. Look west from the picnic area to see flatirons (steeply tilted rock faces) on East Bourland Mountain. The road again slices the hard novaculite 12 miles south of Marathon, where nearly vertical beds of rock are displayed in a roadcut.

Intricate folds in the white Caballos and dark Maravillas outcrops to the southeast of Marathon graphically illustrate the twisted-up, intensely deformed character of the Marathon Uplift.
—Satellite image courtesy Google Earth, 2025

Flatirons of Caballos Novaculite on the mountains west of the highway, 9 miles south of Marathon. (30.0826, -103.2686)

Santiago Peak, an intrusion seen for miles around, is visible on the west side of the highway. (29.9317, -103.2608)

About 20 miles south of Marathon, the surrounding landscape undergoes a fundamental geologic change. US 385 crosses south from the erosional window that looks into the Marathon Uplift and enters the noneroded realm of Cretaceous limestone overburden punctuated by volcanic intrusions. To the east is the jagged profile of the Tinaja Mountains, where the hard Caballos Novaculite holds up the ridge. Farther south, the Eocene to Oligocene-aged intrusions of Santiago Peak and other mountains loom on the skyline, surrounded by flat-lying Cretaceous limestones of the Boquillas Formation of the Terlingua Group. More intrusions dominate the landscape in Big Bend National Park. See the separate section beginning on page 277 for information about the park.

TX 54
VAN HORN—US 62/US180
55 miles

The distinctive skyline profile of Threemile Peak lies to the west of TX 54 north of Van Horn. The mountain has a protective cap of hard Permian Hueco Limestone, whereas the lower slopes are outcrops of extremely ancient sandstones and conglomerates of the Van Horn Sandstone, deposited before most life forms were developed on Earth. These sediments are Cambrian in age, between 519 and 515 million years old. They include zircon grains dated to 520 million years old, which means a volcanic rock or intrusion of that age was eroded so quickly that zircon crystals from it ended up in the sandstone.

To the east is a broad, flat valley, called the Salt Basin, which is a down-dropped segment of crust between two uplifted blocks of the Basin and Range. In this low area, windblown sands and alkali flats are prevalent. Alkali flats are dry, white lakebeds

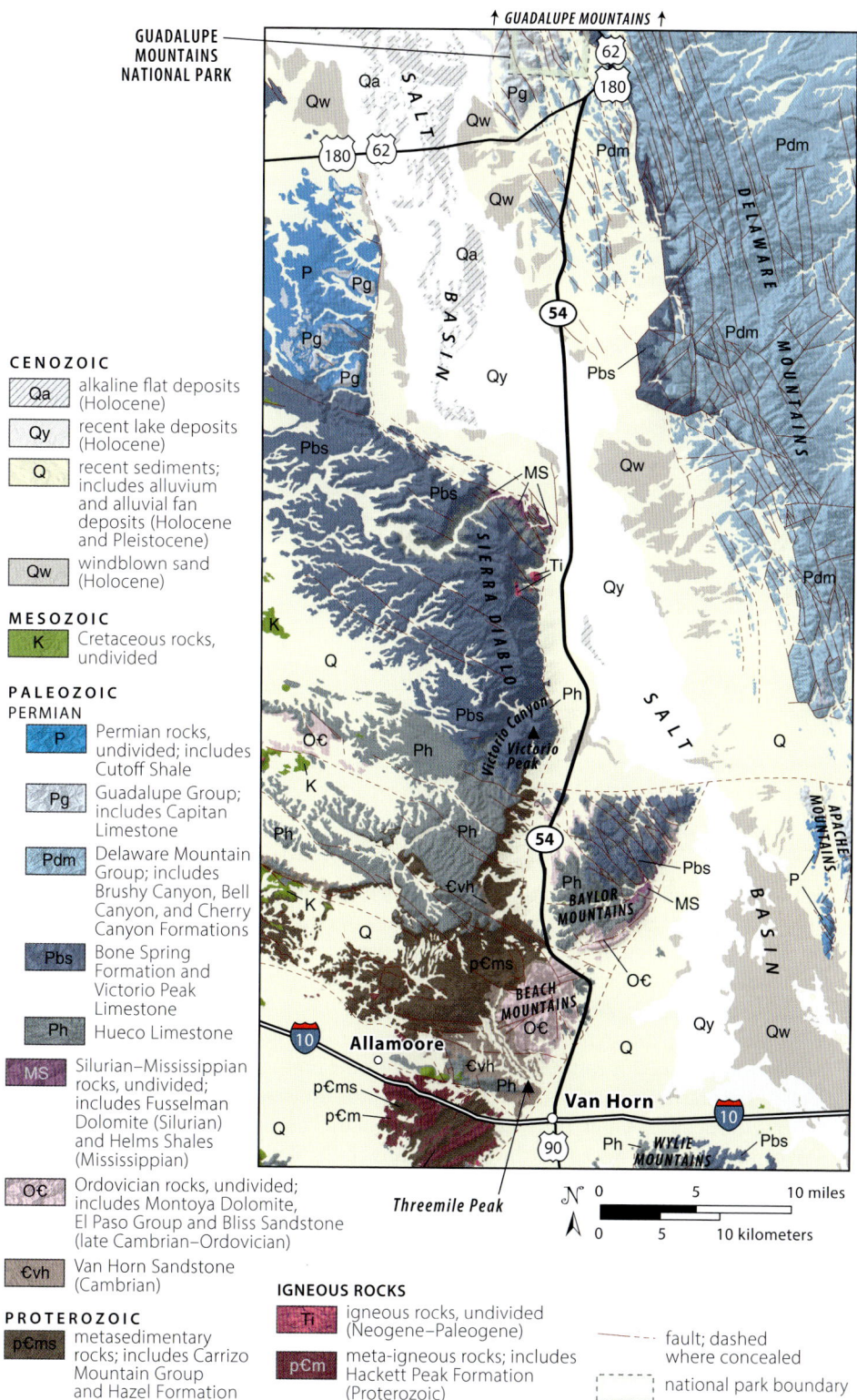

↑ GUADALUPE MOUNTAINS ↑

GUADALUPE
MOUNTAINS
NATIONAL PARK

SALT BASIN

DELAWARE MOUNTAINS

SIERRA DIABLO

SALT BASIN

APACHE MOUNTAINS

Victorio Canyon

Victorio Peak

BAYLOR MOUNTAINS

BEACH MOUNTAINS

Allamoore

Van Horn

WYLIE MOUNTAINS

Threemile Peak

CENOZOIC

Qa	alkaline flat deposits (Holocene)
Qy	recent lake deposits (Holocene)
Q	recent sediments; includes alluvium and alluvial fan deposits (Holocene and Pleistocene)
Qw	windblown sand (Holocene)

MESOZOIC

K	Cretaceous rocks, undivided

PALEOZOIC

PERMIAN

P	Permian rocks, undivided; includes Cutoff Shale
Pg	Guadalupe Group; includes Capitan Limestone
Pdm	Delaware Mountain Group; includes Brushy Canyon, Bell Canyon, and Cherry Canyon Formations
Pbs	Bone Spring Formation and Victorio Peak Limestone
Ph	Hueco Limestone
MS	Silurian–Mississippian rocks, undivided; includes Fusselman Dolomite (Silurian) and Helms Shales (Mississippian)
OꞒ	Ordovician rocks, undivided; includes Montoya Dolomite, El Paso Group and Bliss Sandstone (late Cambrian–Ordovician)
Ꞓvh	Van Horn Sandstone (Cambrian)

PROTEROZOIC

pꞒms	metasedimentary rocks; includes Carrizo Mountain Group and Hazel Formation

IGNEOUS ROCKS

Ti	igneous rocks, undivided (Neogene–Paleogene)
pꞒm	meta-igneous rocks; includes Hackett Peak Formation (Proterozoic)

⎯ ⎯ fault; dashed where concealed

⎯ ⎯ national park boundary

N

0   5   10 miles

0   5   10 kilometers

Geology along TX 54 between Van Horn and US 62/US 180.

where salts have been concentrated by evaporation and poor drainage. The hills in the distance to the east are the Apache Mountains, part of the uplifted block on the east side of the basin.

About 5 miles north of Van Horn, TX 54 closely skirts the northern end of the Beach Mountains to the west, which exposes a dome of Ordovician and Cambrian rocks. Old Proterozoic rocks are exposed along the western flank of the mountain. The road turns abruptly west to pass between the Beach Mountains and the Baylor Mountains to the north.

The northern edge of the Beach Mountains. The upper ridge is Ordovician Montoya Dolomite; the lower slopes are thin-bedded, sandy dolomites of the Ordovician El Paso Group. (31.1640, -104.8420)

The Baylor Mountains, on the right (east) side of the road across from the Beach Mountains, are also a fault block, capped by the Permian Hueco Limestone. Ordovician and rare Silurian rocks are exposed on the eastern flank. Silurian rocks are not preserved nor seen in outcrop in many places throughout the western United States, so these outcrops of the Silurian Fusselman Dolomite are sought after by geologists interested in the Silurian Period.

After passing through the narrow gap between the Beach Mountains and the Baylor Mountains, TX 54 turns north and flanks the high, sheer, impenetrable front wall of the Sierra Diablo for the next 25 miles. It is little wonder why it was named Diablo, the devil. The high cliff face near the top of the range, the Hueco Limestone of Permian age, can be traced northward for many miles. Beneath it are beautiful talus slopes that form "skirts" for the mountain. The talus nearly covers the thousands of feet of older sandstones and conglomerates of the Cambrian Van Horn and Proterozoic Hazel Formations. These reddish, sandy beds are exposed, however, at the very south end of the Sierra Diablo.

Follow the Hueco Limestone ledge northward and notice that it gets lower and lower. As you reach the mouth of Victorio Canyon (across from the entrance to the Blue Origin Launch Site to the east), the Hueco Limestone is at road level and younger Permian rocks are on the skyline. The Sierra Diablo is uplifted along a fault, but it is also tipped toward the north.

The Permian-aged Hueco Limestone forms a high wall at the southern end of Sierra Diablo, with red Proterozoic to Cambrian sandstone and conglomerate beds exposed below (at left) but covered by talus northward (at right). (31.2117, -104.8513)

About 20 miles south of US 62/US 180 (35 miles north of Van Horn), the face of the Sierra Diablo takes an abrupt turn to the west away from the road. On the corner of the range, low down on the slopes, another set of rare Silurian-aged rocks is exposed.

TX 54 from here to the junction with US 62/US 180 traverses the middle of the Salt Basin. To the east, the Permian rock layers in the Delaware Mountains are magnificent. These horizontal layers of sandstone, along with limestone and shale, can be traced for miles. They were deposited into the sea at the edge of the Delaware Basin. Some of the shale beds were laid down on the seaward slopes below the great Permian reef that is now exposed on the cliffs in Guadalupe Mountains National Park. About 5 to 7 miles south of the junction with US 62/US 180, TX 54 crosses low hills where sandstones of the Permian Brushy Canyon Formation (Delaware Mountain Group) rise above the salt flats. As the road nears US 62/US180, views to the north show the imposing cliff walls of El Capitan and Guadalupe Peak in the Guadalupe Mountains. Guadalupe Peak is the highest point in Texas, rising to an elevation of 8,751 feet above sea level.

TX 118
Study Butte—Alpine
79 miles

The southern half of TX 118 is on 80-million-year-old limestone terrains of the Pen and Aguja Formations, punctuated by a large number of Eocene-aged intrusive volcanic rock masses. North of Study Butte for about 10 miles, TX 118 is literally studded on either side by intrusive masses that have punched upward through the surrounding Cretaceous rocks. Erosion of the softer limestone has left them exposed as hills and mountains. East of Study Butte is Maverick Mountain, 2 miles north of town is Bee Mountain on the west and Indian Head Mountain east of the highway, and 4 miles north of town is Willow Mountain standing east of the road. All four are

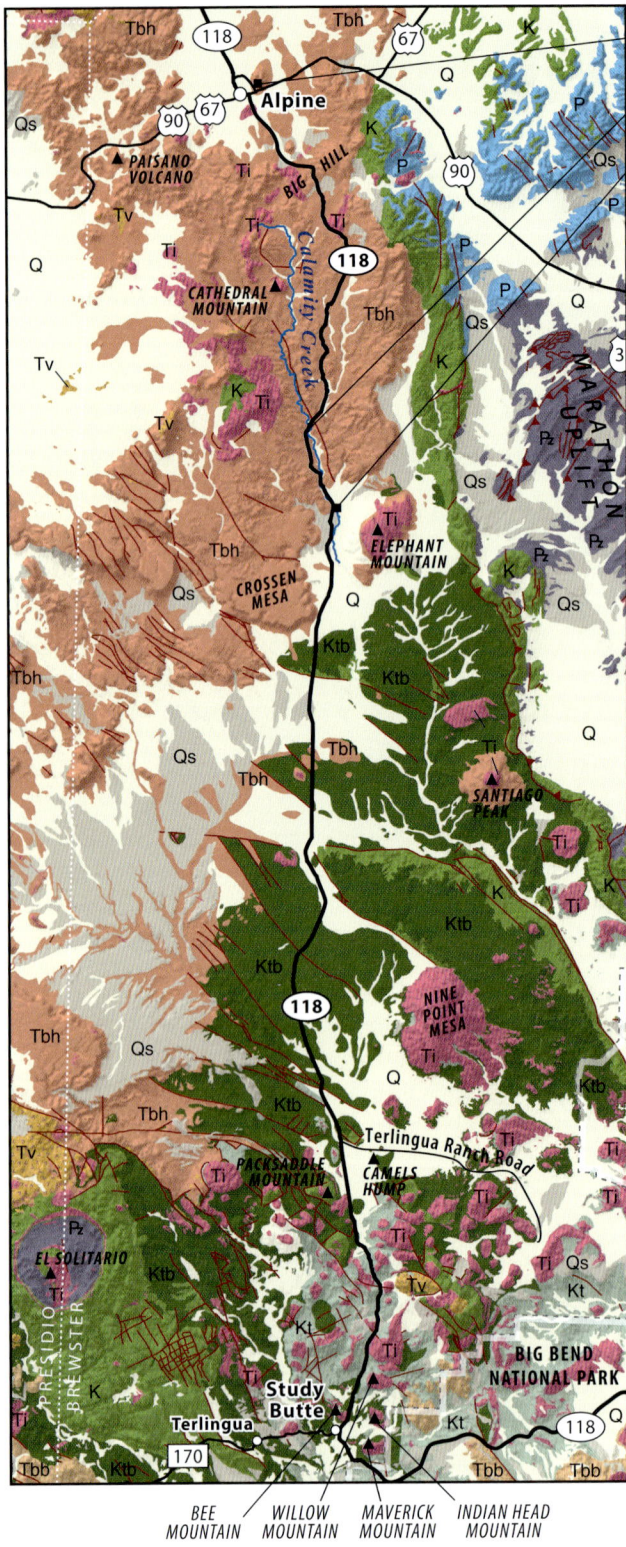

Museum of the Big Bend at Sul Ross State University

ancient soil between lava flows

picnic area at Calamity Creek

CENOZOIC

Q — recent sediments; includes alluvium and alluvial fan deposits (Quaternary)

Qs — older sedimentary deposits (Pleistocene)

VOLCANIC and IGNEOUS ROCKS

Tv — volcanic rocks, undifferentiated (Neogene and Paleogene)

Tbh — Buck Hill Group; includes Decie Member of the Duff Formation, Crossen Trachyte, and Tascotal Formation (Eocene-Oligocene)

Tbb — Big Bend Park Group (Eocene–Oligocene)

Ti — igneous rocks; includes stocks, vent rocks, laccoliths sills, and dikes (Neogene and Paleogene)

MESOZOIC
CRETACEOUS

Kt — Tornillo Group; includes Aguja and Boquillas Formations

Ktb — Terlingua Group; includes Pen Formation

K — Fredericksburg, Washita, Trinity, and Coahuila Groups, undivided; includes Santa Elena Limestone of the Washita Group

PALEOZOIC

P — sedimentary rocks, of the Marathon Uplift, undifferentiated (Permian)

Pz — older Palerozoic sedimentary rocks, undifferentiated (Ordovician–Pennsylvanian)

thrust fault

normal fault

national park boundary

county boundaries shown as white dotted lines

0   5   10 miles

0   5   10   15 kilometers

Labels on map:

Tbh, Alpine, Paisano Volcano, Qs, Q, Tv, Ti, Big Hill, Calamity Creek, Cathedral Mountain, K, Crossen Mesa, Elephant Mountain, Marathon Uplift, Pz, P, Santiago Peak, Ktb, Nine Point Mesa, Terlingua Ranch Road, Packsaddle Mountain, Camels Hump, El Solitario, Presidio, Brewster, Study Butte, Terlingua, Big Bend National Park, Bee Mountain, Willow Mountain, Maverick Mountain, Indian Head Mountain, Tbb

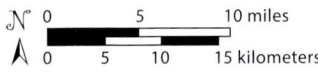

Geology along TX 118 between Study Butte and Alpine.

typical examples of intrusive rock bodies in this area. Willow Mountain is particularly striking because of its vertical rock joints. The mountain mass is a small quartz syenite intrusion, which cooled beneath the surface. As it cooled and contracted, vertical, uniformly spaced cooling cracks, or joints, formed. Weathering removed soft rock, enhancing the joint pattern. Willow Mountain is probably the best display of columnar rock joints in Big Bend country.

North of Willow Mountain, watch for low mesas of thin-bedded, flat-lying Cretaceous rocks near the road. About 14 miles north of Study Butte, look for the Packsaddle Mountain sign (elevation 4,661 feet). The mountain is cored by a dark-colored volcanic intrusion, but you can readily see light-colored, upturned beds of Early Cretaceous rocks on the flanks that were bent at a high angle as the mass pushed upward.

Willow Mountain, an intrusion of quartz syenite that cooled underground and was subsequently uncovered by erosion, is located 4 miles north of Study Butte. Spectacular vertical columnar joints formed as the mass cooled and contracted. (29.3434, -103.5307)

Packsaddle Mountain, 14 miles north of Study Butte, is a dark-colored central intrusion surrounded by inclined, light-colored Cretaceous beds. (29.5167, -103.5393)

Camels Hump, to the east from the Terlingua Ranch turn, is aptly named because it has two humps. North of Camels Hump is an obvious anticline, an up-arched fold, in Cretaceous rocks. About 3 miles north of the Terlingua Ranch turn, the long, flat-topped landform to the east is Nine Point Mesa. It is capped by an Eocene sill of igneous rock about 1,000 feet thick. As you drive north of Nine Point Mesa and crest a small limestone rise, a dominant peak comes into view to the east. This landmark, seen for many miles along TX 118, is Santiago Peak, a large intrusive mass (elevation 6,521 feet).

The northern half of TX 54 traverses Eocene-aged volcanic rocks of the Crossen Trachyte, part of the gigantic Davis Mountains volcanic field. The trachyte, pronounced "tray-kite," is an extrusive lava composed mostly of alkali feldspar. The rock caps Crossen Mesa, a high, dark, flat-topped mesa to the northwest about 50 miles north of Study Butte (30 miles south of Alpine). The 35-million-year-old lava flow is broken by columnar joints. Crossen Mesa is an excellent example of inverted topography where a former low area that lava flowed into is now a topographic high, thanks to erosion removing the surrounding, less-resistant rock. To the east is Elephant Mountain, another mesa capped by the same flow. A picnic area north of it provides good views.

The lava flow that caps Crossen Mesa flowed through a topographic low area. It now stands tall because erosion removed the softer, surrounding rock. (29.9788, -103.5718)

North of the picnic area, a series of curves exposes basalt flows, ash-deposits, and intrusions. About 4 miles north of the picnic area (about 22 miles south of Alpine) is a wonderful roadcut on the west side of the highway, where two dark lava flows are separated by a bright-red to purple layer. The surface of the lower (older) lava flow was exposed to weathering, which oxidized (rusted) the iron within, producing the red color. A poorly developed soil horizon formed on the older lava surface. You can even see weathered blocks of the lower flow mixed into the red layer. Later, the younger (upper) flow covered the red soil.

The distinctive profile of Cathedral Mountain to the west shows up for miles. It is another intrusive mass, exposed by erosion for all to see. Look for it to the west from south of Calamity Creek Road and also from the US border checkpoint 15 miles south of Alpine.

In a roadcut approximately 22 miles south of Alpine, a younger lava flow overlies a reddish, poorly developed soil layer that developed on the surface of the older lava flow at the bottom. (30.0991, -103.5964)

On top of Big Hill, about 6 miles south of Alpine, are excellent roadcuts of dark-gray to green basalt that weathers brown. Note the large crystals scattered among the finer groundmass. From Big Hill, TX 118 drops down to the north toward Alpine. The town seems appropriately named because it stands 4,481 feet above sea level at the entrance to the Davis Mountains, commonly called the Texas Alps. Sul Ross State University, in Alpine, hosts a small museum with displays on the geology and history of the area.

TX 118
ALPINE—FT. DAVIS—I-10
76 miles

For the first few miles north of Alpine, TX 118 crosses a flat, alluvial plain. Off to the left (west) are mesas composed of Oligocene volcanic flows. Mitre Peak, shaped like a bishop's pointed mitre hat, is an intrusive mass exposed by erosion. It stands tall, peaking above the low mesas in the foreground, because it is harder than the eroded rock around it. The best view of it is from the junction with Ranch Road 1837, a little more than 1 mile north of the Jeff Davis–Brewster county line. The flat mesa to the north of the junction is capped by a dark-brown ledge of Oligocene rhyolite and ash-flow tuff. It occupied a valley when it first erupted but now stands high because of inverted topography. Its associated talus slope indicates much weathering is accomplished by gravity in this desert country, as rocks break off cliff faces and tumble down.

Northwest of the Ranch Road 1837 junction, TX 118 begins to climb and weave its way among eroded volcanic hills. Look for the intrusive peak of Barillos Dome on the

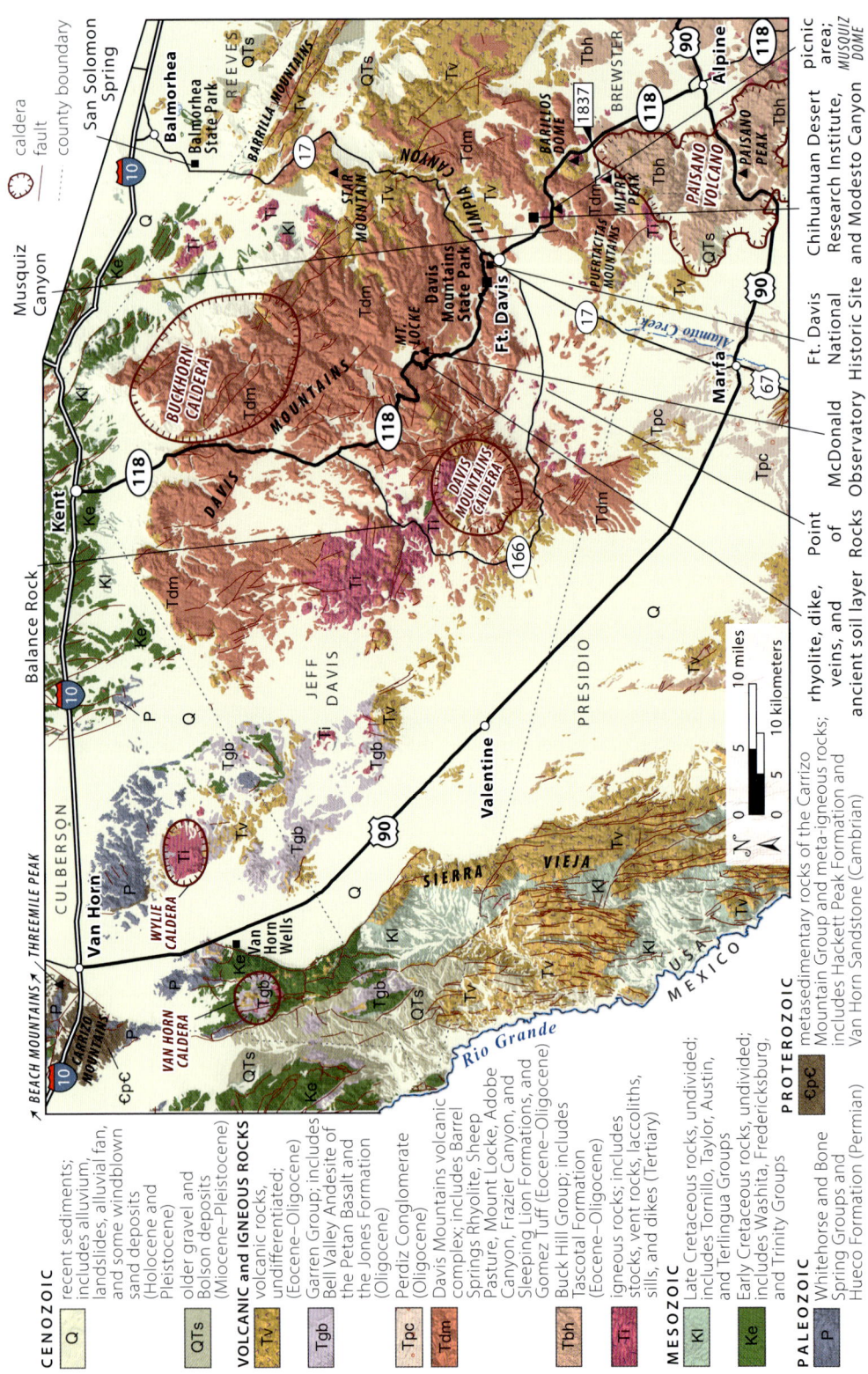

Geology along TX 118 between Alpine and I-10 and along US 90 between Marfa and Van Horn.

NORTH   SOUTH

PAISANO VOLCANO   **Alpine**

to Ft.   *MUSQUIZ*   *BARILLOS*   *MITRE*   *PAISANO PEAK*
Davis   *DOME*   *DOME*   *PEAK*
←

vertically   ▨ intrusive   ▢ light-colored extrusive volcanic rocks;   0   5 miles
exaggerated   rocks   rhyolite, ash-flow tuff, and lava
                                    ▢ dark-colored extrusive volcanic rocks;   0   5 kilometers
                                      basalt, tuff, and lava

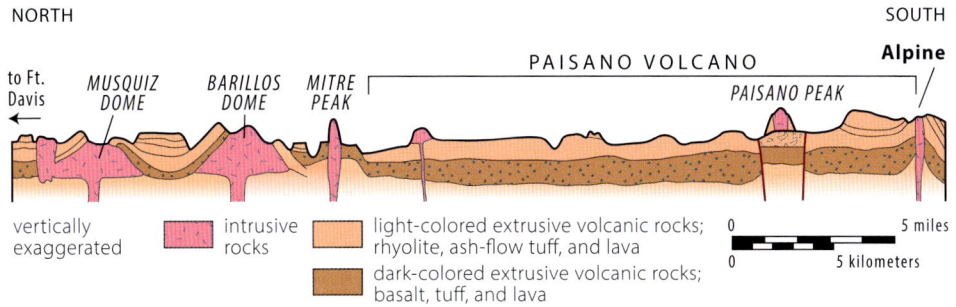

Cross section along TX 118 between Ft. Davis and Alpine.

Mitre Peak, an intrusive mass shaped by erosion, stands tall northwest of Alpine. (30.4808, -103.7474)

left (west). The road swings around the dome to give a good three-dimensional view of the central intrusive core surrounded by upturned younger rocks on the flanks.

At 2 miles north of the Ranch Road 1837 intersection, pay close attention to the large roadcuts on the right. A dense, hard basalt flow rests on a gray ash bed, separated by a very distinct contact. Notice how the top of the ash layer appears to be baked; this is indicative that the older ash bed was deposited first, then a subsequent (and very hot) lava flow surged into this area, baking the top of the ash. In addition, look for voids in the basalt. These former gas bubbles tried to escape the molten lava but got trapped in the flow as it began to cool.

A picnic area, 5.4 miles northwest of the basalt outcrop (or 7.4 miles northwest of Ranch Road 1837) and just past the historic Calamity Creek Ranch, offers views of Eocene-aged intrusions into the Puertacitas Mountains to the south. Note the typical rounded shapes of the intrusions, with their chaotic joint and spheroidal weathering patterns.

Upon leaving the hills and entering a flatter terrain are signs for the Chihuahuan Desert Research Institute. This site, established in 1973 and open year-round, protects 507 acres of semidesert grasslands and features geologic attractions such as Modesta Canyon with its year-round spring, and Clayton's Overlook, a dome with scenic panoramic views of the surrounding landscape. A geologic exhibit can be found on top of the overlook, and a geologic timeline display is in front of the visitor center.

At Ft. Davis is Ft. Davis National Historic Site, where a frontier military post protected travelers on the San Antonio–El Paso Trail from 1854 to 1891. The cliff behind the old fort, standing like a medieval castle, is a 36-million-year-old rhyolite ash-flow tuff known as Sleeping Lion Formation. The vertical fractures are well-developed columnar joints.

Blocks of spheroidally weathered intrusions can be seen 0.4 mile east of the picnic area 17 miles west of Alpine. (30.5270, -103.8193)

Cliffs behind Old Ft. Davis are the 36-million-year-old Sleeping Lion Formation, a rhyolite ash-flow tuff with well-developed columnar jointing. (30.5985, -103.8920)

Excellent exposures of pink to gray Sleeping Lion rhyolite can be seen near the entrance to Davis Mountains State Park. (30.5998, -103.9328)

TX 118 turns westward into Limpia Canyon 1 mile north of the town of Ft. Davis. The canyon cuts into the ash-flow tuff of the Sleeping Lion Formation, and the road follows along the base of the brown cliffs. At the entrance to Davis Mountains State Park are pink and red roadcut exposures of weathered, gray rhyolite, also part of the Sleeping Lion Formation. Look for big feldspar crystals floating in a matrix of very fine-grained rhyolite. The large crystals reveal that the magma was already starting to crystallize at depth before it was erupted to the surface, where the rest of the melt cooled rapidly. Around the parking lot near the state park campground are large blocks of local rocks where you can closely examine fresh rock faces and see the tremendous variety of colors, textures, and compositions of Davis Mountains volcanic rocks.

West of the state park, TX 118 traverses the heart of the Davis Mountains, called by some the Texas Alps, though you won't see any peaks that look anything like the Matterhorn. The Davis Mountains are, however, a gigantic volcanic area that had its origin 39 to 35 million years ago when western North America erupted in a violent fury of volcanism. Volcanic rocks of Eocene to Oligocene age are found in a wide belt from Mexico to Montana, and the Davis Mountains are one of the largest volcanic centers in this chain. Here you will find rhyolite lava flows, with their columnar-jointed palisades and widespread, thick ash-flow tuffs. Dark, iron-rich basalt lavas are present but are thin and not-so-common. The magma that formed these rocks either flowed out or was blasted out of two main volcanic centers: the Paisano volcano west of Alpine and the Buckhorn caldera northwest of Ft. Davis.

As you drive across the Davis Mountains on TX 118, note the many layers of horizontally bedded volcanic rocks that compose this vast eruptive system. The scenic hills surrounding the road are erosional imprints over the flat volcanics—in other words, you are not seeing individual volcanic cones but rather a system of hills and valleys cut through the flat volcanic layers by stream erosion. The roadside park with gazebos, about 6 miles west of the Davis Mountains State Park headquarters, is a good place to take in the scenery.

Rolling hills, valleys, tree-covered slopes, and pleasant vistas are seen for several miles. Occasional volcanic outcrops pop out on some hillsides and in a few roadcuts, but the vegetation on the weathered, rounded slopes covers much of the rock. Rainfall in the Davis Mountains is much higher than the surrounding desert country. The mountains cause westerly winds to rise over them, and as the air climbs, it cools and drops its moisture. The temperate vegetation reflects this increased moisture supply and contrasts sharply with the desert plants in the lowlands surrounding the Davis Mountains.

About 3 miles uphill from the gazebos (and 1.7 miles downhill from the observatory turnoff), where TX 118 slices diagonally across the flanks of Mt. Locke, is a wonderful roadcut, complete with its own parking area and scenic overlook. The Barrel Springs Rhyolite, exposed in the cut, is part of a bold, natural, Oligocene-aged

In a fascinating roadcut on the flank of Mt. Locke, a large, tan-colored rhyolite dike cuts the Barrel Springs Rhyolite. The dike has contorted flow bands that formed as it was emplaced while cooling. Also note the dipping, whitish flow bands in the rhyolite. (30.6651, -104.0318)

Between the time an ash-flow tuff was deposited and a rhyolite lava flow covered it, a poorly developed soil formed. The soil had time to weather, thus creating a sharp, red contact. (30.6651, -104.0318)

outcrop that slashes across the mountain flank. The sharply defined base of this lava flow, visible from the lower end of the pullout, rests on an ash-flow tuff with a reddish soil zone.

Near the uphill-end of the roadcut is a vertical, white brecciated vein that cuts the lava beds. Note how the walls of the vein are hard and very glossy, whereas the center is a bit coarser and more crystalline. The walls cooled fast, while the center cooled somewhat slower and had more time to form larger crystals.

On the uphill end of this roadcut are veins that cut the rhyolite. A 5-foot-4-inch geologist for scale. The close-up shows the brecciated nature of the vein's interior, the harder, more solid edge, and a Swiss Army Knife for scale. (30.6651, -104.0318)

A close-up look at the rhyolite outcrops west of Mt. Locke reveal several zoned feldspar crystals. (30.6758, -104.0706)

The road climbs the flanks of Mt. Locke, where the University of Texas McDonald Observatory is located. Several of the astronomical domes can be seen from the highway. The largest one belongs to NASA, and the next largest one houses the 107-inch McDonald telescope. The visitor's center is open daily, but you need reservations for public viewing of the telescope.

Watch for weathering effects in roadcuts west of Mt. Locke and the observatory turnoff. You can see spheroidal weathering where jointed blocks of lava weather to form rounded outcrops that look almost like boulders. You can also see exfoliation where curving layers of rock weather off the bouldery forms like onion peels. Some of the outcrops also display large feldspar crystals, many of which are zoned because the composition of the newly forming crystal structure changed as the composition of the magma changed. The large crystals formed in a magma chamber below ground, then the fine-grained, surrounding rock erupted and chilled so fast that only small crystals could form.

The road descends toward the west and the junction with TX 166, which skirts the south and west sides of the Davis Mountains. Several locally well-known sites—including Balance Rock, composed of rhyolite, and Point of Rocks, featuring large house-size boulders of granite—can be seen off TX 166.

From the TX 116 junction northward to Kent and I-10, note the low topography and desert vegetation in contrast to the grass and trees higher on the Davis Mountains. The road passes through a canyon-like stretch between mesas of dark volcanic rocks, where a series of 37-million-year-old Eocene lava flows stack up to nearly 1,000 feet thick. Near Kent and I-10, yellow-tan, flat-lying, thin beds of limestones, siltstones, and mudstones line the surrounding low mesas. These Cretaceous rocks were here before the Davis Mountain lavas pushed up through them and flowed out over them, forever changing the landscape of West Texas.

FARM ROAD 170
Study Butte—Presidio
67 miles

Farm Road 170, which follows the Rio Grande for much of the way, is one of the most scenic drives in the entire state of Texas. Dark volcanic mountains brood over lush, green riverside trees, while blazing white, intricately carved ramparts watch over the Rio Grande. The road soars over high passes to offer grandiose views, then plunges to river depths, where enveloping cliffs reveal their innermost geologic secrets.

Between the town of Study Butte and Terlingua, the road weaves through excellent roadcut exposures of the Boquillas Formation, yellow-tan, flaggy beds of mostly limestone, laid down in shallow seawater in Late Cretaceous time. About 1 mile west of the TX 118 junction, the road crosses Terlingua Creek, which flows south and empties into the Rio Grande at the mouth of Santa Elena Canyon (see Ross Maxwell Scenic Drive in Big Bend National Park section). Both Mesa de Anguila and the canyon walls can be seen on the southern skyline, whereas to the north, peaked hills of dark, volcanic intrusive bodies contrast sharply with the light-colored, layered limestones at road level.

About 4 miles west of Study Butte and, barely clinging to life like a tenacious desert plant, is the former mining town of Terlingua, once the mighty producer of one-fourth of the country's supply of mercury. Tailings, abandoned stone houses, and old mercury mine workings abound. Mining began at Terlingua in 1894 and continued under Chisos Mining Company auspices from 1902 to 1946. Howard E. Perry, whose mansion is preserved in Terlingua, was the owner from 1898 to 1942, when Brown & Root Construction Company took over. During its heyday, two thousand people worked in the mines and processing plant. Mercury has many uses, but fulminate of mercury was used in wartime blasting caps and bullet primers. Mercury is also a catalyst for making chlorine and was used in thermometers, switches, and gold extraction. However, as wartime applications decreased and as environmental knowledge has shown mercury to be poisonous to humans, animals, and plant life, the

Remnants of mercury mining around Terlingua, as viewed from FM 170. (29.3120, -103.6230)

Geology along Farm Road 170 between Study Butte and Presidio.

CENOZOIC

Q — recent sediments; includes alluvium, alluvial fan, and windblown sand deposits (Holocene and Pleistocene)

QTs — sedimentary bolson deposits (Miocene to Pleistocene)

VOLCANIC and IGNEOUS ROCKS

Tr — Las Burras Basalt Member of the Rawls Formation; includes Sauceda Lavas (Miocene–Oligocene)

Tfr — Fresno Formation and Santana Tuff (Oligocene)

Tmr — Morita Ranch Formation (Oligocene)

Tpc — Perdiz Conglomerate (Oligocene)

Tbh — Buck Hill Group; includes Duff, Pruett, and Tascotal Formations, and Mitchell Mesa Rhyolite (Eocene-Oligocene)

Tbb — Big Bend Park Group; includes Mule Ear Spring Tuff Member of the Chisos Formation (Eocene–Oligocene)

Ti — igneous rocks; includes stocks, vent rocks, laccoliths, sills, and dikes (Tertiary)

MESOZOIC
CRETACEOUS

Kt — Tornillo Group; includes Aguja and Javelina Formations

Ktb — Terlingua Group; includes Boquillas and Pen Formations

Kw — Washita Group; includes Santa Elena Limestone, Buda Formation, Del Rio Clay, and Sue Peaks Formation and Del Carmen Limestone of the Fredericksburg Group

Ktr — Trinity Group; includes Presidio, Shafter, and Glen Rose Formations

PALEOZOIC

MDO — Caballos Novaculite and Maravillas Chert (Ordovician–Mississippian) and Tesnus Formation of the Morrow Group (Mississippian–Pennsylvanian)

Oϵ — sedimentary rocks, undivided (Late Cambrian–Ordovician); includes Woods Hollow Shale, Fort Pena, Marathon Limestone, and Dagger Flat Sandstone

national park boundary

state park boundary

intermittent creek

normal fault

county boundaries shown as white dotted lines

0   5   10 miles

0   5   10 kilometers

CHINATI MOUNTAINS

CIENEGA MOUNTAINS

BREWSTER

PRESIDIO

BIG BEND RANCH STATE PARK

BOFECILLOS MOUNTAINS

BIG BEND NATIONAL PARK

Study Butte

Terlingua

Lajitas

Presidio

Ft. Leaton State Historic Site

Solitario Basin

SOLITARIO PEAK

EL SOLITARIO DOME

Fresno Creek

Terlingua Creek

Rio Grande

Alamito Creek

Main Park Road

Bofecillos Rd.

RED PLATEAU

MESA DE ANGUILA

SANTA ELENA CANYON

Comanche Creek

LAJITAS MESA

SANTANA MESA

flatirons along the Terlingua monocline

hoodoos

West Contrabondo Trailhead

picnic area with teepee shelters

Arenosa Group Camping Area

THREE DIKE HILL

hoodoos

USA / MEXICO

uses of mercury today are limited to mostly thermometers and barometers. Mining activity at Terlingua shut down in the early 1970s.

The mercury mineralization at Terlingua occurs as red veins of cinnabar, the mineral name for mercury sulfide. The veins cut through the surrounding limestone beds and volcanic rocks. Hot groundwater, carrying the mercury in solution, penetrated upward through cracks, faults, and fissures. As it rose, the solution cooled, leaving behind the cinnabar precipitate in fractures and small cavities in the limestone. The crushed ore was heated in a furnace to release pure mercury vapor, which was then liquified in large condensers in a process similar to the way water vapor condenses on a cold windowpane. The mercury was collected and sold in flasks, each weighing 76 pounds.

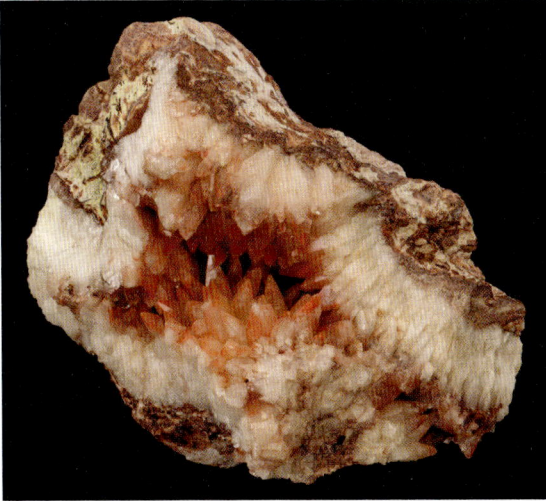

Calcite crystals found at Terlingua are stained red from cinnabar, the primary mercury ore mineral. —Photo courtesy of Rob Lavinsky

The gray-weathered, solid limestone cliffs of the Reed Plateau, south across the highway from Terlingua, are the Cretaceous-aged Santa Elena Limestone, just a bit older than the thin-bedded limestone of the Boquillas Formation at road level.

Between Terlingua and Lajitas, the road crosses several Cretaceous-aged limestone units including the Boquillas, Pen, Santa Elena, and Buda Formations. After passing thick, white limestone beds of steeply dipping Buda about 3.5 miles west of Terlingua, the topography begins to open into a broad valley, where large alluvial fans slope toward the Rio Grande to the south. Look to the right (north) 1.3 miles to the west of the Buda roadcut to see flatirons of Cretaceous-aged Boquillas limestone bordering the south flank of a large fold called the Terlingua monocline. This monocline marks the southern and southwestern edge of the larger Terlingua Uplift, a remnant of Laramide deformation.

Tilted sections of bedded, yellow-tan limestone of the Boquillas still greet you around Lajitas, located at the base of Lajitas Mesa to the north. Comanche Creek, a dry wash most of the year, joins the Rio Grande at Lajitas. Across the Rio Grande at Lajitas, notice how the large mesa surface curves downward toward the river, giving the distinct impression that rocks can bend, fracture, and break if pressure is applied slowly, rather than quickly, in the Earth's crust.

A fundamental geologic change takes place at Lajitas. Between Study Butte and Lajitas, the dark-brown volcanic piles and mounds are eroded intrusive bodies that drove up into, but not through, the surrounding sedimentary rock cover. West of Lajitas, however, the volcanic rocks seen from the roadside are mainly extrusive: they were either blasted from or flowed from a volcano. On the flanks of Lajitas Mesa, north of Lajitas, are wonderful exposures of dark lava flows and white ash deposits of the Chisos Formation (Big Bend Park Group). The road west of Lajitas traverses spectacular rock scenery carved from these volcanic eruptions that occurred between 47 and 32 million years ago. In roadcuts 1 to 2 miles west of Lajitas, you can see white

The white ash was deposited over this dark lava bed, 1 mile west of Lajitas. Air bubbles at the top of the lava indicate gas was being released to the surface as the flow cooled. (29.2640, -103.8040)

Eocene to Oligocene volcanics of the Chisos Formation on Lajitas Mesa at Lajitas. (29.2667, -103.7828)

ash beds lying on dark lava flows. Small faults that offset individual layers of ash are quite clear in these exposures.

Just west of the Presidio County sign is the east entrance to Big Bend Ranch State Park, the largest state park in Texas. FM 170 runs along the southern edge of the park, providing access to various trails and scenic sites. Five to six miles west of Lajitas, the road swings northward in a loop away from the river. At the head of the loop, where the highway dips down to cross the bridge over Fresno Creek, flaggy, light-tan to yellow limestone beds of the Boquillas Formation stripe the creek's vertical walls. The short access road to the West Contrabondo Trailhead, east of the bridge, provides great views of the surrounding landscape.

From the intersection of the access road to West Contrabondo, look at the mountain to the west to see a spectacular face of alternating dark lava and white ash. The thick, black lava layer has been offset because the front of the mountain collapsed. Geologists that have studied this feature interpret it as a gravity slide; however, the cause as to how or when it occurred remains a mystery. About 1.5 miles west of Fresno Creek is a small pullout on the north side of the road where a better view of the slide can be seen.

Equally spectacular hoodoos in eroded ash deposits are exposed along the road next to the river about 4.5 miles west of the Fresno Creek bridge. This ash is associated with the Oligocene-aged Mule Ear Spring Tuff, one of the products of the immense eruptions from the Chisos Mountains. An up-close look at this outcrop reveals holes and pits where crystals, softer than the surrounding ash, have been removed by the selective etching by sparse rainwater.

On the cliff above the picnic area with teepee shelters (just west of the Madera Creek bridge), a lava flow displays well-formed columnar joints. As the lava solidified, it shrank, and these contraction cracks developed from the top to the bottom of the flow.

An impressive gravity slide, which has displaced a section of dark lava flow, can be seen in a mountain side 1.5 miles west of Fresno Creek. (29.2859, -103.8660)

White ash-flow tuff eroded into hoodoos 10 miles west of Lajitas. (29.2873, -103.9087)

Between the picnic area and La Cuesta river access, FM 170 climbs onto the flanks of Santana Mesa to avoid a steep, narrow section of the Rio Grande, formed by a hard intrusive volcanic body. The climb provides a vantage point for marvelous views of the ragged volcanic scene to the west. The intrusive body flanks the east side of the big hill, but extrusive flows and ash compose the rest of Santana Mesa. Eroded lava hills form an almost endless variety of shapes west of Santana Mesa, as the Rio Grande's valley becomes wider and wider.

One mile northwest of the Arenosa Group Camping Area are more hoodoos and a balanced rock on the west (river) side of the road. These impressive features were eroded into alluvial fans composed of sandstone and conglomerates clasts eroded from volcanic rocks of the Cienega and Chinati Mountains to the north. A parking area and short trail provides access to the site.

Beginning about 2 miles west of the the hoodoo parking lot, the highway begins to traverse immense alluvial fans that slope from the northern skyline mountains toward the Rio Grande. About 2.5 miles northwest of the hoodoo parking lot is the west entrance sign to Big Bend Ranch State Park. At first, stopping here seems pointless until you turn around and look to the east. On a hillside are three distinct, dark stripes that give Three Dike Hill its name. In this one hill, four separate volcanic eruptive events are recorded. The tan rocks at its base are from the Fresno Formation, a series of trachyte lava flows that erupted from Fresno volcano around 32 million years ago. Next, around 27.5 million years ago, the Las Burras Basalt erupted, forming the horizontal black bands above the Fresno. Finally, the Leyva Canyon Formation erupted around 27.3 million years ago, forming the buff-colored cliff above the Las Burras. The three black dikes, as well as the black rocks that cap the hill, are the Sauceda Lavas, dating to 27.1 million years ago. At this time, basalt magma rose along numerous cracks and erupted as lava flows on the surface, creating the broad, flat landscape across Big Bend Ranch.

The Cienega Mountains, a sizeable Oligocene-aged intrusion, is seen to the north from Alamito Creek, and ahead in the distance north of Presidio are the Chinati Mountains, formed from several intrusions and piles of 30-million-year-old Oligocene extrusive rocks. Ft. Leaton State Historic Site, on the eastern edge of Presidio, speaks of border battles in the tumultuous times of the mid-1800s. The visitor center is also the stepping point for the magnificent, but treacherous, journey to El Solitario.

View west from the high point on FM 170 to the Rio Grande and surrounding volcanic terrain. (29.2970, -103.9479)

Three Dike Hill tells the story of four separate volcanic events that occurred here over a 5-million-year timespan. (29.3806, -104.1191)

EL SOLITARIO DOME IN BIG BEND RANCH STATE PARK

El Solitario, Spanish for "hermit" or "the lonely one," forms the only large, circular structure in the Big Bend region. Although the origin of the Solitario Dome has been argued by a few geologists to be a meteor crater, most geologists now accept the Solitario Dome was uplifted by a magmatic intrusion. Near the end of the Eocene, about 36 million years ago, magma began to rise toward the surface but solidified as dikes and sills that cut the older, preexisting rocks. Another large volcanic pulse at 35.4 million years ago filled ancient thrust faults from the Ouachita-Marathon mountain-building event, then began to inflate upward like a blister to create an intrusion called a laccolith. This lava bent the overlying rocks upward into a 10-mile-wide dome. The uplift also brought folded rocks of the old Ouachita Mountains near to the surface. During this uplift, some magma made its way to the surface and erupted violently, blasting as much as 6 cubic miles of ash and pyroclastic material into the air and creating a caldera. Much of the eruptive material then fell back into the caldera, along with debris from the caldera walls and slightly later eruptions that filled the caldera. When a laccolith intrusion erupts in this manner, it is called a laccocaldera.

Since its formation, erosion has sculpted and shaped the laccocaldera. Cretaceous-aged limestones, deposited when sea level rose and formed the Western Interior Seaway, form the sharp ridges and flatirons that can be seen from several miles away. A 400-foot-high rim of upturned Cretaceous limestones isolate the Solitario from outside view, which, in addition to its remoteness, probably accounts for early Spanish settlers naming this fascinating geologic feature El Solitario. In addition, the Paleozoic-aged Caballos Novaculite, the same formation exposed near Marathon, can be seen within El Solitario and creates a distinct layer in the folds created during the Ouachita-Marathon mountain-building event. El Solitario thus contains the most

Satellite image of El Solitario showing the distinctive ring structure of a magmatic intrusion called a laccocaldera. —Satellite image from Landsat/Copernicus, May 2024

The limestone flatirons surrounding El Solitario formed when magma pushed upward to create a dome. (29.4074, -103.8581)

southwestward exposures of rocks contorted and folded by the Ouachita event when Africa collided with North America to form the supercontinent Pangea.

To reach El Solitario, stop at the Ft. Leaton State Historic Site office east of Presidio to obtain a special vehicle pass and maps to reach the site. A high-clearance, four-wheel-drive vehicle is necessary to access the region because the two-track roads are extremely rough.

Big Bend National Park

The idea of creating a national park in the Big Bend country began when American troops were stationed in the area while Pancho Villa held sway in northern Mexico. Talk continued for thirty years until Big Bend National Park was officially established June 12, 1944. Ross A. Maxwell, a field geologist, became the park's first superintendent.

While colorful border towns, isolated ranches, mercury mining, and Native American lore provide a rich cultural history, Big Bend National Park is best known for preserving unique natural phenomena, both geological and biological. It contains an outstanding section of the Chihuahuan Desert, with plant and animal species that aren't found elsewhere. It is also a mixing zone where Rocky Mountain species from the north meet Mexican highland species from the south. Biologic zones climb from wet, moist floodplains of the Rio Grande through vast tracts of dry Chihuahuan Desert upward to the cool, moist elevations of the Chisos Mountains, where pine forests predominate.

The geology of Big Bend National Park strikes the visitor in an overwhelming display of topography, odd erosional forms, volcanic remnants, fossil beds, and sheer cliffs of clearly exposed stratigraphy. This great geologic diversity, all painted in rich, earthy colors, becomes even more captivating to one who wants to read the rocks for their meaning.

82 miles from Maverick Entrance to Alpine

Maverick Junction (entrance)

42 miles from Persimmon Gap Entrance to Marathon

Persimmon Gap Ranger Station (entrance)

Dog Canyon Trailhead

Dagger Flat

Qs

Ti

Kc

Kc

Terlingua Creek

Tv

Nine Point Mesa

Ti

385

Qs

Tv

Qs

CHALK DRAW FAULT

SANTIAGO MOUNTAINS

Pz

Pz

Pz

Qs

DESERT

Qs

CHIHUAHUAN

118

Ti

Kc

Ti

Ti

Ti

Ks

Dagger Mountain

Tv

Kc

Ks

2627

Rio Grande

USA MEXICO

Solitaro Peak

Pz

Ti

Corazon Mountains

Qs

Ti

Rosillos Mountains

Ti

Main Park Road

SIERRA DEL CARMEN

Kc

Kc

Christmas Mountains

Ks

Paint Gap Hills

Tbp

Tbp

Ti

Ks

Grapevine Hills

Ti

Study Butte

Terlingua

Ti

Ks

West Road

Ti

Panther Junction

QTs

Ti

Tornillo Flat

170

Qs

Ks

Tv

Basin Rd

12

Ti

Tornillo Creek

Old Ore Road

East Road

Fossil Discovery Exhibit

Lajitas

Tv

Ti

Qs

Ross Maxwell Scenic Drive

Old Maverick Road

MESA DE ANGUILA

CHISOS MOUNTAINS

Nugent Mountain

Ti

Ks

Boquillas Canyon

50 miles from Lajitas to Presidio

Tv

Castolon

Qs

SIERRA PONCE

Ti

Ks

Rio Grande Village

tunnel at milepost 19 through Santa Elena Limestone

Terlingua Creek

BIG BEND NATIONAL PARK

Kc

River Road

Santa Elena Canyon

TERLINGUA FAULT ZONE

Ks

DUGOUT WELLS FAULT ZONE

SIERRA DE CHINO

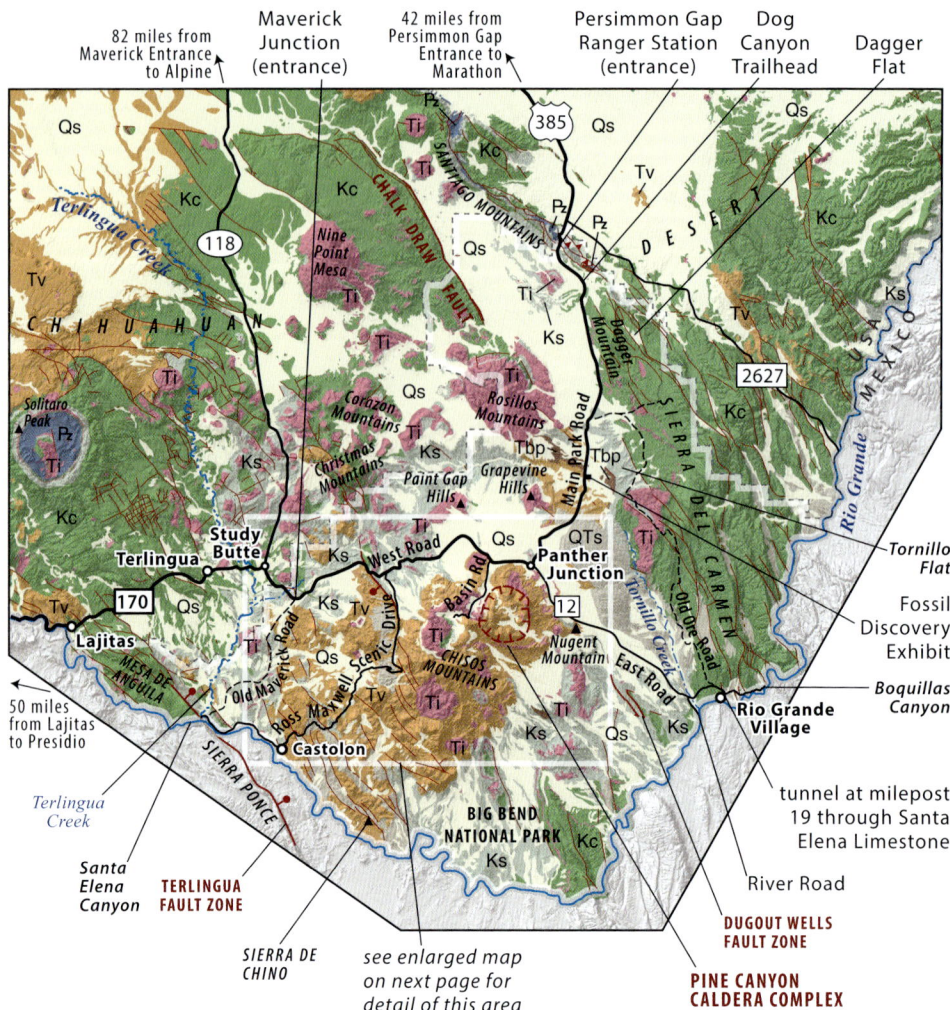

see enlarged map on next page for detail of this area

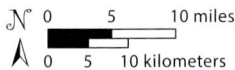

PINE CANYON CALDERA COMPLEX

N

0   5   10 miles

0   5   10 kilometers

fault

thrust fault; teeth on overriding plate

normal fault; ball and bar on downthrown side

caldera

national park boundary

CENOZOIC

Qs	sediments and sedimentary rocks, undifferentiated (Quaternary)
QTs	sedimentary deposits; includes Tertiary basin fill (Miocene–Pleistocene)
Tbp	Black Peaks Formation (Eocene–Oligocene)

VOLCANIC and IGNEOUS ROCKS

| Tv | volcanic rocks, undivided; includes Burro Mesa Rhyolite, South Rim Formation, and Chisos Group of the Big Bend Park Group (includes Mule Ear Spring Tuff Member) (Eocene–Oligocene) |
| Ti | intrusive rocks, undivided (Neogene and Paleogene); includes Government Spring laccolith |

MESOZOIC

| Ks | sedimentary rocks of the Tornillo Group (includes Aguja and Javelina Formations) and Trinity Group (includes Maxon and Glen Rose Formations) (Cretaceous) |
| Kc | carbonate sedimentary rocks, undivided; includes Santa Elena Limestone of the Washita Group, Pen and Boquillas Formations of the Terlingua Group, with some Trinity and Fredericksburg Groups (Cretaceous) |

PALEOZOIC

| Pz | older sedimentary rocks, undivided |

Geology of Big Bend National Park. Area in the white box is on the facing page.

Maverick Junction (entrance)

Rosillos, Corazone, and Christmas Mountains

Government Spring laccolith

SIERRA DEL CARMEN

Croton Peak

Ti

Tv

Ti

118   Kc

Study Butte

Dogie Mountain   Kc

Maverick Mountain   170

Kc   Ks

TORNILLO BASIN

Lone Mountain

Gano Springs Road

385

Main Park Rd

Panther Junction

Dawson Creek

Old Maverick Road

Ross Maxwell Scenic Drive

BURRO MESA FAULT

the Window

Vernon Bailey Peak   Pulliam Peak

Panther Peak

Green Gulch   Panther Peak

Wright Mountain

Basin Rd

Tv

QTs

Ks

Ti

Carter Peak   Panther Pass   the Basin

Pummel Peak

East Road

12

Tule Mountain

Burro Mesa

Casa Grande Peak

Lost Mine Peak

Tv

Nugent Mountain

Ward Mountain   Kc

BIG BEND NATIONAL PARK

Ti

QTs

HAYES RIDGE   Ti

Ks

Ti

QTs

SOUTH RIM

Ti

Qs

Chilicotal Mountain

Tv

Goat Mountain

SIERRA QUEMADA

CHISOS MOUNTAINS

Tbp

Ti

Tuff Canyon

Mule Ear Peaks   DELAHO BOLSON

Tv

QTs

Elephant Tusk

Ks

Ti

U.S.A.

MEXICO

Castolon

Cerro Castellan

Ti

Qs

Rio Grande   volcanic spine   SIERRA DE CHINO

PINE CANYON CALDERA COMPLEX

N   0   5 miles

0   5 kilometers

CHISOS MOUNTAINS

SOUTHWEST

SIERRA DE CHINO   DELAHO BOLSON   SIERRA QUEMADA   SOUTH RIM   HAYES RIDGE   PINE CANYON CALDERA COMPLEX   TORNILLO BASIN

NORTHEAST

SIERRA DEL CARMEN

8,000
4,000
sea level
-4,000 feet

TKs   Tv   Ts

K   Ti   Ti   K

Pz   Pz

2x vertical exaggertion

Ts	basin fill (Pliocene–Pleistocene)
Ti	intrusive rocks (Oligocene)
Tv	sedimentary-volcanic rocks, includes Chisos and South Rim Formations (Eocene–Oligocene)

TKs	Black Peaks Formation (Late Cretaceous–Eocene)
K	Cretaceous sedimentary rocks
Pz	Paleozoic sedimentary rocks

0   5 miles
0   5 kilometers

—— fault

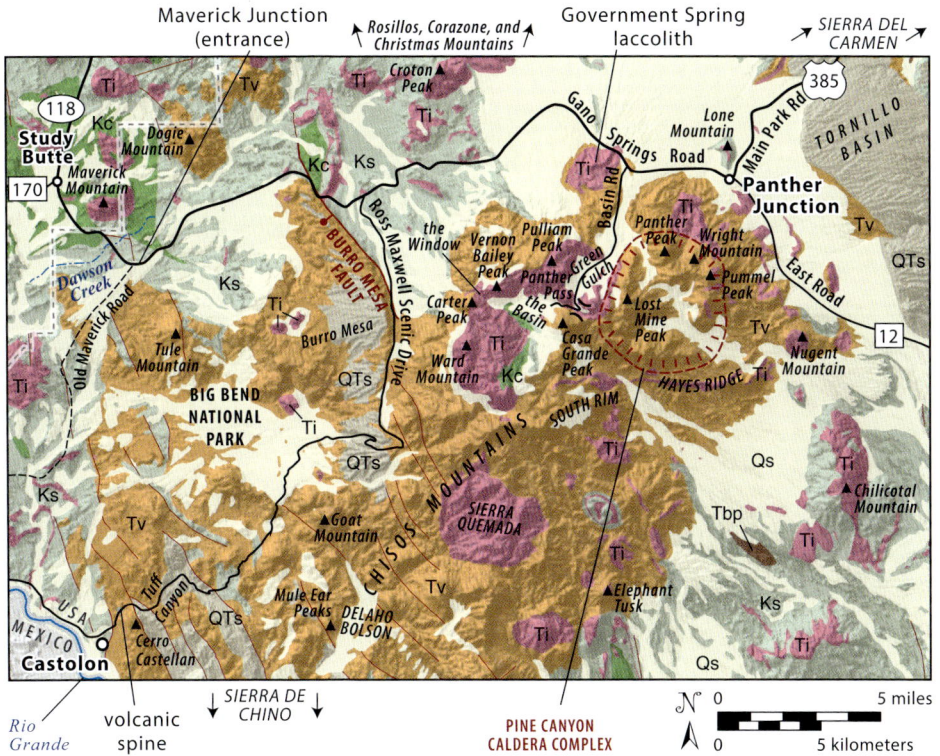

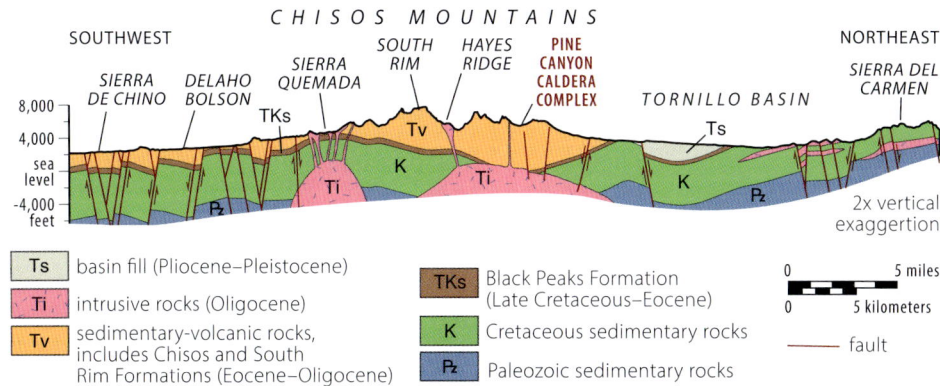

Cross section of the Chisos Mountains in Big Bend National Park. —Modified from Turner and others, 2011

Big Bend stands astride the geologic imprint of two major mountain ranges: the Ouachitas of late Paleozoic time and the younger Rockies. Rocks of the Ouachita Mountains come to the surface along the northern edge of the park. At Persimmon Gap, for example, Paleozoic-aged rocks of Ouachita origin are thrust over Cretaceous limestones. The inexorable push of the Rocky Mountains, responding to plate-crunching motion along the western margin of the North America continent, is dramatically preserved in the northwest-oriented faults, folds, and block mountains forming the Sierra del Carmen on the east side of the park and Mesa de Anguila on the west side.

After the uplifting of the Ouachita Mountains in Paleozoic time, about 270 million years ago, the range was eroded, and thick piles of Mesozoic sandstones, shales, and especially limestones were laid down over the old range. Carbonate deposits composed mostly of the shells of sea-dwelling organisms piled up in the elongate Western Interior Seaway, which extended from the Arctic to the Gulf of Mexico in Cretaceous time. These thick sections of stratigraphy were dramatically elevated thousands of feet above sea level when the new mountain-building episode at the end

STRATIGRAPHIC COLUMN OF THE BIG BEND AREA			
GEOLOGIC AGE		**ROCK NAME**	**ROCK DESCRIPTION**
CENOZOIC / PALEOGENE / OLIGOCENE	Big Bend Park Group	Burro Mesa Rhyolite	rhyolite
		South Rim Formation	rift basin sediments, lava, and flow breccia
		Mule Ear Spring Tuff member of the Chisos Formation	tuff
EOCENE		Chisos Formation	volcanic rocks including lava, ash, tuff, conglomerate, and sandstone
PALEOCENE	Tornillo Group	Black Peaks Formation	mudstone, sandstone, and conglomerate
MESOZOIC / CRETACEOUS		Javelina Formation	varicolored nonmarine mudstone, fossil wood, and dinosaur bone
		Aguja Formation	shoreline dark mudstone, sandstone, coal, fossil wood, and dinosaur bone
	Terlingua Group	Pen Formation	dark marl, marine mudstone (weathers to yellow), contains concretions
		Boquillas Formation	marine flaggy limestone, chalk, and marl
	Washita Group	Buda Formation	white limestone; carbonate shelf
		Santa Elena Limestone	thick limestone beds; deeper shelf
	Trinity Group	Glen Rose Formation	thick limestone beds, marl, dolomite, sandstone, mudstone, conglomerate; deeper shelf
		unconformity	
PALEOZOIC / PENNSYLVANIAN / MISSISSIPPIAN	Morrow Group	Tesnus Formation	sandstone, shale, chert, and conglomerate
ORDOVICIAN	Cincinnatian Group	Maravillas Chert	dark gray to black chert

Stratigraphic column of important rock units in Big Bend National Park.

of Cretaceous time created the Rocky Mountains. The cliffs at Santa Elena Canyon and Boquillas Canyon tell this story of limestone deposition and subsequent uplift.

On the heels of Rocky Mountain compression came a period of apparent continental relaxation when western North America began to pull apart. Normal faults broke the crust and huge volumes of hot, molten mantle material shot upward into these new weaknesses. The magma filled cracks and pushed upward, shoving at the overlying sedimentary rocks, sometimes reaching the surface to erupt lava, liquid, and gas in violent volcanic episodes that must have made Mt. Saint Helens look like a birthday candle. Other vast amounts of hot rock never breached the thick limestone overburden but caused subterranean masses of various sizes and shapes, some even squeezing laterally between layers of the sedimentary column. This volcanic burst in Big Bend occurred simultaneously with other eruptions in New Mexico and Colorado, leaving behind vast volumes of lava, ash, and debris. The Davis Mountains north of Big Bend National Park, for example, are entirely built of volcanic material from this great episode.

The volcanic rocks in Big Bend consist of three main units: the 46- to 33-million-year-old Chisos Formation; the 32-million-year-old South Rim Formation, which erupted from the Pine Canyon caldera; and the Burro Mesa Rhyolite, at 29.4 million years old, the youngest major eruptive unit in the park.

Much of the topography in Big Bend National Park is due to faulting in Miocene time that lifted and dropped great blocks of rock to form mountains and intervening basins. This activity is part of the broader pattern of Basin and Range extension that stretched the crust from eastern California to West Texas. The central part of Big Bend, called the Sunken Block or Tornillo Basin, is one of these down-dropped basins. The Sierra del Carmen on the east and the Sierra de Santa Elena on the west (in Mexico) form the great blocks on either side of the Tornillo Basin. The main period of this extension was from about 26 to 10 million years ago. The down-dropped basins are filled with sediment shed from adjacent highlands. Basin-fill sediments are exposed in eroded banks of Tornillo Creek.

After all the tectonic activity ended and the pyrotechnics of volcanic eruptions and intrusions ceased, Big Bend relaxed and was quiet for millions of years. During this quiescent period, the unending forces of water and gravity worked their incessant magic. During the Pleistocene ice ages, Big Bend was not as dry as today. When the region was wetter, trickles became rivulets, and streams turned into boulder-chunking torrents. Thousands of cubic miles of stones, rocks, and sand grains have been carried out of this region in the last few million years—a short period, really, in the cosmos of geologic time.

Erosion has not worked with an even hand over the landscape, however. If it did, we would see a flat surface. Instead, mountains, canyons, valleys, and mesas stand in unequal stature all across the face of Big Bend. The rocks have something to say about how fast they erode. Soft sediments erode quickly as their loose particles move under the pattering of raindrops. Hard basalt layers, in contrast, mightily resist even the abrasion of landslide scraping or boulder-laden river pounding. But, not forever, and even hard basalt layers and tough limestone ledges eventually weather away.

THE NORTH ROAD (MAIN PARK ROAD)
PERSIMMON GAP—PANTHER JUNCTION
27 miles

The northern park entrance is in Persimmon Gap, a low-elevation pass through the northwest-trending Santiago Mountains. This pass has been the location of a road or trail for hundreds of years. First known as the Comanche Trail, it was then followed by army camel trains, cowboys, freighters, Texas Rangers, stagecoaches, and ore wagons. Today's travelers drive through the pass in air-conditioned cars on smooth pavement. Stop at the Persimmon Gap Ranger Station to see geologic exhibits and maps of the Persimmon Gap area and Big Bend National Park.

The Santiago Mountains at Persimmon Gap have a complex structure. Paleozoic rocks, similar to those seen near Marathon, are thrust-faulted—older rocks over younger—in Persimmon Gap. This thrusting occurred during the Ouachita mountain-building event, beginning about 300 million years ago. Then, during Rocky Mountain building event, about 70 million years ago, these rocks were again thrust-faulted but this time over younger Cretaceous rocks. The Cretaceous rocks are internally thrust-faulted as well.

You can see some of this thrusting in a single hill to the east about one-quarter mile north of the entrance station (milepost 27). The Cretaceous Glen Rose Formation, at the top of the hill, is part of a large fold that includes the Paleozoic thrust fault in the older Tesnus and Maravillas Formations. The entire package was then thrust faulted over younger Cretaceous rocks of the Boquillas, Pen, and Aguja Formations, exposed at the base of the hill.

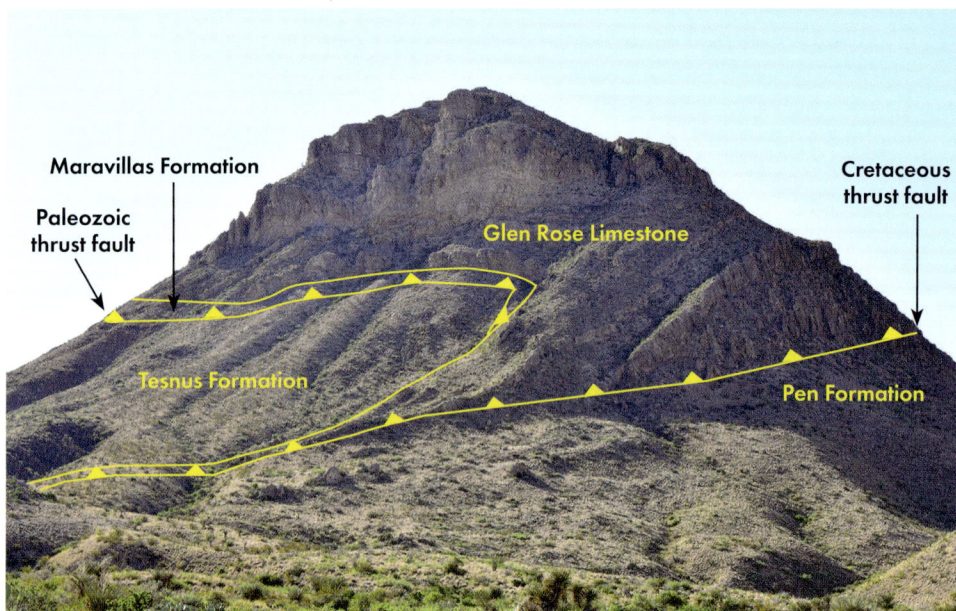

The mountain to the east at Persimmon Gap preserves two thrust faults associated with the Ouachita and Laramide mountain-building events. (29.6630, -103.1718)

A large rockfall along the wall to the east of the Main Park Road near the Dog Canyon Trailhead. (29.6227, -103.1430)

South of Persimmon Gap, the ridges and walls of rock to the east are Cretaceous limestones, elevated in blocks along northwest-oriented normal faults. These ranges are part of the Sierra del Carmen, which extends from Persimmon Gap southward to the Rio Grande. As you drive along, especially near the turn for Dog Canyon Trail, look for a patch of white rockfall in the Cretaceous walls to the east, where a big section of limestone gave away and tumbled to the talus slope below. Multiply this process thousands of times over thousands of years to understand how mighty mountains eventually come tumbling down.

To the west, Cretaceous rocks are punctuated by dark-brown, weathered, intrusive rocks of Oligocene age. Weathering has removed rock from atop and around these hard intrusive bodies, and they now stand higher than the surrounding countryside. The Rosillos Mountains predominate on the skyline to the southwest. Note the brown color and massive, rounded, but internally structureless appearance of these mountains, indicating the rocks are part of a large, intrusive igneous mass, one of but many such bodies in the Big Bend area.

Dagger Mountain, a large dome to the east a couple of miles south of the Dog Canyon Trail, is an anticline, where beds of Santa Elena Limestone have been arched up all around the dome. Dagger Mountain was probably uplifted by an intrusive body, which is still buried. East of Dagger Mountain is Dagger Flat, where mines exploited ore in the Boquillas Formation. When sills of 32-million-year-old gabbro, an intrusive igneous rock similar in composition to basalt, intruded the Cretaceous limestones, the magmatic fluids altered the carbonate rocks, creating mineralized deposits called skarn. The deposits include the minerals wollastonite, garnet, pyroxene, and prehnite. Mines exploited low concentrations of copper, lead, zinc, and molybdenum ores.

About 15 miles south of Persimmon Gap, the road enters colorful flatlands and badlands terrain known as Tornillo Flat. Here, brightly colored, purple, white, gray,

View to the south from Tornillo Flat, with the Chisos Mountains on the skyline. In middle distance are badlands of the Late Cretaceous to Eocene Black Peaks Formation, capped by a flat river terrace. (29.4678, -103.1378)

and tan shales and sandstones of the Black Peaks Formation were laid down in stream channels and floodplains in the Paleocene and Eocene Epochs. Particularly noticeable are flat terrace surfaces where erosion has not yet created badlands. In addition to being colorful, the Black Peaks Formation also shows evidence of one of the greatest natural events the world has ever seen. Rocks contained in the Black Peaks were laid down during and after the mass extinction event that occurred about 66 million years ago at the Cretaceous-Paleogene (K-Pg) boundary. See the discussion about the event in this book's introduction. In the Black Peaks Formation to the east of the Fossil Discovery Exhibit is one of three known areas within Big Bend National Park where the K-Pg boundary can be observed. The others are in the Javelina Formation between the western park entrance and Maverick Entrance Station and 9 miles off Park Road 12 (The East Road) along Glenn Springs Road.

These colorful rocks are also famous for their treasure trove of fossil bones of primitive animals that roamed these parts from the Cretaceous through the Eocene. You can see some of their actual remains at the Fossil Discovery Exhibit, about 19 miles south from Persimmon Gap (or 8 miles north of Panther Junction). The side road heads eastward off the highway to a parking lot, where cross-bedded sandstones and pebble conglomerates deposited in stream channels are exposed. The varicolored shales were deposited in swampy, muddy areas between the channels. In this ancient environment lived the ancestors of today's mammals—dawn horses, gazelle-like camels, and mouse-sized animals. Other creatures look somewhat familiar, but seeing the illustrations at the exhibit makes you wonder how these creatures could possibly be related to anything living today. Their names are also equally unfamiliar—*Ptilodus, Titanoides, Coryphodon*—but the illustrations of these animals give you at least a feel for the fauna that lived in Big Bend from the Late Cretaceous through the Eocene.

Between the fossil exhibit and Panther Junction, the Grapevine Hills lie to the west. These hills are the eroded remnant of a mushroom-shaped igneous intrusion called a laccolith. As the road approaches Panther Junction, the skyline of the Chisos Mountains is dead ahead and the little hill, standing alone to the west next to the highway, is appropriately named Lone Mountain. It is capped by a horizontal layer of

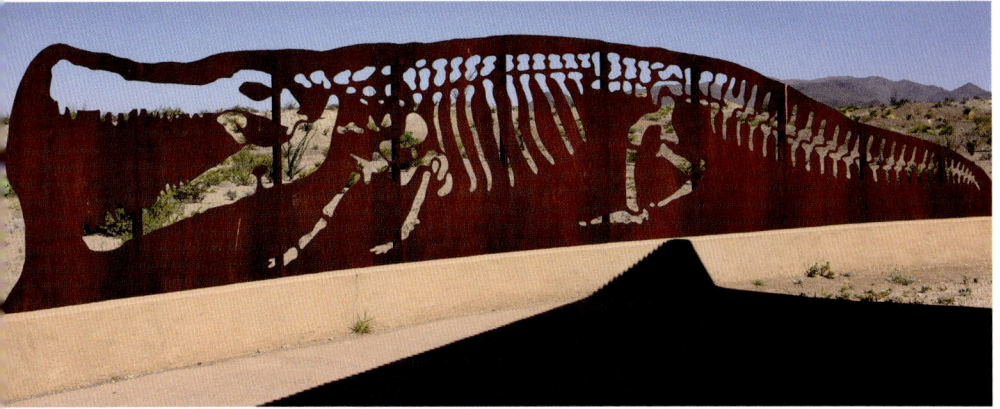

The life-sized, steel cutout of Deinosuchus, the 35-foot-long dinosaur killer of the day, is a must-see at the Fossil Discovery Exhibit. (29.4191, -103.1375)

Nearly complete skeletons of the giant flying reptile Quetzalcoatlus northropi have been found in the Javelina Formation near the Discovery Exhibit. These animals had a wingspan as wide as 40 feet and, when on the ground, stood as tall as a giraffe. —Artwork courtesy of Tama Huguchi

igneous rock, a sill that injected laterally between shale layers of the Cretaceous Aguja Formation. This hard rock protects the soft shale from erosion. At the park headquarters in Panther Junction are books, maps, pamphlets, and displays on the geology, as well as the flora, fauna, and history of Big Bend National Park.

THE EAST ROAD (PARK ROAD 12)
PANTHER JUNCTION — BOQUILLAS CANYON
23 miles

Southeast of Panther Junction, Park Route 12 traverses alluvial fan deposits of sand and gravel that slope away from the Chisos Mountains. The gravelly sediment was eroded from the Chisos Mountains and carried outward in bursts of watery flow by streams swollen by downpours following desert thunderstorms. Views to the southwest are particularly good of the upper volcanic crags—Wright, Panther, and Pummel Peaks—that compose the distinctive skyline of the Chisos Mountains.

About 4 miles southeast of Panther Junction, and south of the road, Nugent Mountain stands out as an outlier of the Chisos Mountain complex. This hard mass of intrusive igneous rock was exposed as the softer, surrounding rock eroded away. Directly south of Nugent Mountain, and to the west as you continue south, is Chilicotal Mountain, a body of intrusive rock surrounded by shales and sandstones of Cretaceous age. *Chilicote* is the Spanish name for the coral bean bushes that grow on the mountain. Look also for the characteristic shape of Elephant Tusk, a peak farther south.

On the east side of the road between Nugent Mountain and River Road are gentle, rather shapeless hills and tan-brown sand and conglomerate deposits. These stream deposits were dissected by later streams, leaving high-standing terraces. Look for flat

Nugent Mountain, a resistant igneous intrusion of the Chisos Mountains. (29.2739, -103.1585)

Sandstone of the Aguja Formation (Cretaceous) at the junction of East Road and River Road. (29.1903, -103.0157)

terrace surfaces low in the distance at about milepost 13. We are in the heart of the Sunken Block, the down-dropped basin drained by Tornillo Creek. Ahead on the skyline is the massive limestone wall of the Sierra del Carmen, an uplifted block.

Where River Road turns to the south (milepost 16), roadside exposures of yellowish to brown sandstone of the Aguja Formation display cross-bedding. Silicified wood and dinosaur bones have been found in this formation of Cretaceous age.

About a quarter mile east of the River Road junction, the road crosses lower Tornillo Creek on the River Bridge. You might wonder why such a substantial bridge is built across this dry wash. Flashfloods are common in the desert following thunderstorm downpours, and water has actually flowed over the bridge here!

On the east side of the bridge is a spectacular change in geology where one formation contacts another. The yellow-weathered clays of the overlying Cretaceous Pen Formation form a sharp contact with the white limestone of the Boquillas Formation, which lines the road as it climbs up from the bridge. Abundant marine fossils of large clams can be found in the limestone. Look, too, for a small vertical dike that cuts across the overlying clay deposits a few hundred yards from the bridge. The road traverses through more white limestone of the Boquillas Formation before reaching the Old Ore Road, a route used by mule and pack trains in the early 1900s to transport ore from mines in Mexico to the railroad station in Marathon.

East of the Old Ore Road junction, the white Boquillas limestone gives way to brown, thicker-bedded limestone of the Buda Formation. Hard, brown-weathered beds of the Cretaceous Santa Elena Limestone form the hills and are the rocks through which the tunnel (near milepost 19) was bored.

All the limestone beds from the tunnel to Boquillas Canyon and Rio Grande Village are the Santa Elena Limestone of Cretaceous age. At the Boquillas Canyon junction, look north to see an anticline, or upward fold, in beds of the Santa Elena.

The road to Rio Grande Village crosses terrace gravels deposited from an older stage of the Rio Grande, and the village is located on the low-lying, modern-day floodplain of the river.

Boquillas Canyon Road cuts into the Sierra del Carmen, a range composed of northwest-trending blocks of rocks impressively uplifted along linear faults that also

An igneous dike cuts the Cretaceous Pen Formation, one-third mile east of the Tornillo Creek bridge. (29.1937, -103.0056)

Tunnel through Santa Elena Limestone of Cretaceous age. The limestone ridge on the skyline is the Sierra del Carmen. (29.2021, -102.9788)

A thick section of Cretaceous Santa Elena Limestone frames the entrance to Boquillas Canyon. (29.1942, -102.9238)

trend to the northwest. As the Sierra del Carmen was uplifted, the Rio Grande cut downward, forming Boquillas Canyon. The road to the Boquillas Canyon parking lot crosses Santa Elena Limestone for a while, then passes some well-exposed road-cuts of stream gravels on the way down to the parking lot. To see the amazing canyon mouth, hike the moderate, 1.5-mile-round-trip Boquillas Canyon Trail. Along the way, look for round potholes on the limestone bank of the river where Native Americans ground their corn. Banks of sand farther on attest to higher flow levels of the Rio Grande. You are standing on a gravel bar composed of cobbles when you can finally see the canyon mouth. Cobble-sized abrasives such as these is how a stream like the Rio Grande can cut such a slash in solid rock, given enough time. Look for ripples in the wind-piled stack of sand to the left of the canyon. And, if you look closely, you will see a nearly vertical fault off-setting beds of limestone inside the canyon on the right-hand wall (looking downstream). The Santa Elena Limestone in Boquillas Canyon is the same rock formation that forms much of the cliff walls at Santa Elena Canyon on the other side of the park.

THE BASIN ROAD
WEST ROAD—CHISOS BASIN VISITOR CENTER
7 miles
See map on page 279.

Basin Road turns south from the main highway (Gano Springs Road) 3 miles west of Panther Junction. Immediately, the road begins to climb, traveling on the graveled surface slope of the alluvial fan that splays outward from the Chisos Mountains. On Pulliam Peak, to the right of the road as you head south, look for the profile of Alsate's face as he lies on his back, looking skyward. Native American legend says that when

Alsate—an Apache leader—was killed, the land shook and his face appeared on the side of the mountain.

The upper parts of the Chisos Mountains to the left, east of the road, are mainly extrusive igneous rocks—lava flows, ash deposits, and broken rock (breccia) that spewed out of volcanoes from 46 to about 33 million years ago in Eocene time. The lower slopes are composed of intrusive igneous rock that was pushed upward into the overlying older volcanic and sedimentary rocks about 32 million years ago in Oligo-cene time but never reached the surface. The mountains to the right, west of the road, are massive bodies of these intrusive igneous rocks, part of the South Rim Formation of the Pine Canyon caldera complex. Pulliam and Vernon Bailey Peaks and Ward Mountain in the Chisos Mountains are part of the same large, intruded igneous body.

As the road enters Green Gulch, the main north-draining wash of the Chisos Mountains, you begin to see solid rock on either side of the highway. Look for dikes that cut the rocks on the lower mountain slopes. You will also get a view of talus slopes forming skirts of eroded rock, sand, and gravel on the lower slopes beneath steep cliff walls. Weathered cracks and joints give the brownish cliffs of intrusive igneous rocks the appearance of castles.

At about 4 miles down the road, watch for the sign that says you have just reached 5,280 feet—1 mile above sea level. Casa Grande Peak, the biggest castle-like battle-ment, looms up ahead. The road twists and winds upward through green shrubs and conifer trees, far different from the cactus, yucca, and ocotillo in the desert below. The apex of the road is at 5,800 feet at Panther Pass, which is the drainage divide between Green Gulch and the Basin ahead. The trail leading to Lost Mine Peak begins from here.

View to the south at milepost 3 in Green Gulch. Pulliam Peak is to the far right, Casa Grande at center right, and Lost Mine Peak to the left. (29.2944, -103.2689)

View looking west from Chisos Basin toward the Window, the gap where Oak Creek exits the Basin. Ward Mountain lies to left. Carter Peak is immediately left of the Window. (29.2754, -103.2936)

After negotiating a couple of genuine hairpin turns, the downward trek of the road takes you into the heart of the Chisos Mountains—the Basin. Gnarled and jagged peaks 2,500 feet above the valley floor surround you now at every point of the compass. With all the volcanic rocks around, you might think the Basin is a crater or caldera, but it is not. The Basin is the pure product of differential erosion, with the harder intrusive rocks of Pulliam and Vernon Bailey Peaks and Ward Mountain standing high, while the softer rocks eroded. When the protective lava cover, seen as a remnant on Casa Grande Peak, was penetrated by erosion, softer rocks below were swiftly eroded away and carried out of the Basin through the gap in the mountain facade known as the Window. All the water and sediment carried off the slopes and peaks into the Basin go out through the window.

The meaning of the word *Chisos* is interesting. It is generally accepted locally that the word means "ghost" or "spirit," referring to the mystical aspect of the Chisos Mountains. An Apache word *chish-ee* means "people of the forest," and Native Americans living in the Chisos Mountains at the time of Spanish contact were mountain people. Maybe they were thus called *chivos*, meaning "goat" or "mountain goat." *Chisos* is probably the composite result of these near-match words.

THE WEST ENTRANCE ROAD
(GANO SPRINGS ROAD AND PANTHER JUNCTION ROAD)
PANTHER JUNCTION—STUDY BUTTE

24 miles
See map on page 279.

From the park headquarters at Panther Junction, you can see Lone Mountain sitting lonely above the alluvial fan plains to the north. It is capped by a resistant igneous sill, a sheet of magma that intruded between layers in the Cretaceous rocks. Look south for magnificent views of the Chisos Mountains, with Pummel, Wright, and Panther

Peaks forming the three highest crags—all part of the central pile of volcanic rocks of the Chisos Mountains.

North of Panther Junction, the park road crosses sloping alluvial outwash emanating from the Chisos Mountains. At Government Spring, just west of the turn-off to the Basin, the road crosses a mound of volcanic intrusive rock known as the Government Spring laccolith. This mushroom-shaped body of lava intruded between layers of the overlying sedimentary rock, domed them up, and then cooled in place. It was later exposed by erosion. As the road heads northwest, the skyline ridge to the north is the Paint Gap Hills, more brown volcanic rocks. Croton Peak is at the west end of this range. The Rosillo Mountains, Corazon Mountains, and Christmas Mountains, east to west, form the distant skyline to the north. West of Croton Spring Road is a little sentinel of yellow sandstone (Aguja Formation) over gray shale, standing to the right of the road between mileposts 11 and 12.

A lone pinnacle of Aguja Formation sandstone stands alone in the desert 11.2 miles west of Panther Junction. (29.3332, -103.3663)

At the turn to Castolon on Ross Maxwell Scenic Drive (milepost 13), Burro Mesa lies to the southwest. Lava flows cap the mesa, forming a resistant layer to erosion. The rhyolite lava was extruded 29.4 million years ago and is called the Burro Mesa Rhyolite. The Burro Mesa fault, a down-to-the-southwest normal fault, lies at the edge of the mesa, separating the volcanic rocks from Cretaceous rocks to the east. Dissected terrace and fan deposits, cut by stream erosion, produce badlands to the north of the road, just west of Ross Maxwell Drive junction. Nearby Dogie Mountain, also north of the road, is a volcanic intrusion.

As you drive along the flats west of Burro Mesa, you are treated to many fine views. Tule Mountain, looking like a ship's prow, is an obvious local landmark to the southwest. The cliff-forming layer at the top is a member of the Chisos Formation named the Tule Mountain Trachyandesite, an extrusive igneous rock with little to no quartz. Below the hard cap are Chisos tuffs. Maverick Mountain, another volcanic intrusion,

Tule Mountain, a distinctive landmark, is capped with a lava flow. (29.3007, -103.4842)

is the large mountain to the north of the highway. In the distance, dead ahead to the southwest, is the formidable wall of Sierra Ponce in Mexico and Mesa de Anguila in Texas. Santa Elena Canyon is the gap in the wall through which the Rio Grande flows.

To the north and west of the highway, knobby, chaotic, disjointed-looking, dark-brown volcanic peaks stand above dissected fans and badlands, colorfully exposed in hues of yellow, white, and tan. The colorful rocks are part of the Cretaceous-aged Javelina and Aguja Formations, in which dinosaur bones and petrified wood are preserved.

West of the old Santa Elena Canyon Road (Old Maverick Road) junction, the highway heads north and drops down into the badlands where Dawson Creek has cut into colorful Cretaceous-aged clay beds. Notice the terrace levels—flat tops—and erosional forms in the badland terrain.

The road climbs out of the badlands near Study Butte, an old mining town. The town is pronounced "Stoody," after a local doctor. Note the red and black mine tailings from the Brewster County Consolidated Mine piled north of the highway at the outskirts of town. The colors are indicative of the mercury mining that took place here. Unusual yellow, gray, and white hoodoos of erosional origin are visible east of the central town junction and contrast with the dark-brown volcanic rocks north of town.

ROSS MAXWELL SCENIC DRIVE
WEST ROAD—CASTOLON AND SANTA ELENA CANYON

30 miles

See map on page 279.

The turnoff to Ross Maxwell Scenic Drive from the main highway (West Entrance Road) is 13 miles west of Panther Junction. This scenic route honors Ross Maxwell, a geologist and first superintendent of Big Bend National Park. Numerous turnouts and illustrated signs explain the geology along this road.

As you head south, the flat-topped mountain to the right (west) is Burro Mesa. Along the sharp, top edge is the Burro Mesa Rhyolite, a resistant 29.4-million-year-old stack of lava. Vertical cracks in the face of this ledge formed as the lava cooled and shrank. At the base of the steep slope is the Burro Mesa fault along which the mesa was uplifted during Basin and Range faulting in Miocene time.

East of the scenic drive, the Chisos Mountains dominate the skyline, and Ward Mountain is unmistakable out in front. Ward Mountain is part of a larger igneous body that is related to, but slightly older than, the Pine Canyon caldera and South Rim Formation. Ward Mountain is composed of mostly granite and shallowly intruded, flow-banded rhyolite that forms the western half of the Chisos Mountains. Between the road and Ward Mountain are a series of Eocene-aged vertical dikes related to the Ward Mountain intrusions that cut across and rise above the horizontal bedding of the similar-aged Eocene volcanic rocks (the Chisos Formation), producing walls that snake across the topography like the Great Wall of China. Look for them to the east of the road at about 4 miles from the main highway, where signs at a turnout explain their geology.

Between mileposts 10 and 11, after you make a couple hairpin turns into the valley, look north for a spectacular view of the colorful stratigraphy on Burro Mesa. Hard, brown-weathered rhyolite lava flows cap its namesake mesa. Below that, orange, red-brown, yellow, tan, and white beds of ash, tuff, lava, conglomerate, and breccia (all associated with the Chisos Formation volcanics) tell the tale of a series of volcanic eruptions, explosions, and lava flows that must have overwhelmed the landscape toward the end of Eocene time.

Ward Mountain, a mixture of granite and rhyolite, forms the western half of the Chisos Mountains. In the foreground are a series of vertical dikes forming walls. (29.2728, -103.3686)

Mule Ear Peaks, a combination of volcanic tuff and dikes, were used by early travelers for navigation. (29.1664, -103.4357)

South of milepost 12, watch to the southeast for the distinctive, twin-pronged Mule Ear Peaks. Early travelers looked for these landmarks to guide their travel. The peaks are a composite of volcanic tuff and intruded dikes that erosion has molded into the shape of mule's ears. The tuff erupted 33.6 million years ago as part of the Mule Ear Spring Tuff.

A display at the turnout near milepost 15 explains the geology of Goat Mountain. It is composed of volcanic tuff, lava, and breccia of the Burro Mesa Rhyolite that filled in a low area and spilled over the top. These eruptions of lava and ash may have risen along a conduit visible in the mountain center.

At Tuff Canyon, near milepost 20, is another geologic turnout and display. A short walk takes you to an overview, where you will see bright-white tuff, ash, and fragments of the Burro Mesa Rhyolite that were blown explosively from a volcano. In deeper parts of the canyon, the ash lies on dark lava, material that flowed out of a volcano. The contrast is distinctive and striking. The power of erosion has exposed these rocks in the canyon walls. Note how the stream cut easily through the soft tuff (composed of ash) but then met the hard lava, which resists erosion and forms ledges and flat, less erodible surfaces.

Near milepost 21, the ash and lava deposits seem to close in on the road, and you drive through tight places where the rocks can almost be touched out the car window. White ash and black lava are interwoven in a stark landscape of artfully eroded shapes. Watch carefully along the right for what looks like a tree trunk partially exposed on a white slope. It has even been interpreted as a petrified tree trunk but is really a small volcanic spine, a little volcanic neck. The hole on the side looks like a knot where a branch emerged, but it is a cavity where a pumice ball weathered out.

The roadside exhibit between mileposts 21 and 22 discusses Cerro Castellan, another distinctive landmark for early travelers and settlers. The bright-orange, red,

Goat Mountain displays multiple eruptions of the Burro Mesa Rhyolite, with dark lava on top above the tan cliff of tuff. The jumble in the middle is an intrusion. (29.1804, -103.4305)

The striking Cerro Castellan, with its colorful volcanic deposits, is protected from erosion by a spine of rhyolite. (29.1519, -103.4984)

and white breccia, basalt, and ash deposits on the lower slope, part of the older Chisos Formation, contrast with the brown-weathered Burro Mesa Rhyolite that caps the hill. The rhyolite spine protects the underlying material from erosion.

The road drops down onto gravelly river terrace deposits as it enters Castolon, a village on the Rio Grande that was named after a settler who lived nearby. Lt. Echols of the US Army visited Castolon in 1860, and the army had a garrison here during the border troubles of 1914–1916. The trading post is an old cavalry barracks, the two houses were army officers' quarters, and the Park Ranger's office was once a Texas Ranger Station.

For 8 miles upstream from Castolon, the road parallels the Rio Grande, passing abandoned houses of an early settlement called Coyote. Sandstone ridges of Cretaceous Aguja Formation lie on the right side of the road. The rock wall of Sierra Ponce looms to the left on the Mexican side of the river. The road ends at the mouth of Santa Elena Canyon, a narrow slot occupied by the Rio Grande.

At the overlook for Santa Elena Canyon, a display in a small shelter at the edge of the parking area has geologic exhibits. The rock wall through which the Rio Grande has cut its gorge is an uplifted block of Cretaceous limestones. The Terlingua fault is at the base of the massive cliff. You are standing on the downthrown side of the fault, and the uplifted block facing you represents 3,000 feet of movement along this Basin and Range fault.

The canyon separates Sierra Ponce in Mexico from Mesa de Anguila in the United States. The long, thin Mesa de Anguila was likely named for its shape (*anguila* means "eel"), for the freshwater eels found in the river here, though there is also the

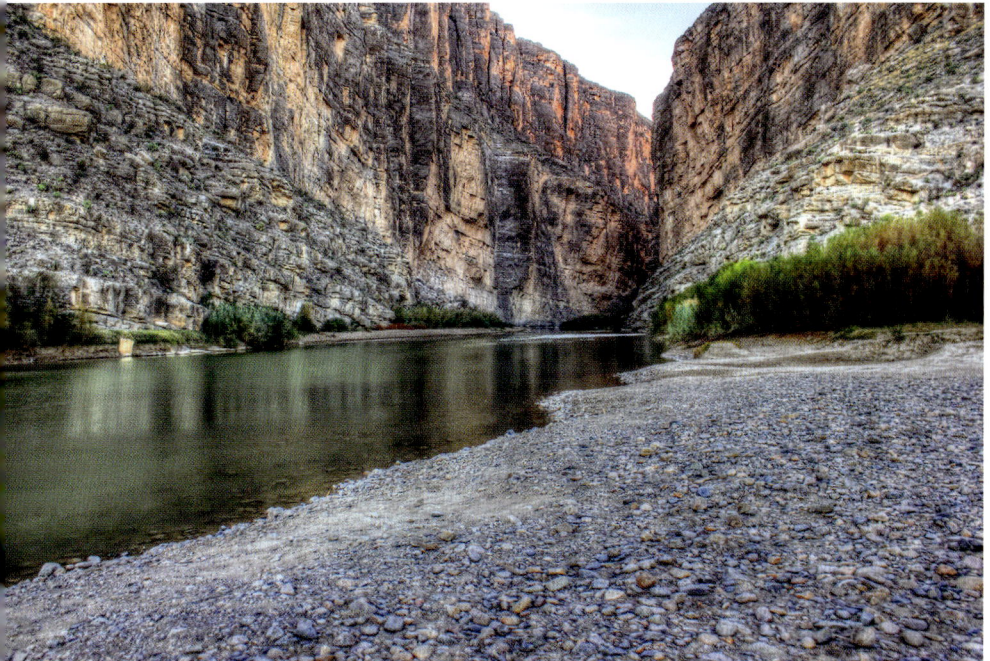

Santa Elena Canyon, one of the most breathtaking views in Texas, is composed entirely of Cretaceous Santa Elena Limestone. —Photo courtesy Wikipedia Commons

suggestion *anguila* is an English misspelling or corruption of the words *aguila* (eagle) or *angulo* (angle or corner). Terlingua Creek, coming from the right (north) flows along the face of Mesa de Anguila and joins the Rio Grande at the mouth of Santa Elena Canyon.

An obvious question is why the Rio Grande cuts across this massive, uplifted block of limestone and doesn't just go around it. The answer is that the Rio Grande (or its ancestor) was here before the block was uplifted. As the block rose along the fault in increments of a few inches to a few feet with each earthquake, the river simply downcut a little faster to accommodate the small change in slope. Repeat this process many, many times over several million years, and you can see how the river cut Santa Elena Canyon little by little, inch by inch—nothing dramatic, just small effects multiplied over millions of years.

HIGH PLAINS

The high, flat, windswept plains of the Texas Panhandle belie a fascinating geologic history, especially where the escarpment-edged borders expose what is hidden beneath the surface. The Texas High Plains is also known as the Llano Estacado, or "Staked Plains" in Spanish, so-called because of the legend of early Spanish explorers setting stakes across the plains to find their way across. The Staked Plains tale is deeply entrenched in Texas mythology, but a more sensible and geologic interpretation is that Llano Estacada means "stockaded" or "palisaded" plains—precisely how the edge of the plains appears when viewed from below the caprock.

This plateau extends from Midland and Odessa north to the Oklahoma border and west to the New Mexico border. At any given location, the topography varies less than a few tens of feet. Though the surface is flat and featureless, it is not horizontal. In the northwest corner of the Panhandle near Dalhart, the elevation is 4,000 feet, but it slopes eastward and southward so that near Big Spring, it is 2,400 feet above sea level.

To understand the genesis of the Texas High Plains, we must look to geology beyond the state borders. In Miocene time, between 23 and 5.5 million years ago, the Rocky Mountains to the north and west of Texas underwent renewed uplift. Erosion of these ranges increased dramatically, causing rivers to deposit a vast apron of gravel, sand, and clay eastward from the mountain front. This wedge of sediments, known as the Ogallala Formation, runs continuously from Wyoming and South Dakota southward through eastern Colo-

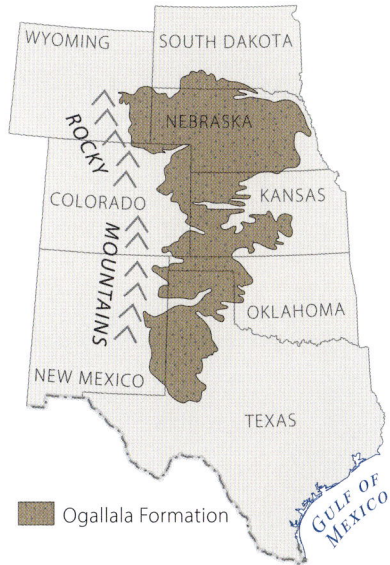

The High Plains is surfaced with the Ogallala Formation, sediments that eroded from the Rockies in Miocene to Pliocene time and extend from Texas north through Nebraska.

rado and into the Texas Panhandle. The wedge slopes eastward from the mountains, creating a gangplank that early settlers traveled up as they headed west, particularly on the Oregon Trail in Nebraska. Over time, streams and rivers spread their erosional load of sediment farther and farther eastward.

For many years, geologists interpreted the sedimentary wedge as a series of coalescing alluvial fans. Modern investigations, however, have shown that the early sediments were deposited by braided streams with interconnected, shifting channels that eventually filled in older valleys. These early sediments were deposited in humid climatic conditions, but as the climate progressively became subhumid to arid, thick

STRATIGRAPHIC COLUMN OF THE HIGH PLAINS			
AGE	**ROCK NAME**	**ROCK DESCRIPTION**	**GRAPHIC COLUMN**

AGE			ROCK NAME	ROCK DESCRIPTION	GRAPHIC COLUMN
CENOZOIC	PLEISTOCENE		Blackwater Draw Formation	sand, fine- to medium-grained quartz, silty, caliche nodules	
			Tule Formation	sand, silt, and clay, gravel at base, caliche nodules	
			Blanco Formation	red and white clays, clayey limestone	
	MIOCENE–PLIOCENE		Ogallala Formation	caliche siltstone conglomerate	
				unconformity	
MESOZOIC	TRIASSIC	Dockum Group	Trujillo Formation	conglomerate, silty shale, sandy shale, sandstone, and shale	
				unconformity	
			Tecovas Formation	mudstone, sandstone, sandy mudstone, and sandy claystone	
				unconformity	
PALEOZOIC	PERMIAN	Quartermaster Formation		red siltstone, gypsum, and red siltstone	
			Alibates Dolomite Member	dolomite, shale, and flint	
		Pease River Group	Blaine Formation	shale, sandstone, gypsum, and dolomite	

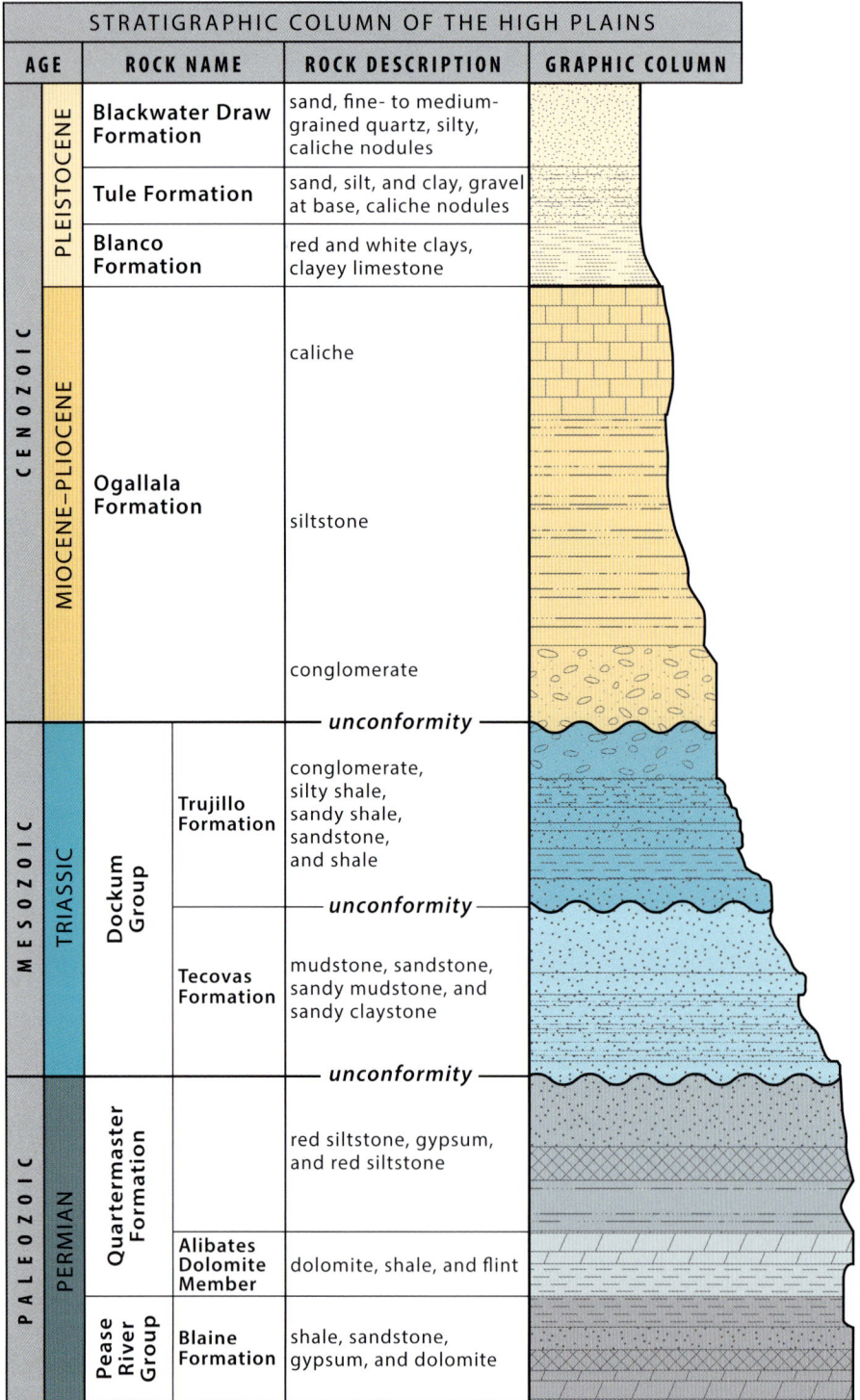

Stratigraphic column of important rock units in the High Plains.

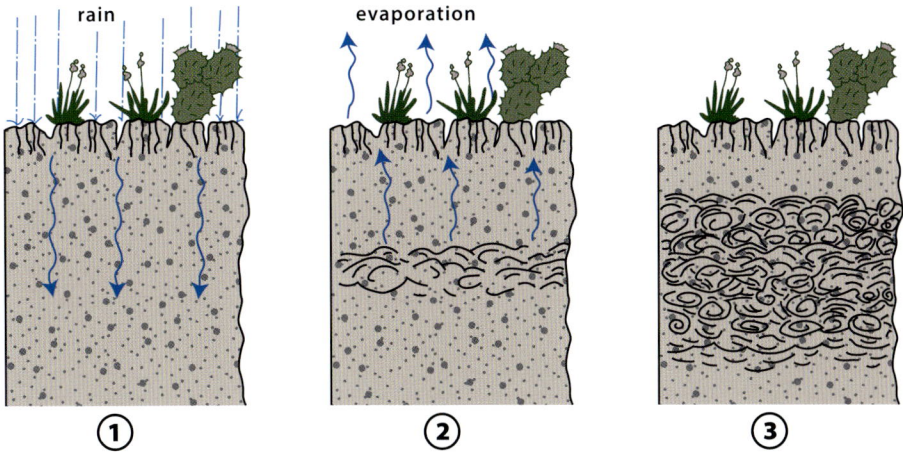

Caliche forms as rainwater dissolves lime (calcium carbonate) and percolates downward through the soil. After the rain, water at the surface evaporates in the dry, sunbaked air, causing the water below to also rise to the surface and evaporate; the lime is left behind as a precipitate, often on the undersides of stones. The perennial repetition of this process builds an impervious, cemented zone of crusts and irregular nodules up to several feet thick beneath the surface.

windblown sand and silt covered the earlier stream deposits. Evaporation of moisture from the upper layers caused calcium carbonate to precipitate and form a hard layer called caliche (pronounced Ka-LEE-chee). The caliche forms a caprock that protects the underlying Ogallala from erosion. The caliche formed in arid conditions with sparse rainfall, much like today's climate on the High Plains.

This cycle of deposition, erosion, and spread of sediment from the mountains continued until the Pleistocene Epoch began. With vast amounts of water on the Earth tied up in continental glaciers that covered much of Europe and North America during the ice ages, sea level correspondingly fell about 400 feet. The mouths of rivers were not only lower but relocated seaward tens to hundreds of miles. At the same time, the cool climate of the Pleistocene produced abundant rainfall, and formerly small streams became raging torrents. Rivers adjust their gradients in response to the sea level (or base level), the volume of sediment load, and the water supplied to them, so the Pleistocene rivers adapted and began to downcut across the sedimentary wedge.

The Canadian River cut an impressive canyon across the northern Panhandle to form the colorful Breaks country. Small streams eroded headward, and the edge of the High Plains was moved little by little westward. Many streams that today seem too puny to effectively carve their large canyons carried more water in the wet Pleistocene Epoch. The Prairie Dog Town Fork of the Red River is an example: it hardly seems capable of carving the spectacular Palo Duro Canyon southeast of Amarillo, but it did when it carried more water. And, of course, this small stream still has tremendous erosive power when its water volume increases dramatically after stormy downpours.

Once the caprock is incised, the underlying rocks are eroded more quickly. The edge of the plains is known as the Caprock Escarpment, which is an impressive cliff in some places and a slight change in elevation in other places. As streams eroded back into their upper drainage divides, a process called headward erosion, the eastern escarpment has moved westward many miles in the last few million years.

View looking east at the caprock edge of the canyon of Timber Creek at the entrance to Palo Duro Canyon State Park. The High Plains surface lies directly beyond the canyon. (34.9800, -101.6909)

Another process is also contributing to the erosion of the High Plains. Parts of the upland are subsiding because of the dissolution of salt layers in Permian rock units hidden beneath the plains. As groundwater reached the salt layers buried 2,000 feet beneath the surface, the salt was dissolved, and the rocks above collapsed into the void, resulting in a corresponding surface trough. Surface drainage, like the Canadian River, was naturally drawn to this low area. The processes of dissolution, salt removal, and continued subsidence are still active, judging from the high salinity (3,000 parts per million) of the Canadian River water entering Lake Meredith. Sinkholes have appeared historically along the Canadian River valley, confirming that salt dissolution continues to be an active local process accompanied by subsidence.

During the last 2 million years, the Pecos River eroded headward, aided by a linear zone of salt dissolution that extended northward into New Mexico. The Pecos captured eastward-flowing streams that had once drained across the Panhandle. With this stream capture, the Llano Estacada was left a high-standing plateau, isolated from its former source, the Rockies. The modern Red, Brazos, and Colorado Rivers originate on the High Plains of Texas, fed by springs emerging from the Ogallala Aquifer.

The white Blanco Formation consists of sedimentary layers eroded from the Ogallala and also clay deposited in shallow, Pleistocene-aged lakes. Fossils of numerous Pleistocene animals have been preserved in the Blanco from a time when North and South America were still joined at the Isthmus of Panama. In addition to native North American animals, such as horses and camels, animals native to South America, such as the great ground sloth and glyptodonts, are fossilized in the Blanco Formation. Herds of grazing mammals, such as mammoths, camels, bison, and rhinos, inhabited these expansive plains because water was plentiful.

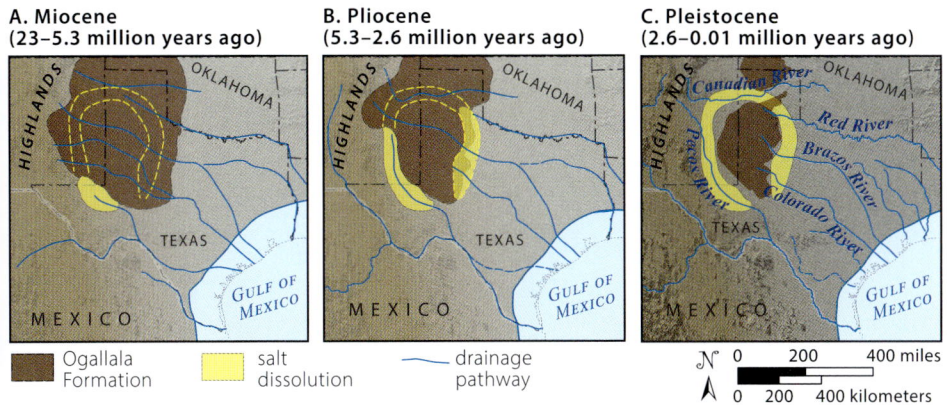

A. Miocene
(23–5.3 million years ago)

B. Pliocene
(5.3–2.6 million years ago)

C. Pleistocene
(2.6–0.01 million years ago)

Ogallala Formation — salt dissolution — drainage pathway

0   200   400 miles
0   200   400 kilometers

Drainage pathways in West Texas and southeastern New Mexico throughout the late Cenozoic changed as rivers followed areas of subsidence caused by salt dissolution beneath the Ogallala. —Modified from Ewing, 2016

Another important stratigraphic unit of the High Plains is the Blackwater Draw Formation, an extensive, windblown sand sheet, which has been called "cover sands" in the past. The lower part of the formation includes the 1.61-million-year-old Guaje ash bed, which erupted from a volcano in New Mexico. It provides a good time constraint for the onset of deposition. Several ancient soil layers within the Blackwater Draw Formation indicate times when the land surface was stable enough for soils to develop. Deposition of the Blackwater Draw Formation likely ended in the Middle Pleistocene Epoch between 300,000 and 350,000 years ago.

Given the flatness and high rainfall of the Pleistocene ice ages, water ponded at the surface, and the region boasted many lakes. Today, a characteristic feature of the Texas High Plains is innumerable round, gray areas called playas or playa lakes. These are the lakebeds of those former ice age lakes. When the weather is dry, they are dusty and usually unvegetated. After a High Plains thunderstorm, however, they turn into ponds. Early pioneers depended on water from these surface ponds for themselves and their livestock because few streams exist on the High Plains surface. Rains didn't always come, though, and the ponds dried up quickly. Rainwater that doesn't evaporate soaks into the underlying, porous sandstones just below the surface to add to the groundwater in the Ogallala Aquifer.

The Ogallala Aquifer is the largest freshwater aquifer in North America. Groundwater in the aquifer is often referred to as "fossil water" because it was added to the aquifer 25,000 to 10,000 years ago. Today, while some recharge of water into the aquifer does occur, in many places it is close to nonexistent. Water of the Ogallala Aquifer has supported agriculture across the Great Plains since the late eighteenth century, and as a result, the level of water in the aquifer has declined as much as 300 feet in places. The twentieth century witnessed an even greater effort to tap groundwater from the Ogallala sands. One study estimated that a volume of water equal to that in Lake Erie was removed from the aquifer by 2010. In recent years, the decline of the water level in the aquifer has slowed as technological developments in irrigation have made water use more efficient.

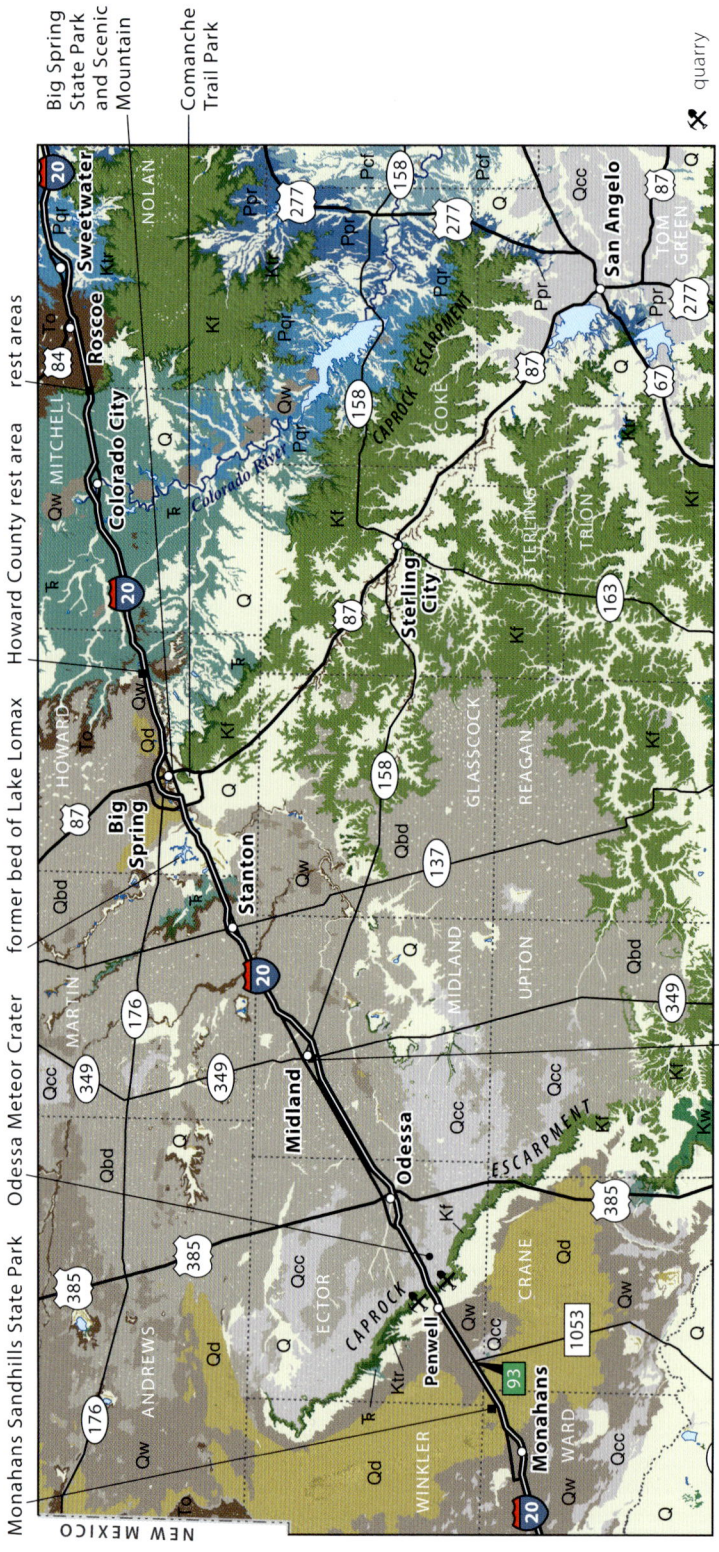

Geology along I-20 between Monahans and Sweetwater.

CENOZOIC (QUATERNARY)

Q — recent sediments; includes alluvium, playa deposits (small dots), gypsite, and older terraces (Holocene and Pleistocene)

Qcc — caliche

Qw — windblown sand, silt, and sand sheet deposits (Holocene)

Qd — windblown dunes (Holocene)

Qbd — Blackwater Draw Formation and windblown cover sand (Pleistocene)

(NEOGENE)

To — Ogallala Formation (middle Miocene to early Pliocene)

MESOZOIC (CRETACEOUS)

Kw — Washita Group

Kf — Fredericksburg Group; includes Edwards Limestone and Segovia Member

Ktr — Trinity Group

(TRIASSIC)

Ŧ — Dockum Group

PALEOZOIC (PERMIAN)

Pqr — Quartermaster Formation, Cloud Chief Formation, and Whitehorse Sandstone

Ppr — Pease River Group

Pcf — Clear Fork Group

✕ quarry

— normal fault

······ county boundary

Big Spring State Park and Scenic Mountain

Comanche Trail Park

rest areas

Howard County rest area

former bed of Lake Lomax

Permian Basin Petroleum Museum

Odessa Meteor Crater

Monahans Sandhills State Park

NEW MEXICO

0   10   20   30 kilometers
0   10   20 miles

N

I-20
MONAHANS SANDHILLS STATE PARK — MIDLAND — SWEETWATER
159 miles

Along the southwest edge of the High Plains is a large area of active sand dunes, created as blowing sand is trapped against the western escarpment by the strong westerly winds that frequently sweep across West Texas. Along I-20, amid the dune field, is Monahans Sandhills State Park, where sand dunes are available for study and play. See the road guide for I-20 on page 210 for information about the dunes.

East of the state park, as you approach FM 1053 (exit 93) and Penwell (exit 101) from the west, look east to see the skyline profile of the High Plains. The southern edge of the High Plains is off to the south. Against this ridge face, held up by flat-lying Cretaceous limestone, the dune sand has been piled over time. I-20 gently climbs up this face. Though no outcrops are exposed near the highway, several quarry operations can be seen where limestone is mined to use for construction gravel and road building.

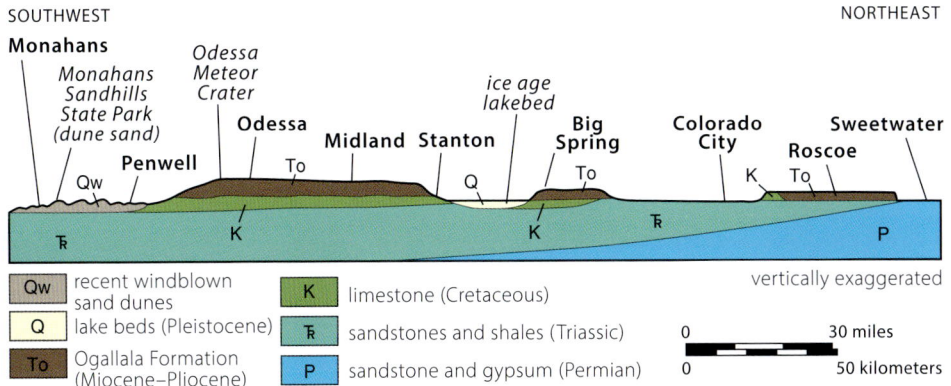

SOUTHWEST   NORTHEAST

Qw	recent windblown sand dunes	K	limestone (Cretaceous)
Q	lake beds (Pleistocene)	Ŧ	sandstones and shales (Triassic)
To	Ogallala Formation (Miocene–Pliocene)	P	sandstone and gypsum (Permian)

vertically exaggerated

0 — 30 miles
0 — 50 kilometers

Cross section along I-20 as it crosses the southern edge of the High Plains between Monahans and Sweetwater.

Exit 93 at FM 1053 offers a great view to the southeast of the Caprock Escarpment that marks the boundary of the High Plains. (31.6723, -102.7044)

Penwell lies at the western edge of the High Plains, and the flatness of the surface is apparent to the east, all the way to Odessa and Midland. The dry, sparsely vegetated countryside is home to three kinds of critters: jackrabbits, pumpjacks, and tough cattle. White rubble in the fields is about all the geology you'll see to tell you of the vast expanse of caliche deposits and gravel of the Ogallala Formation that forms the surface of the High Plains. A few tens of feet beneath the Ogallala surface is another expansive rock unit—flat-lying Cretaceous-aged limestone that you saw in quarries east of Penwell.

In Midland, on the north side of I-20 near exit 135, the Permian Basin Petroleum Museum has a large number of displays on the development of the petroleum industry in this oil-rich region of Texas. The history of early life on the southern High Plains is also presented.

Between Stanton (exits 154–158) and Big Spring (exits 174–179), I-20 crosses a topographically low area that was once the site of an ice age lake called Lake Lomax. It existed when water and rainfall were much more abundant than they are today on the High Plains. The former lakebed makes the terrain even flatter, if that's possible.

ODESSA METEOR CRATER

Between Penwell and Odessa, watch for signs to Odessa Meteor Crater (exit 108), located about 2 miles south of I-20 along Meteor Crater Road. The 550-foot-diameter crater is one of only three documented meteor craters in Texas, the others being Sierra Madera Crater south of Ft. Stockton, and the buried Marquez Crater in Leon County between Houston and Dallas. Geologists think a nickel-iron meteorite from a meteor shower emanating from the asteroid belt between Mars and Jupiter formed this crater about 63,500 years ago. The energy generated by the impact blasted a hole into the Cretaceous limestone bedrock of the High Plains surface and explosively shattered the surrounding rock. Originally, the crater was a round, almost funnel-shaped depression about 100 feet deep. During the past 60,000 years, windblown silt and water-laid sediment have gradually filled the depression nearly to the surface. The crater now appears as a shallow, circular pit with a rubbly rim of limestone defining its outline. The up-tipped rim, the circular shape, and the presence of high-pressure minerals, found in a 165-foot-deep hole drilled in the center of the crater, serve to confirm its meteoric origin.

View across the Odessa Meteor Crater to the building on its far side. (31.7573, -102.4789)

The town of Big Spring, sometimes mistakenly referred to as Big Springs, was named for the single, large spring that once originated between the base of Scenic Mountain and a neighboring hill in the southwestern part of town. The hills are composed of Edwards Group limestones, exposed here at the northern extremes of the Edwards Plateau. The spring was long known to native tribes and was an important assembly point for the Comanche before departing on large-scale raids into northern Mexico during the 1840s and 1850s. Before the arrival of the railroad in the 1880s, the spring had an estimated discharge rate of more than 100,000 gallons per day. The railroad and early town of Big Spring began over drawing from nearby water wells, and the level of the water table dropped below the level of the spring outlet. By the 1920s, the spring was completely dry. Today, the city artificially fills the spring with water from three nearby reservoirs to allow residents and visitors the opportunity to see what the spring may have looked like in the past. To see the spring today, visit Comanche Trail Park in Big Spring and park on Christian Road, then follow a short path to the overlook.

East of Big Spring, I-20 crosses more High Plains surface of Ogallala sand and caliche. About 12 miles east of Big Spring, near the Howard County rest area, the road drops over the rather subdued eastern edge of the High Plains onto red and varicolored beds and soils of the Triassic Dockum Group, exposed by the headward erosion of upland tributaries of the Colorado River.

A narrow belt of Cretaceous limestone, near the rest areas about 10 miles west of Roscoe, forms the western edge of an isolated remnant of High Plains. The numerous windmills alert drivers to this slightly elevated feature. Erosion by the Colorado River separated this remnant from the main body of the High Plains. Roscoe lies on this remnant of the High Plains, and between Roscoe and Sweetwater, the interstate eases downward onto the surface of older Permian rocks. The Caprock Escarpment east of Roscoe is not as impressive as it is in other places, but red bed roadcuts a few miles east of Roscoe are telltale signs you've crossed over into Permian terrain.

I-40
OKLAHOMA — AMARILLO — NEW MEXICO
177 miles

I-40 bisects the Texas Panhandle from east to west, traversing the High Plains in the center, and riding on deeper-eroded, older rocks at each end near the New Mexico and Oklahoma borders. At the Oklahoma border, low hills dominate the landscape. A sheet of sand overlies Permian-aged shales, sandstones, and evaporite deposits of the Blaine Formation. From the Oklahoma border to about milepost 155 (9 miles west of Shamrock), the highway rides along the drainage topography of the North Fork of the Red River, which lies about 2 miles north of Shamrock. The river here has carved into colorful red sedimentary rocks of Permian age. Resistant beds of white anhydrite and gypsum within the Permian sequence are responsible for the low ridges and topographic relief southwest of Shamrock.

I-40 climbs a gentle slope as it curves near milepost 155. This incline represents the eastern edge of the Caprock Escarpment, which in other places (farther south) is indeed an abrupt cliff edge. West of the climb (at exit 152, for example), the highway is on the High Plains surface. The top of the plains is the Ogallala Formation, a vast

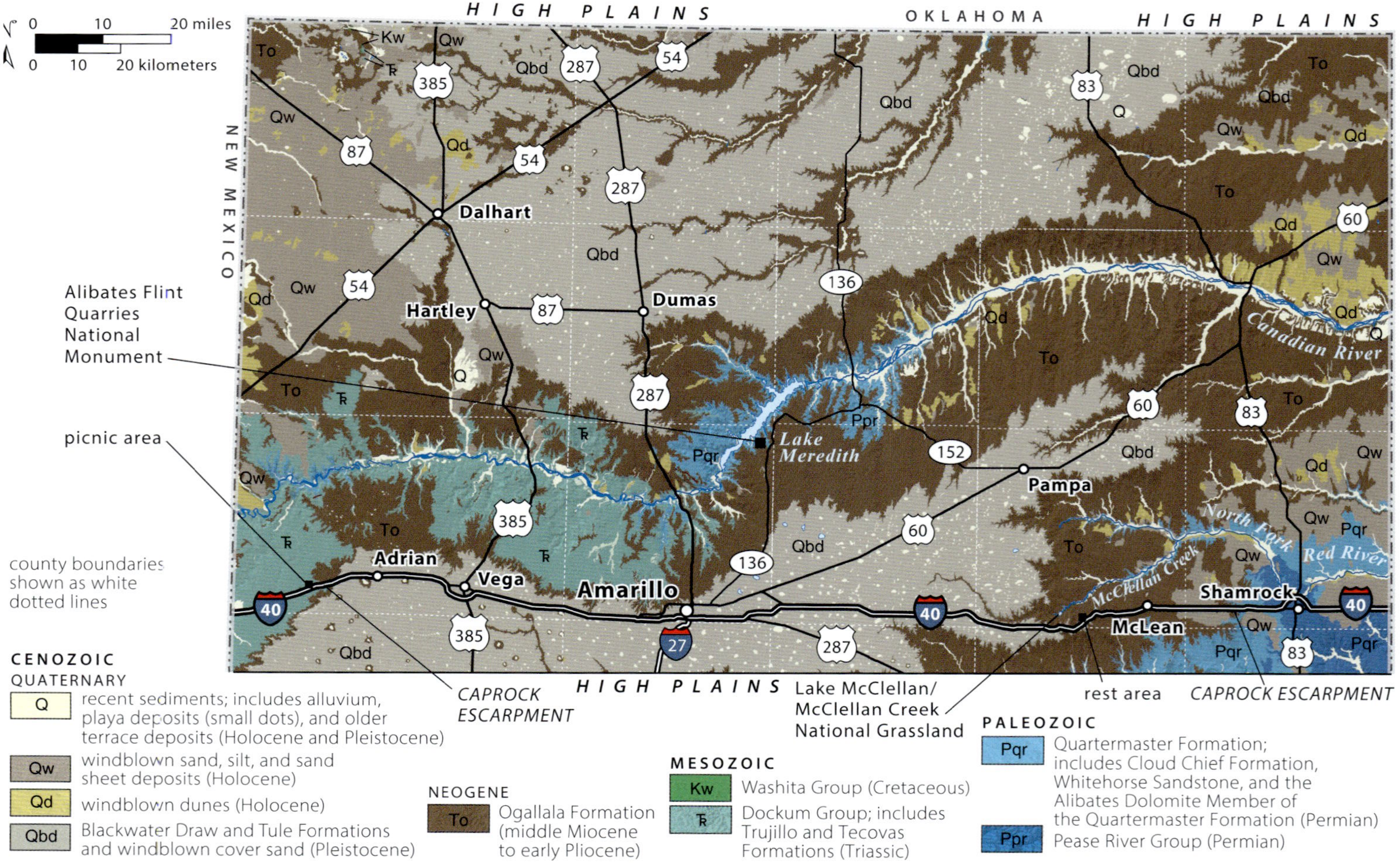
0 10 20 miles
0 10 20 kilometers
HIGH PLAINS
OKLAHOMA
HIGH PLAINS
OKLAHOMA
NEW MEXICO
HIGH PLAINS
Dalhart
Hartley
Dumas
Pampa
Amarillo
Adrian
Vega
McLean
Shamrock
Lake Meredith
Canadian River
North Fork Red River
McClellan Creek
Alibates Flint Quarries National Monument
picnic area
county boundaries shown as white dotted lines
CAPROCK ESCARPMENT
Lake McClellan/ McClellan Creek National Grassland
rest area
CAPROCK ESCARPMENT
CENOZOIC
QUATERNARY
Q recent sediments; includes alluvium, playa deposits (small dots), and older terrace deposits (Holocene and Pleistocene)
Qw windblown sand, silt, and sand sheet deposits (Holocene)
Qd windblown dunes (Holocene)
Qbd Blackwater Draw and Tule Formations and windblown cover sand (Pleistocene)
NEOGENE
To Ogallala Formation (middle Miocene to early Pliocene)
MESOZOIC
Kw Washita Group (Cretaceous)
TR Dockum Group; includes Trujillo and Tecovas Formations (Triassic)
PALEOZOIC
Pqr Quartermaster Formation; includes Cloud Chief Formation, Whitehorse Sandstone, and the Alibates Dolomite Member of the Quartermaster Formation (Permian)
Ppr Pease River Group (Permian)

Geology of the Panhandle along I-40.

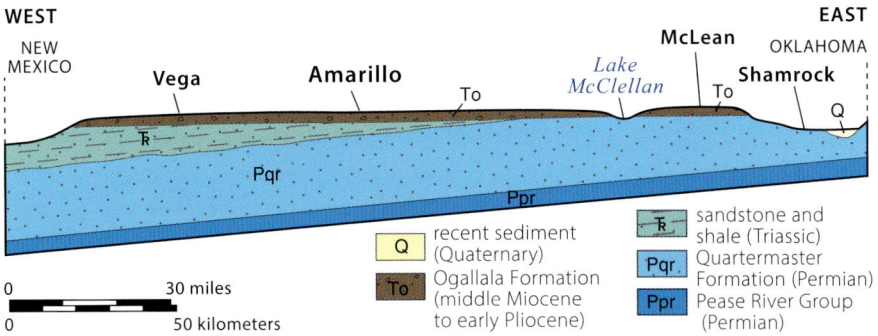

WEST		EAST
NEW MEXICO	McLean	OKLAHOMA

Geologic cross section along I-40 across the northern Panhandle.

Legend:
- Q recent sediment (Quaternary)
- To Ogallala Formation (middle Miocene to early Pliocene)
- Ꞅ sandstone and shale (Triassic)
- Pqr Quartermaster Formation (Permian)
- Ppr Pease River Group (Permian)

sheet of gravel and sand that was carried by rivers and spread eastward across the Panhandle from the Rockies in Miocene to Pliocene time. This sheet formed a continuous wedge-shaped blanket of sediment that lay at the foot of the Rocky Mountain front from Texas to South Dakota. This wedge has subsequently been eroded and dissected but still remains a major geologic feature of North America.

I-40 continues on this flat surface well past Amarillo to within 10 miles of the New Mexico border. The surface is unbroken except for small erosional depressions west of McLean, where the Ogallala has been breached to reveal the underlying Permian rocks around the mostly dry Lake McClellan. The rest area at milepost 130 provides a large, open vista to the north from where the Ogallala can be seen. From exit 128, take FM 2477 north 2 miles to McClellan Creek National Grassland, where the lake used to exist. You can see a small cliff of reddish Quartermaster Formation sandstone just north of the bridge over McClellan Creek.

At the rest stop at milepost 130, the exposed, whitish patches are gravel, sand, and caliche of the Ogallala Formation, the unit surfacing the High Plains. (35.1972, -100.8029)

Beginning at milepost 17, about 5 miles west of Adrian, a marked change to the landscape can be seen to the north. Here, the Canadian River drainage system has eroded through the caprock and into older rocks. The Caprock Escarpment along the western edge of the High Plains is dramatic. At milepost 16, the highway begins dropping west through the Blackwater Draw and Ogallala Formations and into the underlying Triassic-aged Dockum Group sandstones. Great views of this west-facing escarpment and its colorful rock formations can be seen from the rest area at milepost 13 and from Ivy Road at exit 15. Between the rest area and the New Mexico border, the underlying Dockum sandstones give the desertlike landscape an overall orange to red color.

In the McClellan Creek National Grassland, where a shallow lake used to fill the depression, bright-red Permian-aged sandstone of the Quartermaster Formation stands high above the road. (35.2187, -100.8608)

Glowing in the morning sun, the Caprock Escarpment marks the western boundary of the High Plains at exit 15 (Ivy Road). The whitish layer at the top of the escarpment is the Ogallala Formation. (35.2487, -102.7876)

ALIBATES FLINT QUARRIES NATIONAL MONUMENT

From Amarillo, TX 136 leads north to the little-visited Alibates Flint Quarries National Monument and Lake Meredith, a reservoir that fills the breaks of the Canadian River here. TX 136 traverses flat to low, barely rolling countryside typical of the High Plains. Dissection of this surface by tributary drainages to the Canadian River becomes more and more apparent as you drive farther north toward the river. Watch for the turnoff to Alibates Flint Quarries National Monument and Lake Meredith National Recreation Area about 26.5 miles north of Amarillo from the intersection of Amarillo Boulevard (US 60/BR 40) and the Fritch Highway (TX 136). Turn left (west) and follow the signs to the monument. These areas can be visited anytime, but the monument and its flint quarries are only open via park service guided tours.

As you drive into the monument area, red Permian beds are exposed by the erosive action of the Canadian River, which has cut an impressive path across the Texas Panhandle, completely dissecting the High Plains. This segment of colorful erosional topography and canyon country is called the Breaks by local folks because such a break in the vast High Plains surface is noteworthy.

Bright-red siltstone, claystone, sandstone, and white cherty dolomite of the Quartermaster Formation are seen everywhere in the monument from the shoreline level of Lake Meredith up through the dissected hills. These rocks were deposited about 225 million years ago in Permian time. At the tops of the hills and along the road driving into the monument are white sandstone and caliche ledges and blocks of the Ogallala Formation of Miocene to Pliocene age, about 10 to 4 million years old. At the monument, conical, red, haystack-shaped hills are topped by a protective cap of Ogallala sandstone. In colorful contrast, white sandstone blocks rain down the sides of the red hills.

The Alibates Flint Quarries are named for Alibates Creek, in turn named after cowboy Allen "Allie" Bates, who lived in a line camp at the quarry site in the late 1800s. The monument preserves 550 flint quarries and ruins, where for 12,000 years, until about 1870, indigenous peoples dug out multicolored flint to make arrowheads, knives, hammers, and awls for their own use and to trade. The Alibates flint, a fine-grained variety of quartz, displays banded and mottled red, pink, pale blue, pale purple, gray, brown, white, and black hues. It was highly prized and widely traded across Texas and the Great Plains and as far away as Montana and the Great Lakes.

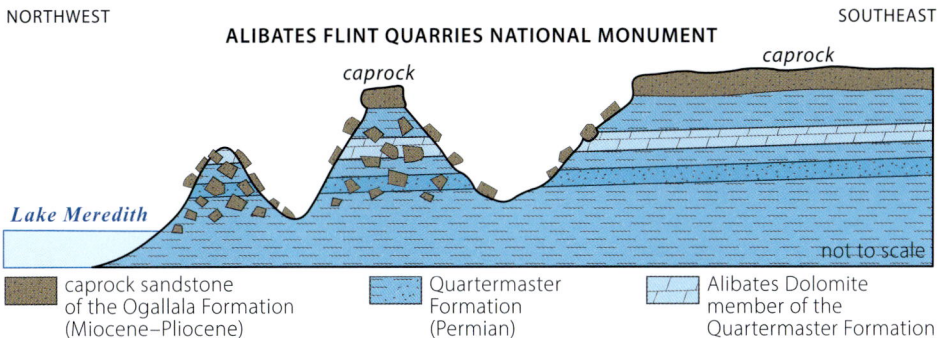

NORTHWEST

ALIBATES FLINT QUARRIES NATIONAL MONUMENT

SOUTHEAST

caprock

caprock

Lake Meredith

not to scale

| caprock sandstone of the Ogallala Formation (Miocene–Pliocene) | Quartermaster Formation (Permian) | Alibates Dolomite member of the Quartermaster Formation |

Geologic cross section at Alibates Flint Quarries National Monument.

The Alibates flint comes from ledges of the Alibates Dolomite, which is part of the Quartermaster Formation. The dolomite has been partly replaced by silica solutions to form chert. Flint is a term applied to dark or highly colored chert.

Note the sharp edges of these 1- to 2-inch-wide flakes of Alibates flint, a stone used for fashioning cutting tools and arrowheads. —Photo courtesy Amir Akhaven

US 84
NEW MEXICO—LUBBOCK—SWEETWATER
212 miles

US 84 from the New Mexico state line southeastward to Lubbock crosses the flat High Plains. Though the landscape has a sound geologic reason for being featureless, it is not a wasteland, by any means, because the brown to reddish soils produce rich crops of cotton, corn, and wheat. Large grain elevators at most of the small towns along this stretch of road attest to this richness. Nonetheless, the town of Levelland, on US 385 about 25 miles south of Littlefield, is well named. A layer of caliche at the top of the Ogallala Formation provides a caprock that resists erosion.

Near the New Mexico border, low topography etched into the Ogallala sandstone lies about 10 to 20 miles south of the highway. Here, ephemeral tributaries in the upend portion of the Brazos River drainage system have carved into the High Plains surface. Although not visible from the highway, low conical hills and knobs of wind-piled sand dunes are visible to the south from the southeastern edge of Muleshoe.

The road passes through several oil fields between Littlefield and Lubbock that produce from Paleozoic reservoirs in the underlying Permian Basin. Thousands of feet of sedimentary rock underlie the High Plains here. Organic material deposited with these sediments is the source of the abundant oil produced today.

Panhandle-Plains Historical Museum
Timbercreek Canyon
Palo Duro Canyon State Park
Tule Canyon; Rock Creek
Caprock Canyons State Park

Muleshoe National Wildlife Refuge/ Muleshoe Depression

Blanco Canyon; Silver Falls Park on the White River

Amarillo
Canyon
Tulia
Silverton
Quitaque
Muleshoe
Littlefield
Lubbock
Levelland
Tahoka
Post
Justiceburg
Lamesa
Snyder
Roscoe
Sweetwater
Big Spring

CAPROCK
ESTACADO
ESCARPMENT
ESCARPMENT
CAPROCK
NEW MEXICO
HIGH PLAINS
LLANO

Lake Mackenzie
Prairie Dog Town Fork of the Red River
White River Lake
Double Mountain Fork of the Brazos River
Colorado River

Lubbock Lake National Historic Landmark
Buzzard Draw/ Sulphur Springs Draw
Yellow House Draw/ Yellow House Canyon
Sam Wahl Recreation Area and Lake Alan Henry

CENOZOIC

QUATERNARY

Q	alluvium; includes, playa deposits (small dots), and older terrace deposits (Holocene and Pleistocene)	
Qcc	caliche	
Qw	windblown sheets deposits of sand and silt	
Qd	windblown dunes	
Qbd	Blackwater Draw and Tule Formations and windblown cover sand (Pleistocene)	

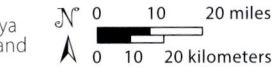

NEOGENE

QTb — Blanco Formation (Pliocene?– Pleistocene)
To — Ogallala Formation (middle Miocene to early Pliocene)

0   10   20 miles
0   10   20 kilometers

county boundaries shown as white dotted lines

MESOZOIC

K — sedimentary rocks (Cretaceous)
Tr — Dockum Group; includes Trujillo and Tecovas Formations (Triassic)

PALEOZOIC

PERMIAN

Pqr — Quartermaster Formation; includes Cloud Chief Formation, and Whitehorse Sandstone, and the Alibates Dolomite Member of the Quartermaster Formation
Ppr — Pease River Group

Geology along US 84 between the New Mexico border and Sweetwater and along US 87 and I-27 between Big Spring and Amarillo.

Between Lubbock and just north of Post, the highway continues to ride the surface of the High Plains through a region that experiences bad dust storms, especially when the wind blows across recently tilled fields. Watch for surface ponds that dot the landscape. Though somewhat subtle, they show where rainfall collects on the flat topography and can infiltrate into the subsurface aquifer.

About 2 miles north of Post, the roadway abruptly drops off the High Plains and into the erosional topography of the Double Mountain Fork of the Brazos River. A

LUBBOCK LAKE NATIONAL HISTORIC LANDMARK

Northwest of Lubbock near the intersection of US 84 and Loop 289 is the Lubbock Lake National Historic Landmark, located in the normally dry Yellow House Draw, a tributary to the Brazos River. Even though it is dry today, in the late 1500s, this was a ten-acre, spring-fed pond called Punta de Agua (Point of Water) by Spanish explorers in the area. This watering hole existed until the early 1900s when over pumping of the aquifer by surrounding farms lowered water levels and caused the springs to go dry. In an effort to reactivate the springs, a project was started that dredged the area around the springs. In the resulting waste piles of rock, two local boys found what looked like a spearpoint and brought it to the West Texas Museum (now the Museum of Texas Tech University) where it was recognized as a Folsom point more than 10,000 years old. Since 1939, archeological studies have been conducted in the area, continuing today.

The geology of the site consists of stream deposits of the Ogallala Formation, overlain by lake deposits of the Blanco Formation, followed by windblown sand deposits of the Blackwater Draw Formation, the youngest unit. During the Pleistocene ice ages, when the climate was cooler and wetter, a stream from the nearby Muleshoe Depression cut a canyon around 50 feet deep through the Blackwater Draw Formation and into the Blanco Formation. Over the years, sediment has filled the canyon with distinct layers, giving archeologists an almost complete record of changing climate and human settlement over the past 11,500 years beginning with the Clovis Culture. Today, visitors to the site can explore the interpretive center to learn more about the Clovis Culture. Outside, life-sized bronze statues of ice age animals that once roamed the area stand tall, and trails from the center lead to important excavation sites.

Around Lubbock Lake National Historic Landmark are bronze statues representing life in the Pleistocene, including this adult and juvenile mammoth. (33.6225, -101.8891)

Just west of where US 84 drops off the Llano Estacado is a large, tan roadcut of Ogallala sediments. (33.2138, -101.4201)

picnic area with great views lies south of the highway at the edge of the eastern escarpment. Where the road drops, roadcuts expose sand and conglomerate of the Ogallala Formation. A view to the west from the highway shows that the Caprock Escarpment, though not very tall, is nevertheless vertical and distinct.

MULESHOE DEPRESSION

White Lake, one of three playa lakes that remain on the salty lakebed of the Muleshoe Depression. (33.9489, -102.7695)

A short side trip south from Muleshoe on TX 214 leads to the Muleshoe National Wildlife Refuge, a wetland area in the Muleshoe Depression. During the Pleistocene ice ages, climate across western North America was cooler and wetter than today, leading to the development of numerous lakes across areas from California to New Mexico with the easternmost of these lakes in the High Plains of Texas. About 20,000 years ago, the Muleshoe Depression was part of a series of lakes connected by streams that drained into Yellow House Draw, which can be seen at nearby Lubbock Lake National Historic Landmark. The enhanced drainage helped carve Yellow House Canyon east of Lubbock. Mammoths, giant bison, camels, dire wolves, and other ice age animals called these lush, grassy plains home. About 14,000 years ago, as the climate warmed and began getting much drier, the lakes mostly disappeared. Today, three small lakes (White Lake, Goose Lake, and Paul's Lake) remain on the salty ancient lakebed. These playa lakes are ephemeral, meaning they don't exist year-round. Water is lost due to evaporation because there is no outlet for the water to escape through. When there is water present, it is usually salty. Just east of these playa lakes are dunes composed of sediment picked up by wind when the lakebeds are dry. Today, the lakes provide an occasional oasis to wildlife and the many birds that migrate along the Central Flyway.

South of Post, the road crosses into the underlying Dockum Group, with its colorful sand and clay beds. The highway continues southeastward for about 20 miles on these varicolored sedimentary rocks of Triassic age. Look for them where US 84 crosses the Double Mountain Fork on the south side of Justiceburg and also along the shore of Lake Alan Henry, an impoundment of the river accessed at Sam Wahl Recreation Area east of Justiceburg. South of Justiceburg, several flat-topped mesas appear on both sides of the road. These mesas are remnants of the High Plains but have been separated from it by erosion.

Three miles south of Post, roadcuts of Triassic Dockum sandstone (top layer) and clay (below) appear on both sides of the highway. The box shows the location of the close-up photo. (33.1583, -101.3456)

Close-up view of cross-bedding in the Dockum sandstone shows that the sand was deposited in a Triassic river channel.

A flat-topped mesa (east of the highway about 9.5 miles south of Justiceburg) is protected by a caprock of more resistant rock layers, while less resistant layers surrounding the mesa have been removed. (32.9519, -101.0846)

US 84 climbs back onto a remnant of the High Plains surface at the rest area 15 miles south of Justiceburg. From Snyder southeastward to I-20 and Roscoe, the road remains on the High Plains surface.

US 87 AND I-27
BIG SPRING — LUBBOCK — AMARILLO
226 miles
See map on page 313.

This stretch of highway between Big Spring and Amarillo is entirely on the flat surface of the Llano Estacado, the staked plains. Undulations in this surface are mainly due to eastward-flowing streams that traverse the High Plains. The headwater tributaries for several of Texas's major rivers such as the Brazos, Colorado, and Red are here. The north-south highway crosses these drainages at nearly right angles, as they flow eastward off the High Plains, to join other streams to become the main rivers that eventually empty into the Gulf of Mexico far to the south.

About 25 miles northwest of Big Spring, small uranium deposits were found in calcrete deposits in both Sulphur Springs Draw and Buzzard Draw. Originally discovered by Kerr-Mcgee Corporation in the mid-1970s, the deposit was recognized in 2017 by the US Geological Survey as the first uranium to be found in calcrete. It is thought that the uranium came from ash deposits within the underlying Dockum Group sediments, and groundwater brought it to and concentrated it in the overlying calcrete of the Blackwater Draw Formation. This Pleistocene-aged unit lies near the surface of the High Plains, so the uranium is suitable for open-pit mining.

Many surface ponds, a unique feature of the High Plains, are seen along the length of US 87 and I-27 between Big Spring and Amarillo. The unvegetated, gray, round,

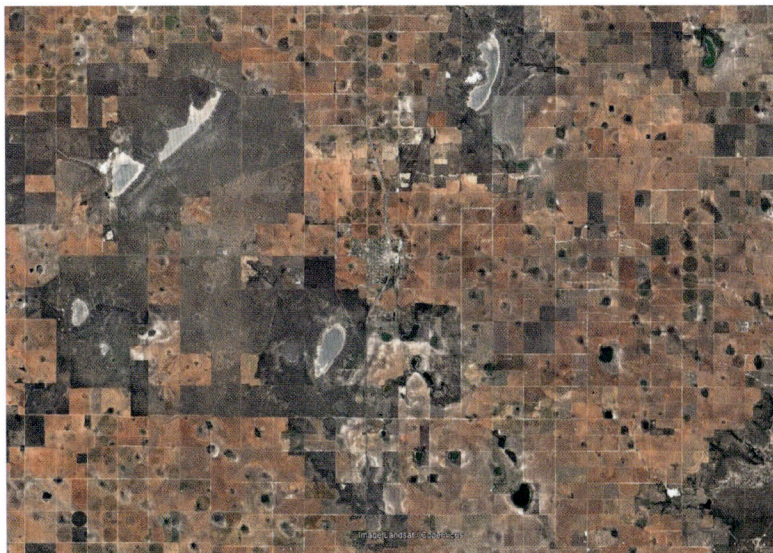

This satellite image shows numerous small surface ponds on the High Plains in the Tahoka area.
—Satellite image from Landsat/Copernicus, July 2024

flat areas fill with water after thunderstorms, then dry up as the ponds evaporate and water sinks into the ground to recharge the underlying Ogallala Aquifer.

About the only rocks to break the flat surface are some low, isolated mesas to the west and northeast of Tahoka (at the US 380 junction). These mesas are remnants of hardened, sandy, shoreline deposits of Cretaceous age that, because of their hardness, survived erosion and now stand above the surrounding, younger Ogallala sand and caliche deposits.

A marked change in vegetation is apparent from south to north on this road. The sparse, dry, mesquite-grazing country around Big Spring contrasts sharply with the lush grass and cropland farther north. The average annual precipitation is not

SILVER FALLS PARK

About 40 miles east of Lubbock, US 82 (TX 114) crosses Blanco Canyon, home to the ephemeral White River, which only contains water during heavy rainstorms. Silver Falls Park, south of US 82 east of the bridge, provides access to rock ledges along the riverbed. Not that long ago, water gushed continuously from the Ogallala Formation at Silver Falls, and the White River was a tree-lined stream with green grass. The White River carved the 34-mile-long canyon deep into the High Plains caprock. Named Blanco Canyon for the stark-white rock of the Blanco Formation in its upper reaches, it is only about 50 feet deep at its head but grows immensely to 6 miles wide and between 300 to 500 feet deep at its mouth. The river cut down through the Pleistocene Blanco Formation and Miocene Ogallala Formation to expose bright-red Triassic-aged sedimentary rocks of the Dockum Group. Numerous artifacts from Francisco Vázquez de Coronado's expedition in 1541 have been found in the canyon, showing that the place was an enticing stop several hundred years ago. Over pumping of the Ogalalla Aquifer dried up the springs to a trickle in the early-twentieth century. Today, only seeps emerge from the boundary between the tan Ogallala ledges and the underlying red Dockum Group.

Ledges of Ogallala Formation formed Silver Falls on the White River before over pumping reduced the flow to just a trickle. (33.6659, -101.1594)

that much different between Big Spring and Amarillo, but the northern Panhandle receives much of its precipitation in the winter in the form of snow, whose water slowly saturates the underlying soil to provide moisture for plant growth. Drying winds and low overall rainfall discourage trees from establishing on the High Plains, except in the low areas along stream drainages.

The town of Canyon is appropriately named because it sits at the headwaters of the canyon cut by the Prairie Dog Town Fork of the Red River, which flows eastward toward Palo Duro Canyon. The town also lies within the basin of a former Pleistocene-aged lake. Do not miss stopping at the Panhandle-Plains Historical Museum, located in Canyon along TX 217 about 2 miles west of I-27. The geologic exhibits are superb and cover geologic time, fossils, the buried Amarillo Mountains, the Ogallala Aquifer, High Plains ponds, Palo Duro Canyon, and oil and gas, in addition to colorful and up-to-date exhibits on history, Native Americans, and firearms.

PALO DURO CANYON STATE PARK

To reach Palo Duro Canyon State Park from I-27, take TX 217 east from exit 106 to Canyon. TX 217 climbs out of town and then heads about 7 miles across the High Plains, where grass and crops lie on either side of the road. The flatness is in sharp contrast to what is encountered ahead. Just before entering the state park, some hint of what is to come is seen in a small side canyon to the right (south) of the highway where ledges of caliche lie over steep-walled cliffs of Ogallala sandstone. But it is not until you have paid your entrance fee and driven to the visitor center overlook that Palo Duro Canyon's size, depth, and color suddenly appear before you. The High Plains are abruptly terminated in colorful cliff walls that drop 700 to 800 feet.

Before becoming a state park, Palo Duro Canyon had a rich history, including 12,000 years of occupation by indigenous peoples. Palo Duro means "hard wood" in Spanish, named for the canyon's junipers from which Native Americans made their hardwood bows. The Spanish explorer Coronado saw the canyon in 1541, and Captain Randolph Marcy wrote of it in 1852, while leading a US army expedition. The Battle of Palo Duro Canyon in 1874 pitted Colonel Ranald Mackenzie's troops against Comanche warriors, ending the Red River War. Only two years later, Charles Goodnight, famous for spearheading Texas-Wyoming cattle drives, began his JA Ranch operation in the canyon. Panhandle citizens' interest in Palo Duro Canyon as a recreation area and led to a failed attempt to make it a national park but succeeded in the creation of the crown jewel of the Texas State Park System in 1934.

Before descending into the canyon, stop at the visitor center located on the Canyon Rim near the first overlook turnoff. A review of the geology and natural history exhibits here set the stage for your trip through the park.

The closeness of the raw rock exposures on the steep descent road to the canyon floor compel you to think about the park's geology. Two main geological stories are written in Palo Duro's colorful walls. The first is a tale of dynamic processes—floods, streams, erosion, down-cutting—those forces that cut the canyon itself out of the body of the High Plains. The second story is of ancient environments, sands, sediments, and long-extinct animals, written in the rocky pages of stratigraphy exposed on the canyon walls.

Park Road 5 crosses the Prairie Dog Town Fork of the Red River just past Hackberry Campground, and it is hard to believe this languid little stream carved Palo

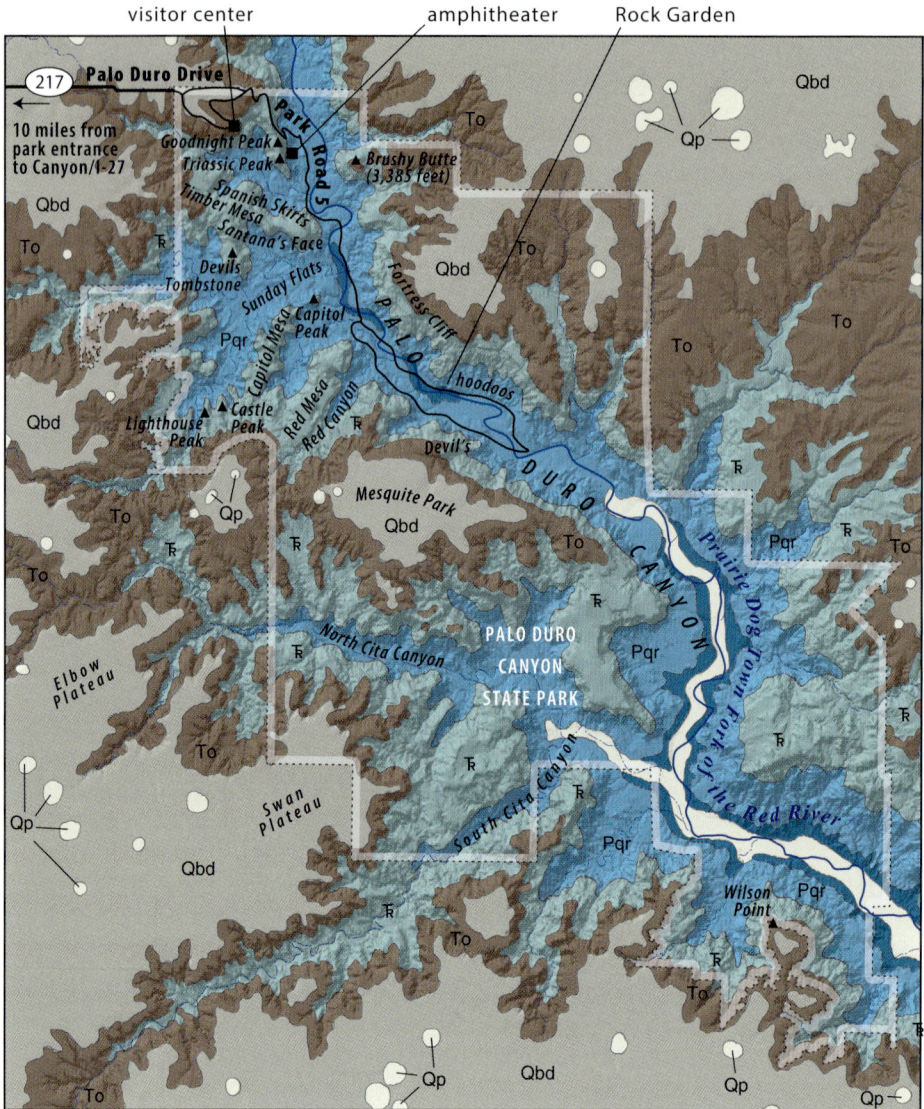

visitor center   amphitheater   Rock Garden

(217) Palo Duro Drive

10 miles from
park entrance
to Canyon/I-27

Goodnight Peak
Triassic Peak
Brushy Butte
(3,385 feet)

Spanish Skirts
Timber Mesa
Santana's Face
Devils
Tombstone
Sunday Flats
Capitol Mesa
Capitol
Peak
Castle
Peak
Lighthouse
Peak
Red Mesa
Red Canyon
Devil's

Fortress Cliff

hoodoos

PALO DURO CANYON

Mesquite Park

North Cita Canyon

PALO DURO
CANYON
STATE PARK

Elbow
Plateau

Prairie Dog Town Fork of the Red River

Swan
Plateau

South Cita Canyon

Wilson
Point

Park Road 5

Qbd
Qp
To
Qbd
To
To
To
To
Pqr
Qbd
Qp
To
To
Qp
Qbd
To
Qp
Qbd
To
To
Pqr
Pqr
Pqr
Pqr
Pqr
Qp
Qbd
Qp
Qp

Geologic map of Palo Duro Canyon State Park.

CENOZOIC

| Qa/Qp | recent alluvium (Qa–Holocene) playa deposits (Qp–Pleistocene) |

| Qbd | Blackwater Draw and Tule Formations (Pleistocene) |

| To | Ogallala Formation (middle Miocene to early Pliocene) |

MESOZOIC

| Ŧ | Trujillo and Tecovas Formations of the Dockum Group (Late Triassic) |

PALEOZOIC
PERMIAN

| Pqr | Quartermaster Formation |

| Pcw | Cloud Chief Gypsum and Whitehorse Sandstone |

⋯⋯⋯ state park
⋯⋯⋯ boundary

N
0   1   2   3 miles
0   1   2   3 kilometers

View to southeast of Palo Duro Canyon from the scenic overlook and visitor center near the park entrance. (34.9793, -101.6912)

Duro Canyon. About 10 million years ago, the Rocky Mountains were given an extra shove upward, with renewed erosion sending deposits of sand, silt, and pebbles eastward across the Panhandle to be deposited as the Ogallala Formation. Then, in the ice ages of the Pleistocene, about 1 million years ago, sea level dropped because much world-wide water was held in continental glaciers. River mouths were now several hundred miles farther offshore and lower. The wetter, cooler climate meant the Pleistocene rivers carried more water. The rivers adjusted to their larger water volume and longer distance to travel by downcutting in their upper reaches. The much bigger Prairie Dog Town Fork of the Red River eroded into the caprock. The fork is still eroding the canyon but at a slower rate. You have to be here after a High Plains downpour to see the creek swell and listen to the boulders thunk together to understand the power of small streams in a flash flood. Over geologic time, little bits of erosion can add up to a lot—even at Palo Duro Canyon!

The layered sedimentary rocks exposed in the walls of Palo Duro Canyon are filled with information that tell a story of conditions on Earth millions of years ago when these rocks were first laid down as fresh sediment. The bright-red rocks at the floor and on the lower slopes of the canyon are the Quartermaster Formation, the main unit exposed. They were deposited at the edge of the seas in Permian time. The wave-formed ripple marks indicate deposition in shallow water. The distinctive white gypsum (calcium sulfate) indicates the periodic drying out of the shallow water. The symmetric ripple marks found in some of the layers also support the interpretation of shallow marine deposition. Spanish skirts, conical exposures of the Quartermaster

AGE (mya)*	ROCK UNIT		DESCRIPTION
recent	caliche		white limy precipitate
MIOCENE– PLIOCENE (10–4)	Ogallala Formation		tan conglomerate and sandstone
TRIASSIC (225)	Dockum Group	Trujillo Formation	gray, pink, white shale, siltstone, and sandstone
			dark gray sandstone
		Tecovas Formation	varicolored (purple, yellow, white, gray, pink, tan, and brown) shale, siltstone, and sandstone
PERMIAN (250)	Quartermaster Formation		interbedded bright-red sandstone, shale, and discontinuous dolomite
			white gypsum
	Cloudchief Gypsum		white, gray, and pink gypsum
	Whitehorse Sandstone		orange-red sandstone and shale

Cross section of rock units in Palo Duro Canyon.

unconformity

Prairie Dog Town Fork of the Red River

mya = millions of years ago

not to scale

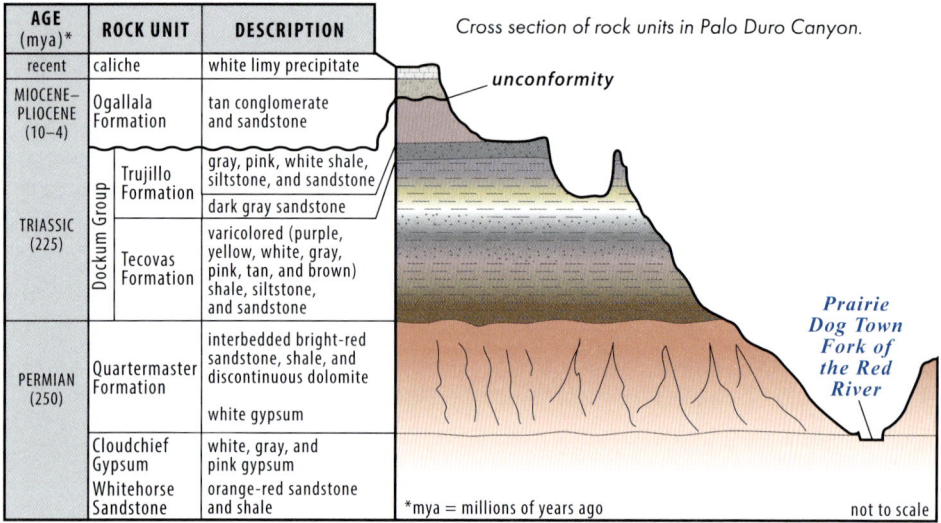

Capitol Peak, viewed from the Lighthouse Trailhead, is composed of two units. The upper and middle section is the Triassic Tecovas Formation, and the lower red slopes are the Permian Quartermaster Formation. (34.9516, -101.6681)

Formation with erosion rills, can be seen from the overlook near the park entrance. These rock formations reminded early Spanish explorers of the skirts women wore.

Above the Quartermaster red beds is a varicolored rock package, consisting mostly of slope-forming shales but also of ledges of sandstone and siltstone. Colors of purple, yellow, white, gray, pink, tan, and brown flicker in the changing light conditions of the passing day. These rocks, which compose the middle slopes of most mesas and hills in Palo Duro, are part of the Dockum Group of Triassic age. The group has two forma-tions in the canyon, the Tecovas and the Trujillo. Tecovas shales are the lower, colorful slopes just above the Permian red beds, whereas the Trujillo is easily recognized by the ledge-forming sandstone that tops many of the canyon's middle-level mesas.

Colorful outcrops of the red Quartermaster Formation are overlain by the purple and tan Tecovas Formation along Park Road 5. The white layers are gypsum. (34.9481, -101.6681)

Mudstones in the Tecovas were laid down in streams and swamps and contain fossils of crocodile-like reptiles (phytosaurs), primitive amphibians, and fish. The multihued colors indicate oxidation and drying and wetting cycles so typical of the varied conditions found in streams and their surrounding swamps. The Tecovas can be seen in the upper part of the Spanish Skirts, on Capitol Peak, and the Devil's Slide.

The Trujillo is easily picked out by the conspicuous sandstone ledge that rests on top of many mesas. Coarse sand and river-type cross-bedding in the Trujillo sandstone tell us it was deposited in an ancient stream bed. While not common, fossils of primitive amphibians, wood, reptiles, and leaves are found in the Trujillo. Santana's Face, the cap on Lighthouse Peak, the Rock Garden, and numerous pedestal caps are all Trujillo sandstone.

At the top of the canyon is the Ogallala Formation, a cliff-forming ledge of tan sandstone, opal-cemented siltstone, and conglomerate, with a ledge of white caliche at the very top. This unit is between 10 and 4 million years old, from Miocene to Pliocene in age. The caliche formed toward the end of a long period of progressive drying, when environmental conditions in the Panhandle changed from a savannah 10 million years ago to a grassland by about 4 million years ago. Conditions became so dry that trees could no longer grow here.

You will note a lot of time is missing between the Trujillo and Ogallala. Either the rocks representing about 200 million years of time were eroded away, or they were never deposited. Whatever the case, the contact line between the multicolored, upper beds of the Trujillo and the lower, tan beds of the Ogallala is an unconformity. The missing time represents most of the Age of Dinosaurs and the Age of Mammals. You are standing on Ogallala rocks, mostly caliche, at the overlook, and you drive down through them on the road to the canyon floor. Fortress Cliff, which forms the east canyon rim, has an impressive wall of Ogallala Formation at its top edge.

Examples of differential erosion—the process in which different rock types respond differently to erosion—are abundantly evident in the canyon. Note the rills and riffles in the red Quartermaster Formation portion of the Spanish skirts, while just above, in the Tecovas Formation, the slopes are rather smooth. Note how the Trujillo sandstone, a harder rock unit than the surrounding mudstones, forms cliffs and ledges on Lighthouse Peak, Santana's Face, and many mesa edges throughout the park. It simply does not erode as fast as the mudstones.

Differential erosion also explains the wonderful pedestals, or hoodoos, seen in many places in the canyon. A hard cap of sandstone protects the pedestal underneath from eroding as fast as the surrounding, unprotected mudstone. Many sandstone slabs are derived from higher up in the cliff walls and have slid downward over the mudstone, which they now protect from erosion. The Rock Garden, at the south end of the park drive, is a jumbled pile of Trujillo sandstone blocks that are the resistant leftovers after the mudstone beneath them was eroded away.

Another erosion process, the local downslope movement of rocks and sediment (mass wasting), creates piles of debris known as talus slopes at the bases of hills, a common feature in the canyon. Rocks fall suddenly or creep down slowly under the influence of gravity.

US 287
AMARILLO — CHILDRESS
117 miles

Between Amarillo and Clarendon, US 287 travels on the High Plains surface, and vistas of flat, immense distances are seen from the car windows. Twenty miles east of Amarillo is the hamlet of Claude, where TX 207 heads south toward Silverton. This scenic road crosses a wide expanse of Palo Duro Canyon east of Palo Duro Canyon State Park. Access to the park, however, is from the west, off I-27, so see the discussion in the previous road guide.

Near Clarendon, watch for tan soils and small outcrop exposures of Ogallala sandstone. Also near Clarendon, you'll notice increased topographic relief as the erosive effects of the Salt Fork of the Red River, which lies a couple miles north of the road, come into play.

The rest area about 4 miles southeast of Hedley sits on the High Plains, but farther southeast the road cuts its way downward through the eastern Caprock Escarpment. Badlands country in the tan sandstone of the Ogallala Formation can be seen to the south of the rest area. Unmistakable red Permian sandstones of the Quartermaster Formation are encountered beneath the Ogallala.

Note red mesas east of Memphis and roadcuts south of town in the distinctly red Permian-aged Quartermaster. A bit farther south of Memphis are remnant buttes that stream erosion has left behind—these, again, are red Permian rocks.

About halfway between Memphis and Childress, US 287 crosses the Prairie Dog Town Fork of the Red River. If water is flowing, it is not hard to understand why early pioneers dubbed this the Red River. The river's sediment load, being composed largely of eroded particles from the surrounding red Permian countryside, impart the river's distinctive color. River bars, channels, and gravel piles are beautifully exposed in the riverbed when water flow is low or nonexistent.

Red Permian-aged Quartermaster sandstone outcrops underlie the whitish-tan Ogallala sandstone about 8 miles southeast of Hedley. (34.7994, -100.5821)

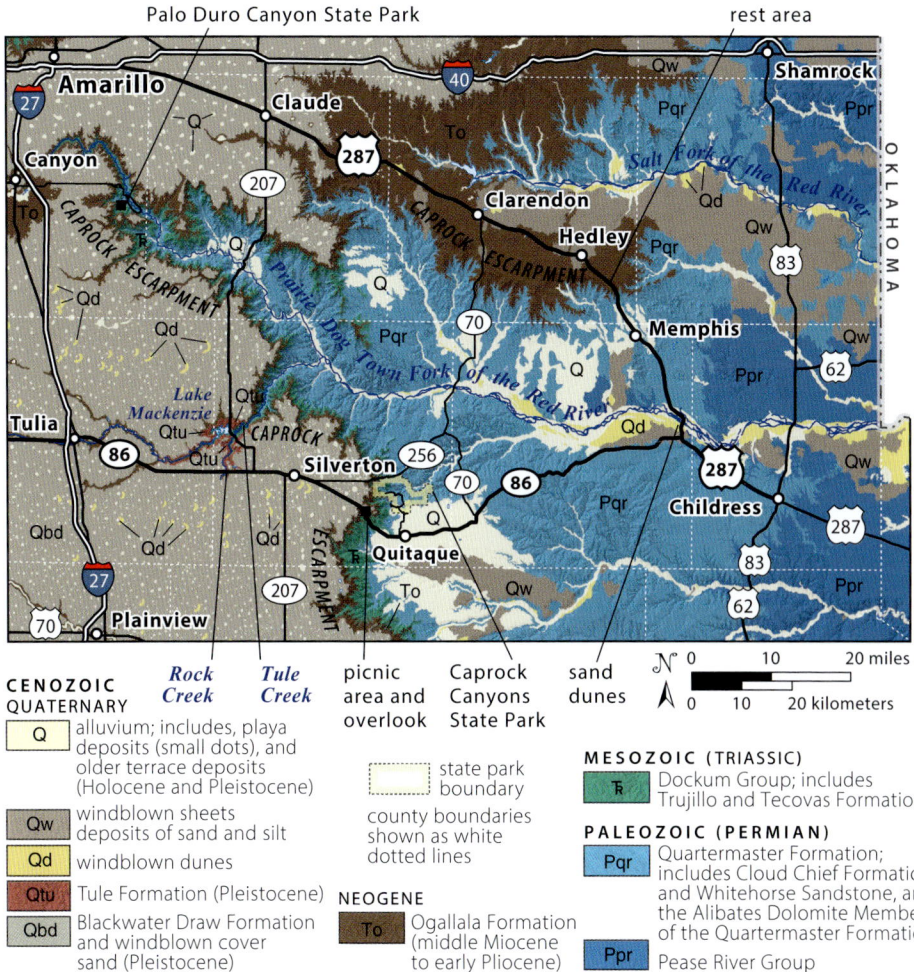

Geology along US 287 between Amarillo and Childress.

CENOZOIC
QUATERNARY

| Q | alluvium; includes, playa deposits (small dots), and older terrace deposits (Holocene and Pleistocene) |

| Qw | windblown sheets deposits of sand and silt |

| Qd | windblown dunes |

| Qtu | Tule Formation (Pleistocene) |

| Qbd | Blackwater Draw Formation and windblown cover sand (Pleistocene) |

picnic
area and
overlook

Caprock
Canyons
State Park

sand
dunes

state park
boundary

county boundaries
shown as white
dotted lines

NEOGENE

| To | Ogallala Formation (middle Miocene to early Pliocene) |

MESOZOIC (TRIASSIC)

| Ṯ͞ʀ | Dockum Group; includes Trujillo and Tecovas Formation |

PALEOZOIC (PERMIAN)

| Pqr | Quartermaster Formation; includes Cloud Chief Formation and Whitehorse Sandstone, and the Alibates Dolomite Member of the Quartermaster Formatio |

| Ppr | Pease River Group |

This remnant butte, composed of Quartermaster sandstone, can be seen in a field south of Memphis. (34.6598, -100.4819)

It's not difficult to figure out how the Prairie Dog Town Fork of the Red River got its name. View from the north side of the US 287 bridge. (34.5730, -100.4370)

About three-quarters of a mile south of the river is a pull-off on the right (west) that offers an opportunity to look at the sand dunes that have been piled up by the wind. With such an abundant sand source lying about in the frequently dry riverbed, it is not surprising the wind would create dunes. All the elements for dune formation are present: wind, a source of sand, and vegetation obstacles to trap the sand.

Between the river crossing and Childress are continuous red soils, an occasional red outcrop, and low, red hills, telling of the presence of Permian-aged sedimentary rocks.

TX 86
TULIA—CAPROCK CANYONS STATE PARK
50 miles

See map on page 325.

From Tulia, follow TX 86 east across the High Plains to reach Caprock Canyons State Park. About 18 miles east of Tulia, TX 86 passes the head of Rock Creek on the left (north) side of the road. Fossils of ice age mammals were collected from this site by paleontologists from Yale University in 1912. Horses, camels, ground sloths, giant short-faced bear, Columbian mammoths, dog-like animals, and pronghorn antelope from this locality date to about 1,000,000 to 700,000 years old.

About 22 miles east of Tulia, take a short 6.5-mile side trip north on TX 207 to reach the canyon of Tule Creek southeast of Lake MacKenzie. The ice age (Pleistocene) Tule Formation sediments in roadcuts directly overlie exposures of the red Trujillo Formation of the Triassic Dockum Group. Look for beautiful, sparkly selenite (gypsum) crystals in the Tule Formation in roadcuts north of the Tule Creek crossing. In addition, the Trujillo Formation forms spectacular buttes and small mesas in and around the canyon.

Great exposures of white Tule Formation directly overlying red Trujillo Formation sandstone, part of the Dockum Group, can be seen upon entering the Tule Creek canyon from the south along TX 207. (34.5390, -101.4238)

East of the TX 207 junction, TX 86 passes through Silverton, and about 11.7 miles east of the TX 207 junction is a picnic area that overlooks the colorful edge of the Caprock Escarpment and the reddened landscape of the Permian plains to the east. From here you can see canyon and badland exposures of the Ogallala sandstone and caliche forming the top edge of the escarpment. Beneath are the varicolored sandstone and shale beds of the Dockum Group, which come in shades of purple, white, gray, tan, red, and pink. At the base of the cliffs, these colorful Triassic beds give way to bright-red sandstones of Permian age that extend eastward for miles.

From the picnic overlook, the road winds downward through the whole panorama to give you a spectacular close-up, car window look at the rocks in nearby hills and roadcuts. The Permian red beds can be seen in hills and soils to the east from the bottom of the escarpment.

CAPROCK CANYONS STATE PARK

Caprock Canyons State Park, located along the edge of the Caprock Escarpment, lies about 3 miles north of Quitaque off Ranch Road 1065. The Ogallala caprock overlies varicolored Triassic-aged Dockum Group rocks that in turn overlie red beds of the Permian-aged Quartermaster Formation. The park road snakes through some of the best close-up outcrops of Permian red beds seen in this part of Texas. White gypsum deposits among the red beds tortuously contorted by extinct geologic forces, are

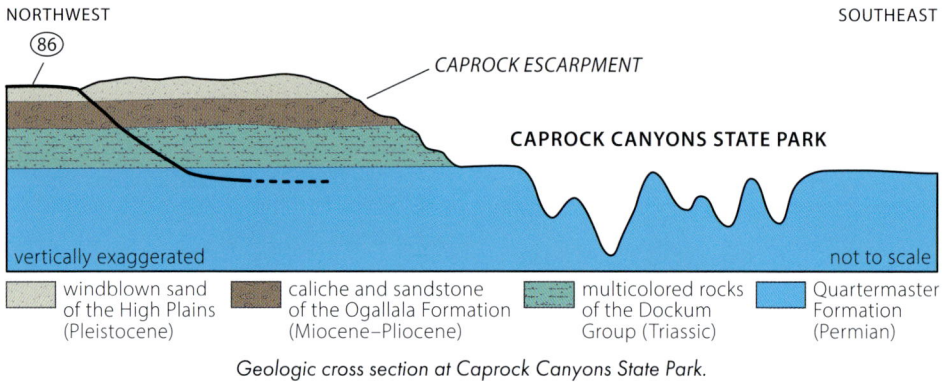

NORTHWEST

86

CAPROCK ESCARPMENT

SOUTHEAST

CAPROCK CANYONS STATE PARK

vertically exaggerated

not to scale

| windblown sand of the High Plains (Pleistocene) | caliche and sandstone of the Ogallala Formation (Miocene–Pliocene) | multicolored rocks of the Dockum Group (Triassic) | Quartermaster Formation (Permian) |

Geologic cross section at Caprock Canyons State Park.

Contorted beds of gypsum within the Quarter-master Formation in a roadcut along the park road. (34.4410, -101.0591)

View of erosional topography in the Quartermaster Formation from the Upper South Prong Trailhead in Caprock Canyons State Park. (34.4414, -101.0930)

View looking northwest from the interpretive exhibit in Caprock Canyons State Park toward the Quartermaster Formation in the foreground and the Trujillo Formation in the background. (34.4202, -101.0600)

spectacular. Gypsum forms when salty water evaporates, so the many layers suggest repeated cycles of wetting and drying that is typical of an arid tidal flat.

A wonderful interpretive exhibit, housed in an open-air rotunda just off the road to the campground, is called "250 Million Years at Caprock Canyons." You'll get more geological understanding in fifteen minutes at this exhibit than most anywhere else. It covers not only geology but also human history and paleobiology, with excellent drawings and displays.

GLOSSARY

algal mat. A tough mat formed in tidal flats where algae grow in spaces between sand grains, binding them together.

alluvial fan. A fan-shaped deposit of stream-deposited sand and gravel found at the base of steep slopes and mountain fronts.

anhydrite. An evaporation mineral, calcium sulfate. It is usually white and is related to gypsum.

anticline. An upward arched fold with the oldest rocks in the center.

aquifer. A body of rock or sediment with good porosity and permeability that holds and readily transmits groundwater.

ash. Rock fragments smaller than 2 millimeters in diameter that have been blasted into the air during a volcanic eruption.

ash-flow tuff. A rock formed from the consolidation and compaction of hot ash flowing from a volcano.

badlands. An intricately carved landscape of soft rock dissected by streams and precipitation runoff.

barrier island. A linear island of sand forming the outer coastline. It is separated from the mainland by a lagoon or sound where finer-grained sediments are deposited.

basalt. A fine-grained, typically dark-colored volcanic rock containing the minerals plagioclase and pyroxene.

batholith. A huge body of light-colored, coarse-grained igneous rock that crystallized beneath the Earth's surface and occupies more than 100 square miles of area at the Earth's surface.

bed. The smallest layer of sedimentary rock, usually a layer from one depositional event, such as a flood or hurricane. A rock with many beds is said to be **bedded** or to contain **bedding**.

bedrock. The solid rock that underlies soil or loose surface material.

bolson. A Southwest desert term for a low area that has no stream drainage outlet.

brachiopod. A marine invertebrate characterized by two bilaterally symmetrical shells. It has lived from the Cambrian to today.

breccia. A rock composed of angular, broken fragments.

bryozoan. An invertebrate characterized by branching, stick-like colonial growth. It has lived from the Ordovician to today.

calcareous. Rocks or materials containing calcium carbonate.

calcite. A light-colored mineral composed of calcium carbonate that is the main constituent of limestone, most marble, and many marine fossils.

caldera. A large volcanic depression formed after a large eruption when land collapses into the emptied magma chamber below.

caliche. A layer of calcium carbonate that forms in soils and sediments in dry regions.

caprock. An overlying rock layer that is unusually hard and protects the underlying layers from erosion. It caps landforms such as mesas, plateaus, and buttes.

carbonate bank. A large sedimentary rock unit with moderate relief that is composed of primarily calcareous deposits. The Great American carbonate bank is an example.

carbonate rock. A sedimentary rock composed mostly of calcite and dolomite.

chalk. A soft, white, earthy limestone of marine origin, formed mainly from the shells of floating microorganisms.

chert. A sedimentary rock composed of quartz crystals too small to see with the naked eye. It forms as a chemical deposit and is found primarily as nodules precipitated from water percolating through limy sediments after deposition. It can also be precipitated directly from water on the seafloor.

claystone. A sedimentary rock of mud or clay. It is also called mudstone or shale.

coastal plain. An area of very low topographic relief next to the coast.

columnar jointing. The parallel fracturing in a lava flow that causes the flow to break into columns.

concretion. A hard ball or odd-shaped mass of mineral matter, usually formed from the precipitation of some minor mineral in the host rock. It forms around a nucleus such as a bone, shell, leaf, or fossil.

conglomerate. A sedimentary rock composed of pebbles, cobbles, and/or boulders eroded from older rocks. The large size implies a vigorous process of deposition, such as that of a mountain stream.

contact. The boundary or surface between two different rock types, formations, or ages of rocks. Contacts may be depositional bedding planes, faults, edges of intrusive bodies, or unconformities.

corals. Marine, bottom-dwelling, mostly colonial animals that secrete an external skeleton composed of calcium carbonate. Their skeletons form major components of modern and ancient reefs.

country rock. The preexisting bedrock intruded by or surrounded by igneous rocks.

crinoid. An echinoderm with a cup-shaped body, numerous feathery arms, and a stalk for attachment to the seafloor. It has lived from the Ordovician to today.

cross-bedding. Sedimentary layers deposited at an angle to the main, horizontal beds.

crust. The outermost layer of Earth. **Oceanic crust**, composed of basalt and rocks of similar composition, ranges from 3 to 6 miles thick. **Continental crust**, composed mainly of lighter-colored, less dense rock—such as granite or gneiss, often with a veneer of sedimentary rock—normally ranges from 20 to 40 miles thick.

crystalline. Said of a rock containing crystals such as igneous rocks, where crystals form from cooling magma.

cuesta. A hill with a gentle slope on one side and a steep slope on the other.

deformation. A general term for the folding, faulting, shearing, extension, or compression of rock.

delta. A body of sediment deposited where a river enters a standing body of water.

deposition. The process of sediment settling out of water or air.

diatom. A microscopic, single-cell plant with silica walls that lives in oceans and lakes.

differential erosion. The selective wearing away of softer rocks so that harder rocks remain forming landscape features.

dike. A tabular body of intrusive igneous rock that cuts across the surrounding country rock.

dip. The downslope direction on an inclined (or dipping) bedding surface.

dissolution. The process of dissolving rock, usually by acidic rainwater or groundwater.

dolomite. A sedimentary rock akin to limestone that contains magnesium as well as calcium carbonate. Dolomite typically forms when fluids moving through buried limestone precipitate magnesium.

entrenched river. A river that cuts down into bedrock, having inherited its course from a previous cycle of erosion on flatter topography.

erosion. A general term for several processes that loosen, dissolve, or weather and then transport earthen materials and thus wear away landscapes.

evaporite. A sedimentary rock typically formed by the partial or total evaporation of brine.

escarpment. The steep face of a ridge or plateau along which the land drops abruptly to a lower level.

exfoliation. A weathering process where scales of rock are peeled off the surface, often resulting in dome-shaped hills.

fault. A fracture in rock where one side moves up, down, or sideways or is offset, relative to the other side.

fault block. A piece of crust bounded by faults.

feldspar. The most abundant mineral group in Earth's crust. Includes plagioclase (calcium- and sodium-bearing) feldspars and alkali (potassium-bearing) feldspars. Feldspar is not as common as quartz in soil or sedimentary rocks because weathering alters feldspar to clay.

fine-grained. Said of sedimentary rocks that have particles that are relatively small, usually averaging less than 2 millimeters in diameter. Also said of igneous rocks with relatively small, hard-to-see crystals.

flatiron. A plate of steeply dipping, resistant rock on mountain flanks that weathers into the triangular shape of a clothes iron.

flint. A dark-colored chert. The term is commonly used to describe arrowhead material.

floodplain. An area adjacent to a riverbed that may lie underwater when the river overflows its banks.

foliation. The parallel surfaces or layers in metamorphic rock caused by the growth, flattening, or dissolving of mineral grains under stress during metamorphism.

formation. The basic subdivision of sedimentary rocks that can be mapped from place to place.

fossil. Any preserved evidence of past life.

fossil hash. A fossil description of the situation in which all of the organic material in an environment falls to the ocean floor and fossilizes, creating a mixture of many different fossils within the rock.

fracking. Artificially fracturing an oil- or gas-bearing rock unit so that the hydrocarbon moves into the well.

glauconite. An earthy, green, iron-rich mica mineral. It forms as small pellets in shallow marine water where sedimentation is slow.

gneiss. A coarse-grained, foliated metamorphic rock characterized by light and dark bands of minerals. It forms during high-grade regional metamorphism.

granite. A coarse-grained intrusive igneous rock consisting of quartz, alkali feldspar, and mica.

groundwater. The subsurface water in the zone of saturation, the area below where rainwater percolates through the soil and underlying rock. Groundwater may include ancient saltwater and brine.

group. A formal rock unit containing two or more formations.

gypsum. A hydrous calcium sulphate ($CaSO_4 \cdot 2H_2O$), a soluble mineral that forms as water evaporates.

headward erosion. The lengthening and cutting upstream of a stream or gully, by rain wash, gullying, or slumping.

hoodoo. The differentially eroded columns or pinnacles of rock that resemble animals or creatures.

hornblende. The most common dark silicate mineral of the amphibole group. It is a constituent of granite, gneiss, and schist.

hydrocarbon. An organic compound consisting of carbon and hydrogen atoms, like oil and natural gas.

igneous. A rock that cooled from molten material, either from magma within the Earth (intrusive or plutonic) or from lava at the surface (extrusive or volcanic).

interbedded. Said of rock units with alternating layers of differing rock types, characteristics, or both.

intrusive igneous rock. A rock that cools from magma beneath the surface of Earth. The body of rock is called an **intrusion**. The magma is said to **intrude** into other rocks of the crust.

island arc. An offshore volcanic arc or linear chain of volcanoes formed along a convergent tectonic plate margin.

joint. A fracture in a rock without displacement.

karst topography. A landscape shaped by the dissolution of soluble rocks and characterized by caves, sinkholes, and underground drainage.

laccolith. A mushroom-shaped igneous intrusion formed when rising magma spreads between layers of sedimentary rock.

lava. Melted rock, or magma, that erupts at Earth's surface.

lignite. Brownish to black coal, between peat and bituminous coal in quality.

limestone. A sedimentary rock composed of calcium carbonate, including the calcareous skeletons of invertebrate fossils.

limy. Describes sediments, soils, or rocks that contain a significant amount of lime (calcium oxide, CaO), often associated with limestone formation.

lithosphere. The rigid crust and uppermost mantle of the Earth.

llanite. A rhyolite rock with large crystals of red feldspar and blue quartz.

longshore drift. The movement of sand along the shore, driven mainly by oblique waves striking the coast.

magma. Molten rock. Termed **lava** where it erupts at the surface of the Earth.

mantle. The part of Earth between the interior core and the outer crust.

marble. A metamorphic rock composed of calcite or dolomite.

marl. A limestone containing a significant amount of mud.

meander. Series of loops, turns, or bends in the course of a river that form as the river swings from side to side across its floodplain.

metamorphic. Said of minerals and rocks with compositions and textures that were changed by heat, pressure, or both. For example, slate is a metamorphic rock that forms when shale is metamorphosed, and marble forms when limestone is metamorphosed.

mica. A group of platy minerals, such as biotite. Micas are common constituents of igneous and metamorphic rocks.

mineral. A naturally occurring chemical element or compound with a characteristic crystal form.

mudstone. A sedimentary rock composed mainly of clay. Unlike shale, it doesn't tend to split into thin pieces.

neck (volcanic). The eroded remnant of solidified lava filling the conduit of an extinct volcano.

normal fault. A fault, created by tensional stress, in which the rock above the fault moves down relative to the rock below the fault.

obsidian. A dark-colored volcanic glass.

olivine. A dark-green silicate mineral rich in iron and magnesium. It is common rock-forming mineral in basalt.

ore. A rock that contains desirable minerals in concentrations that are economic to extract.

oxbow lake. The abandoned, water-filled loop of a stream meander that looks like a U-shaped ox yoke.

oxidation. The chemical process where oxygen is added to minerals or other compounds to form **oxides**. Weathering oxidizes minerals, and burning wood is a type of oxidation.

peat. Partly carbonized plant remains. Peat is an early stage in coal development.

pegmatite. A very coarse-grained, usually granitic igneous rock with crystals at least 1 inch long.

petrified wood. Wood that has been fossilized by silica replacing the wood.

pictograph. An image painted on rock.

playa. An ephemeral lakebed.

point bar. A series of low, arcuate ridges of sand on the inside bend of a river meander loop.

pumice. A porous volcanic rock formed in an explosive eruption when a gas-rich lava solidifies rapidly.

pumpjack. The surface equipment that drives a piston pump in an oil well.

pyroclastic. Fragments of rock violently ejected from a volcano.

pyroclastic flow. A very hot, rapidly moving mix of lava, pumice, ash, and volcanic gas erupted from a volcano.

quartz. One of the most common minerals in the Earth's crust. The most common variety is colorless and clear like glass. Quartz is composed of silicon and oxygen, is the main constituent of most sand, and is common in sedimentary rocks and light-colored igneous rocks.

red bed. A sedimentary layer composed of reddish grains of any size, usually due to coatings or cements of hematite or other iron oxide minerals.

reef. A mound or ridge built along a coastline by lime-secreting organisms, such as corals.

reservoir (for hydrocarbon). Any subsurface, porous, and permeable rock that contains oil or gas.

resistant. Said of a rock or rock outcrop that withstands the effects of weathering or erosion.

rhyolite. A fine-grained, pale-gray to pink volcanic rock containing the minerals quartz, sanidine, plagioclase, and minor amounts of hornblende and biotite.

rift. A linear valley marking where a tectonic plate is being pulled apart by tensional tectonic forces.

ripple, ripple mark. A series of small ridges of sand produced when wind or water moves sediment.

sandstone. A sedimentary rock consisting primarily of sand-sized grains, usually of the mineral quartz.

salt dome. A column of salt that rose upward and deformed surrounding layers of rock.

sand sheet. A large plain of windblown sand that lacks large dunes.

schist. A metamorphic rock that is strongly layered due to an abundance of visible, platy minerals.

sedimentary rock. A solid rock composed of sediment that has been naturally compacted, cemented, or both.

shale. A fine-grained sedimentary rock, composed mainly of clay, that tends to split into thin pieces parallel to its bedding.

shelf. A flat area on the continent margin. It can be dry land or covered by shallow ocean water, depending on sea level.

silica. Silicon dioxide, the compound that composes quartz in all its varieties, including chert.

sill. A tabular body of intrusive igneous rock that is parallel to the surrounding country rock.

siltstone. A sedimentary rock consisting primarily of silt-sized grains, particles of rock larger than clay but finer than sand. It can have texture and composition similar to shale but lacks the thin-bedded, platy appearance and tends to be better cemented.

sinkhole. A surface depression that resulted from the collapse of an underlying cavity.

source rock. An organic-rich sedimentary rock that yields oil or gas when subjected to proper heat, pressure, and time.

speleothem. Any structure formed in a cave by the deposition of minerals from water.

spheroidal weathering. A form of chemical weathering where spherical shells of decayed rock split off to create rounded boulders.

stock. A discordant, irregularly shaped body of intrusive igneous rock with less than 40 square miles exposed at Earth's surface.

subduction. The process of an oceanic plate sinking into Earth's interior beneath another tectonic plate.

subsidence. The sinking or settling of the Earth's surface, often due to subsurface movement or groundwater extraction.

sulfide. A compound containing sulfur. Sulfide minerals often contain valuable metals, such as copper and zinc.

syenite. A coarse-grained intrusive igneous rock consisting of the minerals orthoclase feldspar, lesser to minor plagioclase feldspar, minor augite and hornblende, and no quartz.

syncline. A trough-shaped fold with the youngest rocks at the center.

talus. A pile of rocks that accumulates at the base of a cliff from falling rocks.

tectonic. Referring to large-scale processes affecting the structure of the Earth's crust.

terrace. A flat bench, above and next to a stream, that marks a former, higher stream level.

thrust fault. A fault dipping less than 45 degrees that forms by tectonic forces of horizontal compression. Generally, the rock above the fault moved upward and over the rock below it.

trachyte. A fine-grained igneous rock composed mostly of alkali feldspar.

trilobite. A marine arthropod common in the Paleozoic Era.

tufa. A calcareous rock deposited around hot springs or seeps.

tuff. An igneous rock composed of ash, pumice, and other debris erupted explosively from a volcano.

turbidity current. An ocean current of dense water and sediment that flows rapidly down continental slopes, triggered by earthquakes or an oversupply of sediments.

unconformity. A surface between two rock units representing a gap in geologic time.

vein. A tabular body of minerals precipitated in cracks in rock.

wash. A dry stream bed that only carries flowing water after a heavy rainstorm.

washover fan. A small delta built on the lagoon side of a barrier island by storm waves breaking over the barrier island.

water table. The top surface of the water-saturated part of an aquifer.

weathering. The process by which rocks break down near Earth's surface due to exposure to air, water, and the action of organisms.

wildcatter. Someone who drills an exploratory oil well in an area not known to be an oil field.

xenolith. A foreign rock inclusion, usually in an igneous rock.

REFERENCES

Ambrose, W. A., Flaig, P., Zhang, J., and others. 2020. The Midway to Carrizo succession in the southeastern Texas Gulf Coast: Evolution of a tidally influenced coastline. *Gulf Coastal Association of Geological Societies Journal* 9: 41–75.

Anderson, J. B. 2007. *The Formation and Future of the Upper Texas Coast: College Station.* Texas A&M University Press.

Anderson, J. B., and A. B. Rodriguez. 2008. *Response of Upper Gulf Coast Estuaries to Holocene Climate Change and Sea-level Rise.* GSA Special Paper 443.

Barker, D. S. 2000. *Down to Earth at Tuff Canyon, Big Bend National Park, Texas.* Bureau of Economic Geology, DE0002.

Barker, D. S., and R. M. Reed. 2010. Proterozoic granites of the Llano Uplift, Texas: A collision-related suite containing rapakivi and topaz granites. *GSA Bulletin* 122: 253–264.

Barnes, M. A., Anthony, E. Y., Williams, I., and G. B. Asquith. 2002. Architecture of a 1.38-1.34 Ga granite-rhyolite complex as revealed by geochronology and isotopic and elemental geochemistry of subsurface samples from west Texas, USA. *Precambrian Research* 119: 9–43.

Bickford, M. E., Soegaard, K., Nielsen, K. C., and J. M. Mclelland. 2000. Geology and geochronology of Grenville-age rocks in the van Horn and Franklin Mountains area, west Texas: Implications for the tectonic evolution of Laurentia during the Grenville. *GSA Bulletin* 112: 1134–1148.

Big Spring State Park. 2019. *Guide to Historic Rock Carvings on Scenic Mountain.* Texas Parks and Wildlife.

Boyd, F. M. and C. W. Kreitler. 1986. *Hydrogeology of a Gypsum Playa, Northern Salt Basin, Texas.* Bureau of Economic Geology Report of Investigations No. 158.

Cepeda, J. C., and C. D. Henry. 1983. *Oligocene volcanism and multiple caldera formation in the Chinati Mountains, Presidio County, Texas.* The University of Texas at Austin, Bureau of Economic Geology Report of Investigations No. 135.

Cogan, D., and C. Lea. 2021. *Vegetation Inventory Project: Big Bend National Park.* Natural Resource Report NPS/CHDN/NRR—2021/2275.

Collard, M., Buchanan, B., Hamilton, M. J., and M. J. O'Brien. 2010. Spaciotemporal dynamics of the Clovis-Folsom transition. *Journal of Archeological Science* 37: 2513–2519.

Cross, B. L., and N. Fox. 2011. *Gem Trails of Texas (9th edition).* Baldwin Park, California: Gem Guides Book Company.

Dattilo, B. F., Howald, S. C., Bonem, R., and others. 2014. Stratigraphy of the Paluxy River tracksites in and around Dinosaur Valley State Park, Lower Cretaceous Glen Rose Formation, Somervell County, Texas. In *Fossil Footprints of Western North America,* eds. M. G. Lockley and S. G. Lucas, *New Mexico Museum of Natural History & Science Bulletin* 62: 307–338.

Dickerson, P. W., Hoffer, J. M., and J. F. Callender, eds. 1980. *Trans Pecos Region (West Texas).* New Mexico Geological Society 31st Annual Fall Field Conference Guidebook.

Evans, T. J. 1974. *Bituminous Coal in Texas.* The University of Texas at Austin, Bureau of Economic Geology Handbook 4.

Evans, T. J. 1975. *Gold and Silver in Texas.* Bureau of Economic Geology Mineral Resource Circular No. 56.

Ewing, T. E. 1986. Balcones volcanoes in South Texas: Exploration methods and samples. In *Contributions to the Geology of South Texas,* ed. W. L. Stapp, South Texas Geological Society: 368–379.

Ewing, T. E. 1991. *The Tectonic Framework of Texas: Accompanying Tectonic Map of Texas.* The University of Texas at Austin, Bureau of Economic Geology.

Ewing, T. E. 2016. *Texas Through Time: Lone Star Geology, Landscapes, and Resources.* The University of Texas at Austin, Bureau of Economic Geology Udden Series No. 6.

Finch, W. I. 1996. *Uranium Provinces of North America: Their Definition, Distribution, and Models.* USGS Bulletin 2141.

Finsley, C. E. 1999. *A Field Guide to Fossils in Texas* (3rd edition). Houston: Gulf Publishing.

Finsley, C. E. 1999. *Discover Texas Dinosaurs.* Houston: Gulf Publishing.

Flis, J. E., Yancey, T. E., and C. J. Flis. 2017. Middle Eocene storm deposition in the northwestern Gulf of Mexico, Burleson County, Texas, USA. *Gulf Coastal Association of Geological Societies Journal* 6: 201–225.

Frohlich, C., and S. D. Davis. 2002. *Texas Earthquakes.* Austin: University of Texas Press.

Galloway, W. E., Whiteaker, T. L., and P. Ganey-Curry. 2011. History of Cenozoic North American drainage basin evolution, sediment yield, and accumulation in the Gulf of Mexico basin. *Geosphere* 7: 938–973.

Gray, J. E., and W. R. Page, eds. 2008. *Geological, Geochemical, and Geophysical Studies by the US Geological Survey in Big Bend National Park, Texas.* USGS Circular 1327.

Griffith, G. E., Bryce, S. A., Omernik, J. M., and others. 2004. *Ecoregions of Texas.* US Geological Survey Map, scale 1:2,500,000.

Grimes, S. W., and P. Copeland. 2004. Thermochronology of the Grenville Orogeny in west Texas. *Precambrian Research* 131: 23–54.

Hall, S. A. 2001. Geochronology and paleoenvironments of the glacial-age Tahoka Formation, Texas and New Mexico High Plains. *New Mexico Geology* 23: 71–77.

Harbour, R. L. 1972. *Geology of the Northern Franklin Mountains, Texas and New Mexico.* USGS Bulletin 1298.

Hart, M. B., Yancey, T. E., Leighton, A. D., and others. 2012. The Cretaceous-Paleogene boundary on the Brazos River: New stratigraphic sections and revised interpretations. *Gulf Coastal Association of Geological Societies Journal* 1: 69–80.

Hayward, O. T. 1988. *South-Central Section of the Geological Society of America.* GSA Centennial Field Guide 4.

Henry, C. D. 1998. *Geology of Big Bend Ranch State Park.* The University of Texas at Austin, Bureau of Economic Geology Guidebook 27.

Henry, C. D., and W. R. Muehlberger, eds. 1996. *Geology of the Solitario Dome, Trans-Pecos Texas: Paleozoic, Mesozoic, and Cenozoic Sedimentation, Tectonism, and Magmatism.* Bureau of Economic Geology, Report of Investigations No. 240.

Henry, C. D., Kunk, M. J., Muehlberger, W. R., and W. C. McIntosh. 1997. Igneous evolution of a complex laccolith-caldera, the Solitario, Trans-Pecos Texas: Implications for calderas and subjacent plutons. *GSA Bulletin* 109: 1036–1054.

Henry, C. D., Price, J. G., Rubin, J. N., and others. 1988. Widespread, lavalike silicic volcanic rocks of Trans-Pecos Texas. *Geology* 16: 509–512.

Holliday, V. T. 1985. Archaeological geology of the Lubbock Lake site, Southern High Plains of Texas. *GSA Bulletin* 96: 1483–1492.

Holliday, V. T., Kring, D. A., Mayer, J. H., and R. J. Goble. 2005. Age and effects of the Odessa meteorite impact, western Texas, USA. *Geology* 33: 945–948.

Johnson, A., Boyd, C., and A. Castaneda. 2011. Lower Pecos Rock Art Recording and Preservation Project. Archaeological Institute of America Site Preservation Program, p. 1-7.

Kasmarek, M. C., Gabrysch, R. K., and M. R. Johnson. 2010. *Estimated Land-surface Subsidence in Harris County, Texas, 1915-1917 to 2001.* USGS Scientific Investigations Map SIM-3097, 2 sheets.

KellerLynn, K. 2008. *Guadalupe Mountains National Park Geologic Resource Evaluation Report.* Natural Resource Report NPS/NRPC/GRD/NRR—2008/023.

KellerLynn, K. 2015. *Lake Meredith National Recreation Area and Alibates Flint Quarries National Monument.* Geologic resources inventory report. Natural Resource Report NPS/NRSS/GRD/NRR—2015/1046.

Kyle, J. R., ed. 1990. *Industrial mineral resources of the Delaware Basin, Texas and New Mexico.* Society of Economic Geologists Guidebook 8.

Land L. and G. Veni. 2018. *Karst Hydrogeology Scoping Investigation of the San Solomon Spring Area: Culberson, Jeff Davis, and Reeves Counties, Texas.* National Cave and Karst Research Institute Report of Investigation 8.

Lee, J-H., and R. Riding. 2022. Stromatolite-rimmed thrombolite columns and domes constructed by microstromatolites, calcimicrobes and spongesvin late Cambrian biostromes, Texas, USA. *Sedimentology* 70: 293–334.

Lehman, T. and J. Schnable, eds. 1992. *Guidebook for the Geology of the Southern High Plains at Caprock Canyons State Park Texas.* Texas Tech University.

Levine, J. S. F., and S. Mosher. 2010. Contrasting Grenville-aged tectonic histories across the Llano Uplift, Texas: New evidence for deep-seated high temperature deformation in the western uplift. *Lithosphere* 2: 399–410.

Love, D. W., Hawley, J. W., Kues, B. S., Austin, G. S., and S. G. Lucas. 1993. *Carlsbad Region (New Mexico and West Texas).* New Mexico Geological Society 44th Annual Fall Field Conference Guidebook.

Lovejoy, E. M. P. 1975. An interpretation of the structural geology of the Franklin Mountains, Texas. In *Las Cruces Country*, eds. W. R. Seager, R. E. Clemons, and J. F. Callender, New Mexico Geological Society 26th Annual Fall Field Conference Guidebook, 261–268.

Lucas, S. G., and D. Ulmer-Scholle, eds. 2001. *Geology of Llano Estacado.* New Mexico Geological Society 52nd Annual Fall Field Conference Guidebook.

Machenburg, M. D. 1984. *Geology of Monahans Sandhills State Park.* Bureau of Economic Geology Guidebook 21.

Maxwell, R. A. 1968. *The Big Bend of the Rio Grande.* Bureau of Economic Geology Guidebook 7.

McAnulty, W. N., Muehlberger, W. W., Rodgers, R. C., and others. 1971. *A Glimpse of Some of the Geology and Mineral Resources Sierra Blanca: Van Horn Country, Hudspeth and Culbertson Counties, Texas.* El Paso Geological Society 5th Annual Field Trip Guide.

McBride, E. F. 1989. Stratigraphy and sedimentary history of pre-Permian Paleozoic rocks of the Marathon uplift. In *The Appalachian-Ouachita Orogen in the United States.* GSA, eds. R. D. Hatcher, W. A. Thomas, and G. W. Viele, The Geology of North America F-2: 603–620.

Mosher, S. 1998. Tectonic evolution of the southern Laurentian Grenville orogenic belt. *GSA Bulletin* 110: 1357–1375.

Mosher, S., Helper, M., and J. Levine. 2008. *The Texas Grenville Orogen, Llano Uplift, Texas.*

Fieldtrip guide to the Precambrian Geology of the Llano Uplift, central Texas. GSA Annual Meeting Field Trip 405.

Muhs, D. R., and V. T. Holliday. 2001. Origin of late Quaternary dune fields on the Southern High Plains of Texas and New Mexico. *GSA Bulletin* 113: 75–87.

Nelson, K. 1992. *A Road Guide to the Geology of Big Bend National Park*. Big Bend Natural History Association.

Nielson, R. L., and C. Barker. 2016. *Geology of the Northern Llano Uplift, Junction to Llano, Texas*. Stephen F. Austin State University Faculty Publications 15.

Paine, J. G., Mathew, S., and T. Caudle. 2012. Historical shoreline change through 2007, Texas gulf coast: Rates, contributing causes, and Holocene context. *Gulf Coastal Association of Geological Societies Journal* 1: 13-16.

Perttula, T. K. 2021. *Ancestral Caddo Mounds and Monuments in East Texas*. Friends of Northeast Texas Archaeology Special Publication No. 60.

Prive, J. G., Henry, C. D., Standen, A. R., and J. S. Posey. 1985. *Origin of Silver-copper-lead Deposits in Red-bed Sequences of Trans-Pecos Texas: Tertiary Mineralization in Precambrian, Permian, and Cretaceous Sandstones*. The University of Texas at Austin, Bureau of Economy Geology Report of Investigations 145.

Reese, J. F., Mosher, S., Connelly, J., and R. Roback. 2000. Mesoproterozoic chronostratigraphy of the southeastern Llano uplift, central Texas. *GSA Bulletin* 112: 278–291.

Sansom, A. 2008. Water in Texas, an introduction. Austin, University of Texas Press.

Sawyer, D. S., Buffler, R. T., and R. H. Pilger Jr. 1991 The Crust Under the Gulf of Mexico Basin. In *The Gulf of Mexico Basin,* ed. A. Salvador, GSA, Geology of North America J (Chapter 4): 53–72.

Shah, S. D., and J. Lanning-Rush. 2005. *Principal Faults in the Houston, Texas Metropolitan area*. USGS Scientific Investigations Map 2874.

Shannon W. M., Barnes, C. G., and M. E. Bickford. 1997. Grenville Magmatism in West Texas: Petrology and Geochemistry of the Red Bluff Granitic Suite. *Journal of Petrology* 38: 1279–1305.

Sharpe, R. D. 1980. *Development of the Mercury Mining Industry: Trans-Pecos Texas*. The University of Texas at Austin, Bureau of Economic Geology Mineral Resource Circular No. 64.

Spearing, D. 1991. *Roadside Geology of Texas*. Missoula, Montana: Mountain Press Publishing Company.

Stafford, K. W., Shaw-Faulkner, M. G., and J. L. DeLeon. 2011. Spring hydrology of Colorado Bend State Park, central Texas. *Stephen F. Austin State University Faculty Publications* 14: 152–159.

Stafford, K. W., and G. Veni. 2018. *Hypogene Karst of Texas*. Texas Speleological Survey Monograph 3.

Stern, R. J., and W. R. Dickinson. 2010. The Gulf of Mexico is a Jurassic backarc basin. *Geosphere* 6: 739–754.

Thomann, W. F. 1980. Ignimbrites, trachyites, and sedimentary rocks of the Precambrian Thunderbird Group, Franklin Mountains, El Paso, Texas. *GSA Bulletin* 92: 94–100.

Turner, K. J., Berry, M. E., Page, W. R., and others. 2011. *Geologic map of Big Bend National Park, Texas*. USGS Scientific Investigations Map 3142, scale 1:75,000, pamphlet.

Veni, G. 2021. *A Cross Section of Central Texas Cave and Karst Management: Show Caves,*

Preserves, and Private Property. National Cave and Karst Research Institute Field Guide 1.

White, J. C., Miggins, D. P., Barker, D. S., and others. 2012. *Magmatism in Big Bend National Park: Recent Studies.* Field Trip Guide for South-Central GSA Meeting.

Wilkins, D. E., and D. R. Currey. 1997. Timing and extent of Late Quaternary paleolakes in the Trans-Pecos closed basin, West Texas and south-central New Mexico. *Quaternary Research* 47: 306–315.

Williams, C. G. 2012. Replanting the (Really) Lost Pines of Texas. *Forest History Today*: 12-15.

Yates, R. G., and G. A. Thompson. 1959. *Geology and Quicksilver Deposits of the Terlingua District Texas.* USGS Professional Paper 312.

Zolensky, M. E., Sylvester, P. J., and J. B. Paces. 1988. Origin and significance of blue coloration in quartz from Llano rhyolite (llanite), north-central Llano County, Texas. *American Mineralogist* 73: 313–323.

INDEX

Page numbers in bold face include photographs

Paul Brandes earned a BS in geology at the New Mexico Institute of Mining and Technology and an MS in geology at Michigan Technological University. He has worked as a geology consultant, an environmental enforcement officer, an exploration geologist, and a professor of geology at a community college. He has contributed photographs to several textbooks and laboratory manuals in geology and is on the management team of mindat.org. He is the author of *Michigan Rocks! A Guide to Geologic Sites in the Great Lakes State* and the photographer of two other books in the Rocks Series: *Texas Rocks!* and *New Mexico Rocks!* He lives and works in Houston when he is not traveling in search of geologic features to photograph.

Dar Spearing wrote the second edition of *Roadside Geology of Texas* and is also the author of *Roadside Geology of Wyoming* and *Roadside Geology of Louisiana*. He whetted his interest in the geology of Texas as a research geologist and exploration manager for Marathon Oil Company. He lived in Grand Lake, Colorado, for many years until his death in 2018. He was excited to help with the third edition. Many of the new color figures are based on ones he drew for the second edition, and some of his colorful descriptions are included as well.